(6) 91 Need for FGF1 & HSPG to induce neurone
    form^n. MAP2 then expressed

DEVELOPMENT OF THE CEREBRAL CORTEX

The Ciba Foundation is an international scientific and educational charity (Registered Charity No. 313574). It was established in 1947 by the Swiss chemical and pharmaceutical company of CIBA Limited—now Ciba-Geigy Limited. The Foundation operates independently in London under English trust law.

The Ciba Foundation exists to promote international cooperation in biological, medical and chemical research. It organizes about eight international multidisciplinary symposia each year on topics that seem ready for discussion by a small group of research workers. The papers and discussions are published in the Ciba Foundation symposium series. The Foundation also holds many shorter meetings (not published), organized by the Foundation itself or by outside scientific organizations. The staff always welcome suggestions for future meetings.

The Foundation's house at 41 Portland Place, London W1N 4BN, provides facilities for meetings of all kinds. Its Media Resource Service supplies information to journalists on all scientific and technological topics. The library, open five days a week to any graduate in science or medicine, also provides information on scientific meetings throughout the world and answers general enquiries on biomedical and chemical subjects. Scientists from any part of the world may stay in the house during working visits to London.

Ciba Foundation Symposium 193

DEVELOPMENT OF THE CEREBRAL CORTEX

1995

JOHN WILEY & SONS

Chichester · New York · Brisbane · Toronto · Singapore

©Ciba Foundation 1995

Published in 1995 by John Wiley & Sons Ltd,
Baffins Lane, Chichester
West Sussex PO19 1UD, England

Telephone: National (01243) 779777
International (+44) (1243) 779777

All rights reserved.

No part of this book may be reproduced by any means,
or transmitted, or translated into a machine language
without the written permission of the publisher.

Other Wiley Editorial Offices

John Wiley & Sons, Inc., 605 Third Avenue,
New York, NY 10158-0012, USA

Jacaranda Wiley Ltd, 33 Park Road, Milton,
Queensland 4064, Australia

John Wiley & Sons (Canada) Ltd, 22 Worcester Road,
Rexdale, Ontario M9W 1L1, Canada

John Wiley & Sons (SEA) Pte Ltd, 37 Jalan Pemimpin #05-04,
Block B, Union Industrial Building, Singapore 2057

Suggested series entry for library catalogues:
Ciba Foundation Symposia

Ciba Foundation Symposium 193
viii + 338 pages, 52 figures, 3 tables

Library of Congress Cataloging-in-Publication Data

Development of the cerebral cortex.
    p. cm. — (Ciba foundation symposium ; 193)
  Editors: Gregory Bock and Gail Cardew.
  Includes bibliographical references and indexes.
  ISBN 0 471 95705 4 (alk. paper)
    1. Cerebral cortex—Growth—Congresses. 2. Developmental
neurology—Congresses. I. Bock, Gregory. II. Cardew, Gail.
III. Series.
    [DNLM; 1. Cerebral Cortex—growth & development—congresses. W3
C161F v. 193 1995]
QP383.D48   1995
612.8′25–dc20
DNLM DLC
for Library of Congress   95-21512
                                                                       CIP

British Library Cataloguing in Publication Data

A catalogue record for this book is available from the British Library

ISBN 0 471 95705 4

Typeset in 10/12pt Times by Dobbie Typesetting Limited, Tavistock, Devon.
Printed and bound in Great Britain by Biddles Ltd, Guildford.
This book is printed on acid-free paper responsibly manufactured from sustainable forestation, for
which at least two trees are planted for each one used for paper production.

Contents

Symposium on Development of the cerebral cortex, held at the Ciba Foundation, London, 29 Nov–1 Dec 1994

This symposium is based on a proposal made by Jack Price

Editors: Gregory R. Bock (Organizer) and Gail Cardew

C. Blakemore Introduction: mysteries in the making of the cerebral cortex   1

C. Walsh and **C. Reid**   Cell lineage and patterns of migration in the developing cortex   21
Discussion   34

J. G. Parnavelas, M. C. Mione and **A. Lavdas**   The cell lineage of neuronal subtypes in the mammalian cerebral cortex   41
Discussion   52

General disussion I   Dendritic differentiation   59
Cell lineage   61

J. Price, B. P. Williams and **M. Götz**   The generation of cellular diversity in the cerebral cortex   71
Discussion   79

P. F. Bartlett, L. R. Richards, T. J. Kilpatrick, P. T. Talman, K. A. Bailey, G. J. Brooker, R. Dutton, S. A. Koblar, V. Nurcombe, M. O. Ford, S. S. Cheema, V. Likiardopoulos, G. Barrett and **M. Murphy**   Factors regulating the differentiation of neural precursors in the forebrain   85
Discussion   94

E. Boncinelli, M. Gulisano, F. Spada and **V. Broccoli**   *Emx* and *Otx* gene expression in the developing mouse brain   100
Discussion   110

General discussion II   Teleology for tangential migration   117
Differentiation factors in culture   124

Z. Molnár and **C. Blakemore**   Guidance of thalamocortical innervation   127
Discussion   140

v

A. Ghosh Subplate neurons and the patterning of thalamocortical connections 150
Discussion 165

J. Bolz, A. Kossel and D. Bagnard The specificity of interactions between the cortex and the thalamus 173
Discussion 186

General discussion III The handshake hypothesis 192

P. Levitt, R. Ferri and K. Eagleson Molecular contributions to cerebral cortical specification 200
Discussion 207

D. D. M. O'Leary, D. J. Borngasser, K. Fox and B. L. Schlaggar Plasticity in the development of neocortical areas 214
Discussion 223

D. J. Price, R. B. Lotto, N. Warren, G. Magowan and J. Clausen The roles of growth factors and neural activity in the development of the neocortex 231
Discussion 244

General discussion IV *In vitro* and *in vivo* evidence for areal commitment of neuron precursors in the ventricular zone of the neocortex 251

N. W. Daw, S. N. M. Reid, X.-F. Wang and H. J. Flavin Factors that are critical for plasticity in the visual cortex 258
Discussion 270

E. G. Jones Cortical development and neuropathology in schizophrenia 277
Discussion 289

G. W. Roberts, M. C. Royston and M. Götz Pathology of cortical development and neuropsychiatric disorders 296
Discussion 316

Final discussion Determination and specification 322

Index of contributors 326

Subject index 328

Participants

P. Bartlett The Walter and Eliza Hall Institute of Medical Research, Royal Melbourne Hospital, Parkville, Victoria 3050, Australia

C. Blakemore (*Chairman*) Department of Physiology, University of Oxford, South Parks Road, Oxford OX1 3PT, UK

J. Bolz INSERM U371 'Cerveau et Vision', 18 avenue du Doyen Lépine, 69500 Bron, France

E. Boncinelli DIBIT, Istituto Scientifico H.S. Raffaele, Via Olgettina 60, 20132 Milano, Italy

T. Bonhoeffer Max-Planck-Institut für Psychiatrie, Am Klopferspitz 18a, D-82152, Planegg-Martinsried, Germany

N. Daw Department of Ophthalmology and Visual Science, Yale University School of Medicine, 330 Cedar Street, PO Box 208061, New Haven, CT 06520–8061, USA

A. Ghosh Division of Neuroscience, Enders 250, Children's Hospital and Department of Microbiology and Molecular Genetics, Harvard Medical School, 300 Longwood Avenue, Boston, MA 02115, USA

G. M. Innocenti Institute of Anatomy, 9 rue Bugnon, CH-1005 Lausanne, Switzerland

E. G. Jones Department of Anatomy and Neurobiology, University of California, Irvine, CA 92717, USA

H. Kennedy INSERM U371, 18 avenue du Doyen Lépine, 69500 Bron, France

L. A. Krubitzer Vision, Touch and Hearing Research Centre, Department of Physiology & Pharmacology, University of Queensland, St Lucia, Brisbane, Queensland 4072, Australia

A. S. Lamantia Department of Neurobiology, Medical Center, Duke University, Box 3209, Durham, NC 27710, USA

P. Levitt Department of Neuroscience and Cell Biology, Robert Wood Johnson Medical School–UMDNJ, 675 Hoes Lane, Piscataway, NJ 08854, USA

L. Maffei Istituto di Neurofisiologica del CNR, Via S. Zeno 51, I-56127, Pisa, Italy

Z. Molnár (*Bursar*) Department of Physiology, University of Oxford, South Parks Road, Oxford OX1 3PT, UK

R. Murray Department of Psychological Medicine, Institute of Psychiatry, De Crespigny Park, Denmark Hill, London SE5 8AF, UK

D. D. M. O'Leary Molecular Neurobiology Laboratory, The Salk Institute, 10010 N Torrey Pines Road, La Jolla, CA 92037, USA

J. G. Parnavelas Department of Anatomy and Developmental Biology, University College London, Gower Street, London WC1E 6BT, UK

D. J. Price Department of Physiology, The University of Edinburgh Medical School, Teviot Place, Edinburgh EH8 9AG, UK

J. Price SmithKline Beecham Pharmaceuticals, New Frontiers Science Park, Harlow, Essex CM19 5AW, UK

P. Rakic Department of Neurobiology, Yale University, School of Medicine, C303 SHM, PO Box 3333, New Haven, CT 06510–8001, USA

G. W. Roberts Department of Molecular Neuropathology, SmithKline Beecham Pharmaceuticals, New Frontiers Science Park, Harlow, Essex CM19 5AW, UK

J. L. Rubenstein Department of Psychiatry, University of California at San Francisco, Neurogenetic Laboratory, 401 Parnassus Avenue, San Francisco, CA 94143, USA

M. Schmutz K-125.11.08, Ciba Geigy Ltd, CH-4002 Basle, Switzerland

C. Walsh Neurogenetics laboratory, Beth Israel Hospital/Harvard Medical School, Alpert 347, 200 Longwood Avenue, Boston, MA 02115, USA

Introduction: mysteries in the making of the cerebral cortex

Colin Blakemore

University Laboratory of Physiology, Parks Road, Oxford OX1 3PT, UK

The human cerebral cortex contains probably about half the neurons in the brain. It is a vast sheet of biological computing power, some 3 mm thick and 1600 cm^2 in area. Each neuron receives up to tens of thousands of synapses. The numbers of interconnections within the cortex and, therefore, computational operations that it could potentially perform are astounding. Perhaps the most formidable task faced by developmental neurobiology is to give an account of how this remarkable neuronal machine is constructed.

The Ciba Foundation symposium that is published in this volume provided a timely opportunity to define the current issues, to air controversies and to arrive at a consensus about at least some aspects of the development of the cerebral cortex. The symposium, held late in 1994, coincided with a significant anniversary in the history of this field. 330 years ago, in 1664, two books were published that stamped an indelible mark on this subject. In France, the French edition of René Descartes's *Traité de l'Homme*, published posthumously, made his revolutionary ideas accessible to a wide audience. Although this book continued to promote Descartes's notion of what would now be called 'interactive dualism', with the fingers of the soul prodding the machinery of the brain at times of voluntary choice or ethical dilemma, the overwhelming message was radically mechanistic:

> *I would like you to consider that all the functions I attribute to this machine, such as digestion . . . nutrition...respiration, waking and sleeping; the reception of light, sounds, odours . . .; the impression of ideas in the organ of common sense and imagination; the retention of those ideas in memory; the inferior stirrings of the appetites and passions; and finally the movements of all the external members . . .; I say that I would like you to consider that these functions occur naturally in this machine solely through the disposition of its organs, no less than the movements of a clock.*

This view, that the physical machinery of the brain can determine even the highest, cognitive functions of the mind remains a tenet of faith in neuroscience. In the context of this belief, the cerebral cortex, given the complexity of its organization, clearly presents the greatest challenge to understanding how such deterministic mechanisms are implemented.

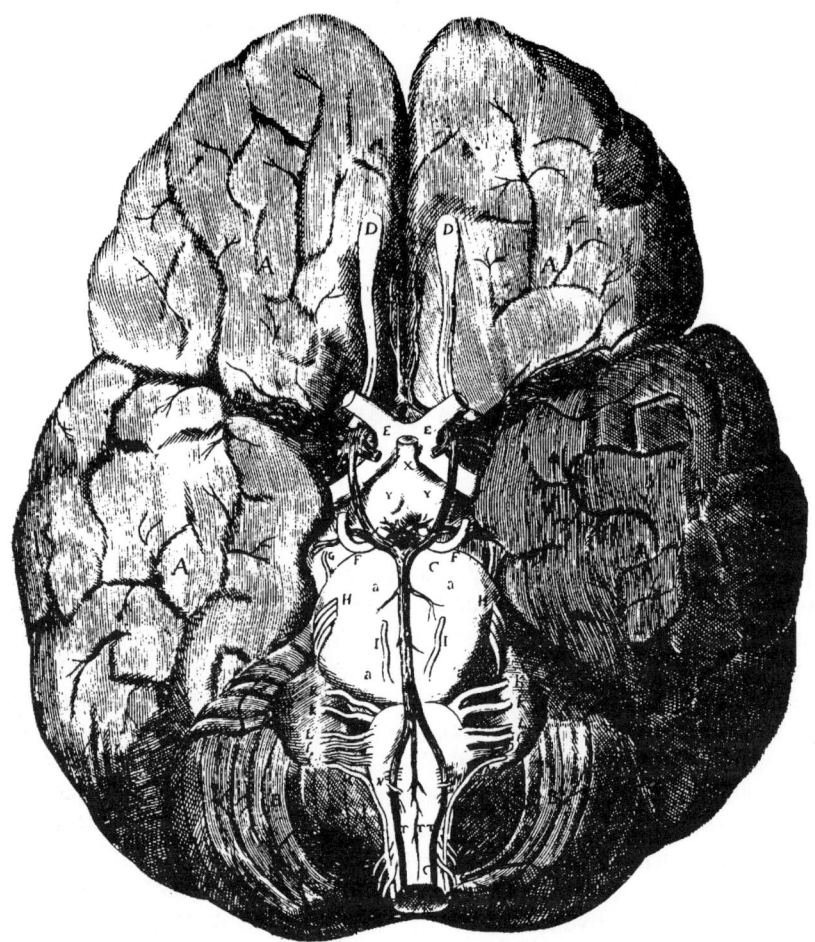

FIG. 1. This fine illustration from Thomas Willis's *Cerebri Anatome* (1664) was probably drawn by Christopher Wren, then Professor of Astronomy in Oxford, who worked with Willis. It shows the undersurface of the human brain, with its enormous cerebral cortex (A).

The other seminal publication in 1664 was the *Cerebri Anatome* of Thomas Willis, the Court Physician who lived and worked in Oxford. Until that book, all the speculations about brain function in western science since the time of Galen (129–199 AD), including Descartes', had focused on the role of the cerebrospinal fluid. Following the Aristotelian emphasis on fluids within the body as the basis of its functions and diseases, special significance was attributed to this clear fluid, then called 'animal spirit', within the ventricles of the brain. The fluid was supposed to contain the mechanisms of sensation, imagination, reason, memory and the control of movement. According to the 'cell doctrine', which remained dominant throughout the Middle Ages and up to the end of the 17th century, these 'mental' processes were thought to take place within the cerebrospinal fluid, in sequence, from the lateral ventricles (the First cell) to the fourth ventricle (the Third cell). The revolutionary contribution of Willis's book (obvious though it seems today) was the proposal that the tissue of the brain—the cells and connections between them—perform the important functions. Moreover, Willis, who, together with his collaborator Christopher Wren, dissected the brains of animals and humans, argued that the cerebral hemispheres, being so clearly enlarged in human beings, must be responsible for those functions that are specially developed in man—perception, thought, intelligence, memory and all voluntary functions (see Fig. 1). That view is still dominant in brain research.

Functional subdivisions

As early as the fifth century BC, Alcmaeon of Croton, one of the founders of neurology, suggested that each major sense has its own territory within the brain fed by a conduit (sensory nerve) from the sense organ, as well as a *sensorium* (or *sensus*) *communis* within which sensory information is integrated. Alcmaeon also considered sensation and thought (or understanding) to be separate processes and proposed that humans alone have the power of understanding (Rose 1994). However, the classical cell doctrine did not recognize subdivisions within the *sensorium communis* of the First cell. Neither did Willis go so far as to suggest that the various functions that he ascribed to the cerebral hemispheres were carried out by different parts of the cortex. Indeed, the notion that the cortex operates as an entity, performing 'mass action', was espoused by such eminent figures as Marie-Jean-Pierre Flourens in the 19th century and Karl Lashley in this century. But it has gradually become accepted that the cortex is divided and subdivided into a variety of functional units, some sharply delineated and segregated, others overlapping.

First, and most obviously, the thickness of the mammalian cortex is split into layers—classically six of them for the neocortex. The layering, which is visible in microscopic sections, even when unstained and to the naked eye for some regions of the cortex, is produced by variation in the types and density of cell bodies and fibres through the cortical depth. There is clear specialization of function among the layers. The major input to the cortex from the thalamus terminates in layer IV; the pyramidal neurons of the supragranular layers usually project to other regions of the cortex, including the opposite hemisphere; those of the lower layers tend to project out of the telencephalon, with distinct cell types sending their axons to such structures

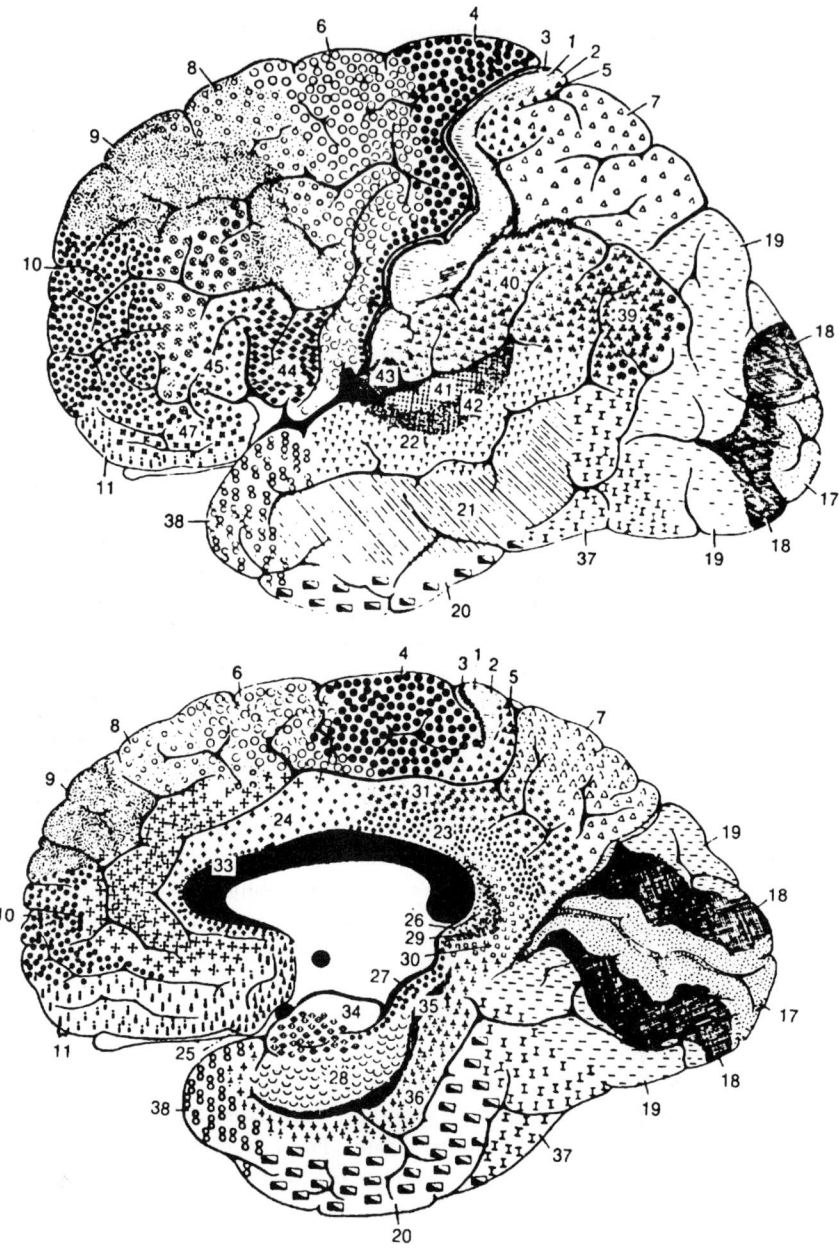

as the thalamus, the claustrum, the corpus striatum, and the brainstem and spinal cord.

Arranged tangentially, orthogonal to the lamination, are a variety of subdivisions and modular arrangements. In particular, the cortex is parcelled into distinct territories or fields, with more or less sharp boundaries between them. After the discredited efforts of the phrenologists, from the end of the 18th century, to assign such functions as 'sublimity', 'benevolence' and even 'Republicanism' to discrete 'organs' across the surface of the brain, the genuine evidence for localization of function in the cerebral cortex grew rapidly and inexorably (see Clarke & O'Malley 1968, Blakemore 1977). Broca's and Wernicke's accounts of areas concerned with different aspects of language; Hughlings Jackson's description of the motor cortex; the discovery of regions for hearing and vision: all these attributions of functional responsibility came originally through the time-honoured approach of inferring normal function from the symptoms of brain damage or disorder in human patients.

The gamut of techniques now available to neuroscience has amply confirmed the doctrine of localization of function in the cortex. Observation of the results of electrical stimulation of the human cortex; behavioural studies of animals after cortical lesions; electrophysiological recording in anaesthetized and alert, trained animals; and now neuroimaging studies on normal humans and clinical patients: all these approaches have complemented the view gained from neuropsychological assessment of the consequences of brain damage, to give us a coherent and convincing picture of the cerebral cortex as a patchwork of distinct functional areas, each committed to participation in the processing of a particular aspect of a sensory input or motor output, the storage of memories or the planning of behaviour. The multitude of highly specific connections between these areas creates a number of parallel, but partly interconnected 'streams' for information processing (see Felleman & Van Essen 1991, Young et al 1994).

At least some of these functional subdivisions coincide with subtle local variations on the common anatomical themes of the cortex. The striking appearance of the stria of Gennari, which characterizes the striate or primary visual cortex, was reported as early as 1782 (Glickstein 1988). Other reports of regional differences in the histological appearance of the cortex accumulated throughout the last century, culminating in Korbinian Brodmann's (1909) remarkable cytoarchitectonic studies of the cortex (Fig. 2). Largely on the basis of only nissl-stained sections and primitive fibre staining, he was able to delineate a large number of distinct fields according to the relative thickness of cell and fibre layers, with sharp boundaries between neighbouring areas (an

---

FIG. 2. Brodmann's (1909) cytoarchitectonic fields of the human brain: (*top*) lateral view and (*bottom*) medial surface of left hemisphere.

achievement that astounds anyone who looks down the microscope at sections of the cortex, even with Brodmann's descriptions to guide them). Moreover, Brodmann and others were sufficiently confident of the absolute characteristics of many of the major areas that they employed the same numbers for what they considered to be equivalent fields in a variety of species. It is a testimony to Brodmann's extraordinary powers of observation that his numbers are still widely employed to describe many of the major divisions of the cortex, from rodents to humans.

Regional fields are not the only form of tangential specialization. The more that individual fields are investigated with sophisticated anatomical and physiological methods, the more that they are seen to be further divided into arrays or mosaics of specialized subdivisions. Within radial columns, neurons generally share some functional property, with physiological characteristics varying from column to column. Other tangentially distributed groupings can overlap the columnar organization.

The most extreme example of tangential modularity (perhaps because it has been most intensively studied) is the primary visual cortex of Old World monkeys, where cells are generally selectively sensitive to the orientation of line stimuli, respond through either eye and are sometimes chromatically sensitive. Neurons in individual radial columns share the same preferred orientation, which shifts systematically from column to column. There is a second set of columns or slabs, the ocular dominance columns, superimposed on the orientation columns, in which cells are dominated by one eye or the other, reflecting the strictly segregated distribution of right-eye and left-eye afferent axons terminating in layer IV (Hubel & Wiesel 1977). Along the length of each ocular dominance slab, in the supragranular layers is a regular pattern of 'blobs', recognized by a high level of expression of the mitochondrial enzyme cytochrome oxidase, within which neurons tend to be chromatically selective and have poor orientation selectivity (Livingstone & Hubel 1984).

The nature of regional differentiation

What is the basis and significance of cytoarchitectonic variation? The local differences in patterns of lamination seen by Brodmann are caused and/or accompanied by a variety of other regional characteristics. The proportions of different morphological classes of cortical neuron vary from field to field (see Peters & Jones 1984). For example, primary sensory areas are characterized by a relatively thick granular layer IV, containing non-pyramidal (stellate) and/or small pyramidal cells, whereas motor areas have a thicker layer V, with distinctive large pyramidal neurons (Betz cells), many of which project into the corticospinal tract.

Although certain aspects of intrinsic circuitry may be similar throughout the neocortex (Douglas & Martin 1991), the regional variations in morphology presumably reflect local differences of microcircuitry as well as the proportions of neurons projecting to different targets. In some cases, the differences in output projections between cortical fields are more extreme. For instance, in advanced mammals such as primates and carnivores, the striate cortex, unlike most other cortical fields, neither sends nor receives interhemispheric connections, nor does it project to the spinal cord.

Even the small modular subdivisions within an individual field can vary in the pattern of their projections. For instance, cells in the cytochrome oxidase blobs of the monkey striate cortex project to particular subdivisions of the second visual area, distinct from those to which the neurons between the blobs project (e.g. Hubel & Livingstone 1987).

Brodmann himself (1909) had no doubt about the meaning of cytoarchitectonic variation:

The specific histological differentiation of cortical areas proves irrefutably their specific functional differentiation.

The most distinctive difference between major cortical areas, and the one that most obviously determines their functional specialization, is the afferent input that they receive from the thalamus. The thalamus transmits nearly all input destined for the cortex. The sense organs (except the olfactory) and major subcortical motor centres each provide input to one or more specific thalamic nuclei, which have well-defined reciprocal interconnections with particular cortical areas. In rodents the entire neocortex receives thalamic input (Caviness & Frost 1980) and even in the adult there is rather precise isomorphism between the topographic arrangement of thalamic nuclei and the cortical territories to which they project (Caviness 1988). In primates too, most if not all areas of the neocortex receive thalamic input, and the main sensory cortical areas are dominated by input from primary relay nuclei in the thalamus. However, corticocortical projections are more important in determining the functional status and specialization of the many non-primary fields that flank the major sensory areas.

Formation of the cortical plate

The basic principles of cortical neurogenesis are similar in all mammals (see Uylings et al 1990). The neurons and glia that will eventually form the cortex originate in the pseudostratified proliferative neuroepithelium that lines the forebrain vesicle. The cells are born over an extensive fraction of prenatal life in eutherian mammals (seven days in rats; about 100 days in humans). At early stages most are produced in the ventricular zone, which lies immediately at the surface of the ventricles. A separate population of progenitor cells, which accumulates further from the ventricular surface to form the subventricular zone, gradually takes over the generation of committed cells.

Newborn neurons migrate towards the margin of the cerebral wall along the processes of radial glial cells to form the cortex (see Rakic 1981). In general, they accumulate in an inside-out sequence, each wave of arriving cells migrating through the existing cells of the cortical plate before stopping at the top of the plate (see Berry & Rogers 1965, Lund & Mustari 1977, Rakic 1981). However, the earliest-generated cells are an interesting exception to this general rule: they accumulate at the outer edge of the cerebral wall as a rather undisciplined layer, the primordial plexiform zone (Marin-Padilla 1971) or preplate, which is then split, by the arrival of true cortical neurons, into the marginal zone (cortical layer I) at the pial surface and the subplate below the accumulating cortical plate (see Fig. 3: nomenclature according to the Boulder committee 1970). Thus, the marginal zone disobeys the general inside-out rule. The cortical plate proper, sandwiched between the two components of the preplate, steadily thickens as subsequent waves of arriving neurons migrate through the existing cells of the subplate and plate and take up their position under the marginal zone (Luskin & Shatz 1985a,b).

Neurons of the preplate are unusual in a variety of ways, all of which suggest that they might perform an important function during development. They express transmitter and receptor systems precociously; they send the earliest descending axons out of the cortex; and a substantial fraction of them die, around the time of completion of cortical neuronogenesis (see Shatz et al 1988). Another indication that cells of the subplate play an important developmental role is the fact that thalamic axons, which arrive early in the process of corticogenesis, accumulate within the subplate layer for a waiting period, lasting between a few days in rodents to several weeks in primates (Lund & Mustari 1977, Rakic 1976, Shatz & Luskin 1986), before they invade the cortex proper. Although in parts of the rodent cortex, some thalamic axons appear to penetrate the lower layers of the true cortical plate soon after they arrive (Catalano et al 1991), there is evidence that at least some thalamic fibres actually form temporary synaptic connections on subplate cells during the waiting period (Herrmann et al 1994, Friauf & Shatz 1991).

The possibility that the early corticofugal projection from subplate cells constitutes a scaffold that helps to guide thalamocortical and perhaps even other corticofugal projections, and that interaction of thalamic fibres with subplate cells is essential for the subsequent innervation of the cortex, are topics of considerable current interest and controversy (see this volume: Molnár & Blakemore 1995, Bolz et al 1995, Ghosh 1995).

What determines areal and modular specialization?

It is not until after the completion of neuronal migration (after birth in rodents) that the obvious regional differences in cortical cytoarchitectonics

Introduction

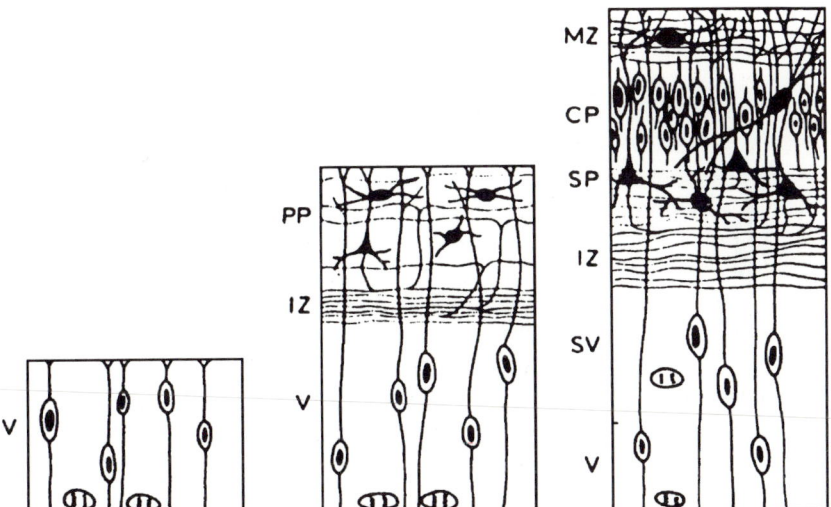

FIG. 3. These three panels illustrate cross-sections of the cerebral wall at different stages in the formation of the cortex. Initially (*left panel*), progenitor cells in the ventricular zone (V) undergo symmetrical division, increasing the area of the neuroepithelium. They move cyclically, up and down, with mitosis occurring close to the ventricular surface (*bottom*). Corticogenesis starts (*middle panel*) with the first asymmetric divisions, producing neurons (drawn as filled profiles) that migrate upwards along the processes of radial glia (see Fig. 4) to the outer edge of the cerebral wall to form the primordial plexiform zone (Marin-Padilla 1971) or preplate (PP). Between the ventricular zone and the preplate is a thickening layer of transverse fibres, the intermediate zone (IZ). At later stages (*right panel*), progenitor cells also accumulate in a second proliferative layer, the subventricular zone (SV). Neurons of the true cortical plate (CP) migrate up and take their place, in an inside-out order, between the outer part of the preplate (the marginal zone, MZ, containing the Cajal–Retzius cells) and the lower part, the subplate (SP), containing the bulk of the early-generated cells. A few presumed preplate cells, with characteristic bipolar or multiform morphology, extend obliquely across the thickening cortical plate. (Modified from Uylings et al 1990.)

become evident. At very early stages the cortex appears more or less homogeneous. How and when is regional specialization determined?

According to Reznikov et al (1984) and Rakic (1988) there is a fate-mosaic or proto-map across the proliferative cells of the ventricular zone: the characteristics of each cortical area are determined by the genetic commitment of the cells delivered to it by the germinal zone, to which it is tied by the migration pathway of the radial glial processes (Fig. 4). This hypothesis fitted well with the prevailing view at that time: that migration is strictly radial and that each radial fascicle or 'mini-column' of neurons visible in the adult cortex is the product of a fixed 'proliferative unit' lying in the corresponding region of the germinal zone (Smart & McSherry 1982, Rakic 1988).

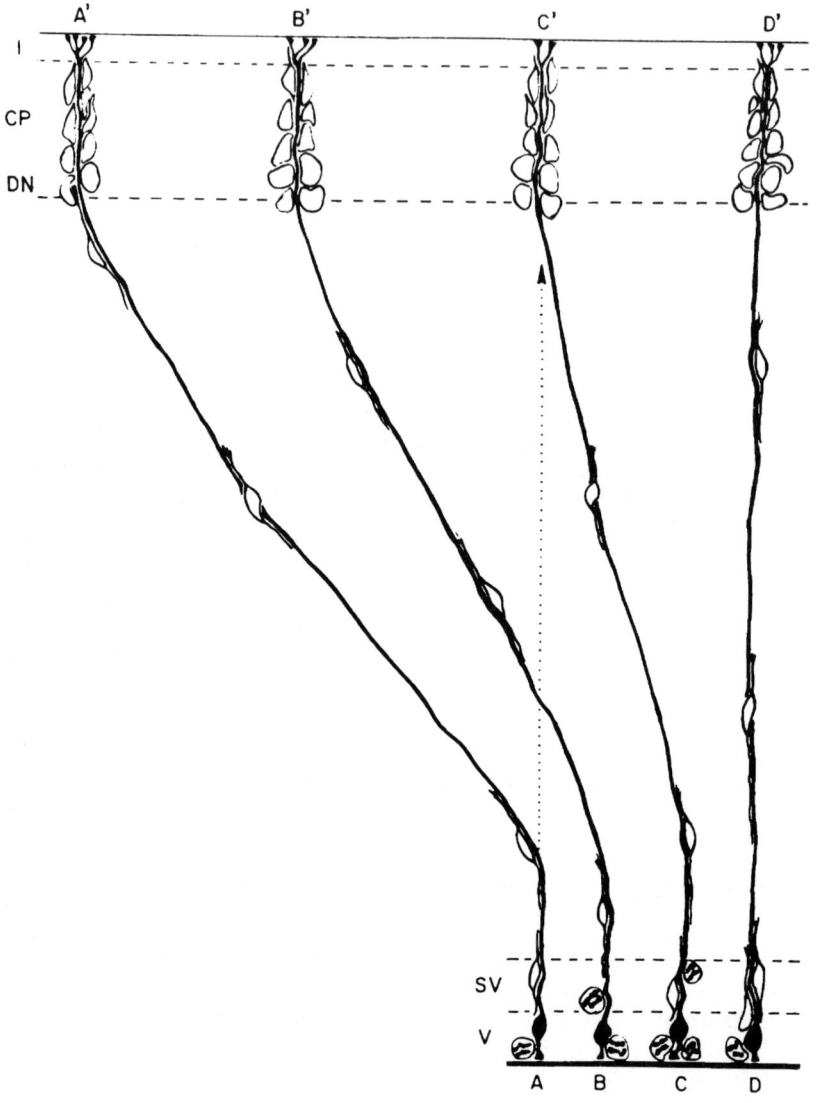

Introduction

Others (e.g. Driesch 1914, O'Leary & Stanfield 1989, O'Leary 1989) have proposed that the cortical plate is initially a homogeneous protocortex—a more or less equipotential sheet, committed only to a basic pattern of laminar organization but with little or no inherent tendency to regional differentiation. According to this theory, areal specificity depends on an extrinsic signal, which triggers regional differentiation and determines the characteristic local discrimination and target-specific outgrowth from the cortex. The most obvious candidate for such an extrinsic signal, at least for the primary sensory areas, is the thalamic innervation, which arrives very early in development and is more or less precisely distributed to the appropriate cortical areas.

Probably neither extreme hypothesis is entirely acceptable. In favour of the suggestion that some aspects of local fate are specified in the neuroepithelium are: (1) McConnell & Kaznowski's (1991) finding that the cortical lamina for which each immature neuron is destined is determined very early during mitosis; and (2) the discovery of selective gene expression in the germinal neuroepithelium at very early stages (see Boncinelli et al 1995, this volume). However, the regions over which specific genes are precociously expressed do not generally correspond to particular functional fields. Moreover, it is now clear that certain progenitor cells can move freely in the germinal zone. By marking clonally related cells with retrovirus-mediated transfer of a histochemical marker gene, Walsh & Cepko (1992, 1993) demonstrated widespread dispersion of individual clones across different functional regions of the cerebral cortex (see Walsh & Reid 1995, this volume). Nevertheless, the numerically dominant pattern of clonal distribution is radial (Tan & Breen 1993) and the radially distributed clones or sub-clones are usually of a single neuronal or glial type (see this volume: J. Price et al 1995, Parnavelas et al 1995, Walsh & Reid 1995).

Trivial support for the protocortex theory comes from the fact that thalamic axons become incorporated within the cytoarchitecture of the area they innervate. Local anatomical features that directly reflect the thalamic input,

---

FIG. 4. Rakic's hypothesis of the formation of 'ontogenetic columns' through migration of neurons along radial glial cells. Progenitor cells in four different 'proliferative units' at locations A, B, C and D in the ventricular (V) and subventricular (SV) zones are connected to corresponding positions (A', B', C' and D') in the cortical plate (CP) by the processes of radial glia. Newly arrived neurons migrate past existing deep neurons (DN) in the plate to take up their position below cortical layer I. Despite the expansion of the cortex, the topographic relationship between cortex and proliferative zone is maintained by the integrity of the glial guides, preventing a mismatch such as that indicated by the dotted line and arrow between A and C'. However, recent evidence suggests that at least some progenitor cells migrate laterally within the neuroepithelium over considerable distances (see Walsh & Reid 1995, this volume), and that migrating neurons can jump from one radial glial fibre to another, thus blurring the correspondence between cortex and proliferative zone. (Reproduced with permission from Rakic 1981.)

such as the segregated thalamic terminal distributions that make up the ocular dominance stripes in layer IV of the primate striate cortex (Hubel & Wiesel 1977) and the 'barrel' specializations in rodent somatosensory cortex, corresponding to the input from individual vibrissae (Van der Loos & Woolsey 1973), obviously depend on the thalamic innervation. It is hardly surprising that such features do not appear until after the invasion of the cortical plate by thalamic axons (e.g. Rakic 1976, Jhaveri et al 1991).

In the monkey, removal of both eyes early in fetal development, which causes a reduction in the number of neurons in the corresponding thalamic nucleus, the lateral geniculate nucleus, results in a corresponding reduction in the size of the striate cortex (recognized by its distinctive cytoarchitecture) and a change in the entire sulcal pattern of the occipital lobe (Rakic 1988).

Perhaps the most dramatic evidence that the thalamic input plays a regulatory role comes from experiments involving transplantation of cortex from one region to another in the rat (see O'Leary et al 1995, this volume). Schlaggar & O'Leary (1991) demonstrated that late-embryonic occipital (putative visual) cortex transplanted into the position of somatosensory cortex in the rat develops barrels histologically indistinguishable from normal. Similar procedures suggested that the distinctive differences in the efferent projections of cortical regions are also influenced by the particular thalamic input that they receive. If an explant of cortex from the occipital region of a late-embryonic rat is transplanted to a more rostral region of the hemisphere of a neonatal host, it adopts the output characteristics of the host site, retaining its projection to the spinal cord (O'Leary & Stanfield 1989).

How might local fate be regulated epigenetically?

An extreme version of the proto-map hypothesis would attribute local differences of cell type and number, laminar distribution of particular transmitter receptors, projection patterns and so on, to patterns of gene expression imposed on the daughter neurons of each region of the germinal epithelium. However, there are a variety of other ways in which local specialization might be achieved through extrinsic influences acting epigenetically.

(l) The first is through the regulation of cell death. There is substantial death of neurons in the cortical plate itself early in development (e.g. Heumann & Leuba 1983). Differential elimination of various types of neurons could obviously lead to local differences in cytoarchitecture and microcircuitry (Price & Blakemore 1985, Finlay & Pallas 1989). Such local variation in cell death could be regulated entirely inherently, as part of the genetic program, but it might also be influenced by transmitters or other factors produced by afferent axons, or by factors transported back from target structures along the axons of cortical cells.

Introduction

(2) The morphological and functional characteristics of neurons in the cortex might also be influenced by substances transported retrogradely from target structures.

(3) Thalamic (or other afferent) axons might, through the production of transmitters or other factors, directly influence gene expression in their targets, leading to structural changes such as spine formation on dendrites (Valverde 1967), or more subtle differences such as the production of channels or receptors by the target neurons.

(4) Initial 'exuberant' axon distribution followed by selective withdrawal of axons from particular target structures appears to occur in many projection systems (see Kind & Innocenti 1990). Moreover, some of this selective pruning of connections appears to be epigenetically regulated, in that it depends on the activity passing through the pathway during the period of elimination. For instance, in immature rats and cats (but interestingly not in Old World monkeys) there are dense interhemispheric projections to and from the body of the striate cortex, which are gradually eliminated after the formation of the cortical plate, and the final pattern of this projection can be influenced by disturbances of binocular vision early in life (see Kind & Innocenti 1990). The withdrawal of the early projection from the rat striate cortex to the spinal cord, reported by O'Leary & Stanfield (1986), is another example of refinement on the basis of selective pruning of an exuberant projection; although it must be said that Oudega et al (1994) have recently reported that the striate cortex lacks a corticospinal projection right from the start.

(5) The characteristics of the target neurons, the distribution of thalamic terminals, the synapses between them, the morphology and function of cortical cells, and the circuitry to which they contribute might all be influenced by activity arriving in the cortex along thalamic afferents and distributed through cortical circuits. There are now innumerable examples of structural and functional plasticity in the neocortex depending on the pattern of afferent stimulation during particular critical times during development. For example, reduction in afferent activity from the vibrissae of a mouse during the first postnatal week (by cauterizing some of the whisker follicles or blocking the inflow of sensory afferentation by cutting the trigeminal nerve) prevents formation of the barrels corresponding to the missing sensory input (Van der Loos & Woolsey 1973, Welker & Van der Loos 1986). And the pattern of ocular dominance stripes in layer IV of the monkey can be modified by selective deprivation of one eye (e.g. LeVay et al 1980, Swindale et al 1981). Synaptic plasticity, often due to long-term potentiation and/or depression (e.g. Artola et al 1990, Fregnac et al 1992), mediated at least in part via the N-methyl-D-aspartate receptor (e.g. Komatsu & Toyama 1988, see Daw et al 1995, this volume), is now recognized as a hallmark of the neocortex. The learning algorithms that such plasticity can support might contribute an essential part to the computational machinery of the cortex, not only during early development, but throughout life.

14

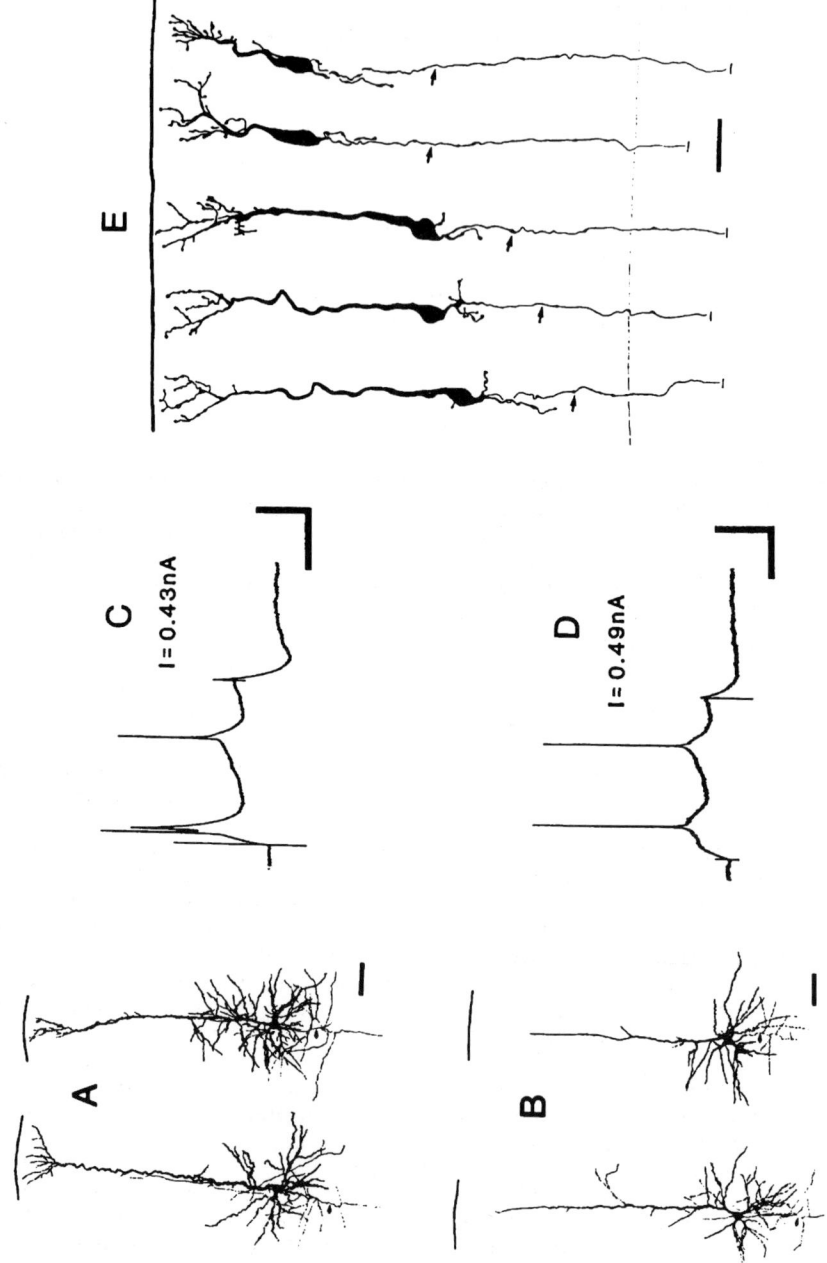

Introduction 15

The magnitude of the problem of selective differentiation

The chapters in this volume show that considerable progress has been made in defining the factors, inherent and epigenetic, that control the differentiation of cortical cells (see McConnell 1989). However, it would be wrong to trivialize the problem that remains. Even before attempting to account for regional differences, the task is no less than to explain how every distinguishable type of cortical neuron is generated. We do not even know how many distinct categories of cells exist, but it is certainly very large. There are at least four major types of morphologically different smooth stellate cells, containing γ-aminobutyric acid (GABA) and presumed to be inhibitory (see Peters & Jones 1984), but with subclassifications according to layer and so on, the number of distinct inhibitory interneurons could easily be as many

FIG. 5. In layer V of the rat striate cortex there are two major types of pyramidal neuron (and similar classes have been described in other species). One type, which projects out of the telencephalon to the superior colliculus etc., has a thick apical dendrite with a prominent terminal tuft, and fires a distinctive burst of action potentials when injected with current. The other type, which projects to other cortical areas, including the opposite hemisphere, has a slender, untufted apical dendrite and produces more-regular trains of impulses. A and B illustrate examples of the soma–dendritic morphology of corticotectal (A) and interhemispheric neurons (B), which were initially back-labelled from their target structures by injection, *in vivo*, of fluorescent microspheres into the superior colliculus and the visual cortex of the opposite hemisphere, respectively. These prelabelled cells were then identified in fixed slices of the cortex and were filled with Lucifer Yellow dye, which was photo-oxidized for the preparation of these camera lucida drawings. These particular examples came from a postnatal day (P) 10 rat, at which stage the morphological differentiation between the two classes is virtually complete. Bars = 50 μm. C and D show typical firing patterns for identified corticotectal and interhemispheric cells, respectively. Neurons in adult rats, back-labelled from their target as described above, were visualized in living slices and impaled for intracellular recording. Each trace shows the change in membrane potential and the action potentials produced by the injection of a 250 ms current pulse (current strength indicated). Note the characteristic two-impulse burst shortly after current onset in the corticotectal cell (C). Bars = 25 mV (*vertical*) and 50 ms (*horizontal*). For the first two weeks of postnatal life, no corticotectal cells exhibit such bursting. The first start to burst at P15 and all of them demonstrate bursting by P21. Although the axons of these two classes of cells appear to grow correctly towards the appropriate target shortly after they have migrated into the cortical plate (some days before birth), the soma–dendritic morphologies do not begin to diverge until about P5. E shows camera lucida drawings of pyramidal-like cells filled in fixed slices from the rat cortex on embryonic day 21, when only layers V and VI are in place below the superficial dense cortical plate of immature neurons. At this stage, when the axons of layer V pyramids have already reached their targets, all the cells have similar soma–dendritic morphology, with thick tufted apical dendrites. Arrows indicate axons. Bar = 20 μm. How might the various steps in the differentiation of these two types of co-generated neurons be regulated? (A, B and E: modified from illustrations in Kasper et al 1994b, and C and D modified from Kasper et al 1994a.)

as 20. There might easily be similar numbers of sharply delineated spiny stellate (non-pyramidal) neurons and even more pyramidal neurons, defined by the lamina in which they lie, their biophysical characteristics, the morphology of their dendrites, their local axon collateral distribution and the targets of their cortifugal axons.

Consider one striking example. In rodents (and probably in many other species) pyramidal cells of layer V of the visual cortex fall into two classes (Fig. 5) that can be distinguished on the basis of a variety of criteria in the adult (see Kasper et al 1994a). One type has a thick apical dendrite with a prominent terminal tuft extending into layer I, it produces distinctive bursts of action potentials when injected with current, and its axon projects out of the telencephalon to the superior colliculus (and often also to the pontine nuclei). The other type has a slender apical dendrite without a tuft, produces more-regular trains of impulses and projects to other cortical areas, including the opposite hemisphere.

The cell bodies of these two types can lie immediately adjacent to each other in layer V, implying that they are co-generated in the germinal epithelium. Moreover, pyramidal neurons of layer V appear indistinguishable when they first migrate into position (all have stout apical dendrites with terminal tufts) and none of them fires bursts. Yet the two types immediately start to express their differential characteristics by spinning out their axons unerringly towards the appropriate target structure three or four days before birth. Soma-dendritic morphology does not diverge until several days after birth, and it is not until 14 days postnatally that the first corticotectal pyramidal cells develop the burst-firing characteristic (Kasper et al 1994b). The main defining properties of these two classes of neuron are presumably specified at a very early stage, but emerge in these distinct steps over a period of some weeks. Moreover, at least some of their properties are epigenetically influenced, because these corticotectal cells in the visual cortex, which in the normal mature animal do not project to the spinal cord, will do so if the cortex in which they reside is transplanted to a more rostral region of the hemisphere, from which layer V normally contributes to the corticospinal tract (O'Leary & Stanfield 1989).

The stimulus selectivities of sensory cortical neurons (such as the preferences for orientation, direction, velocity, spatial frequency and disparity in the primary visual cortex) obviously depend on the sum of their inputs. If one includes those characteristics as part of the criteria for the classification of neurons, the number of distinctly differentiated types would probably rise to thousands or tens of thousands. How can such a rich diversity be specified? This volume shows that much progress has been made, yet we are still a long way from defining the interplay of prespecified and epigenetic influences that determines the fate of each class of cortical neuron and glial cell and, hence, the specialization and function of each cortical area.

References

Artola A, Bröcher S, Singer W 1990 Different voltage-dependent thresholds for inducing long-term depression and long-term potentiation in slices of the rat visual cortex. Nature 347:69–72

Berry M, Rogers AW 1965 The migration of neuroblasts in the developing cerebral cortex. J Anat 99:691–709

Blakemore C 1977 Mechanics of the mind. Cambridge University Press, Cambridge

Bolz J, Kossel A, Bagnard D 1995 The specificity of interactions between the cortex and the thalamus. In: Development of the cerebral cortex (Ciba Found Symp 193) p 173–191

Boncinelli E, Gulisano M, Spada F, Broccoli V 1995 *Emx* and *Otx* gene expression in the developing mouse brain. In: Development of the cerebral cortex (Ciba Found Symp 193) p 100–116

Brodmann K 1909 Vergleichende Lokalisationslehre der Groshirnrinde in ihren Prinzipien dargestellt auf Grund des Zellenbaues. J. A. Barth, Leipzig

Boulder committee: Angevine JB Jr, Bodian D, Coulombre AJ et al 1970 Embryonic vertebrate central nervous system: revised terminology. Anat Rec 166:257–261

Catalano SM, Robertson RT, Killackey HP 1991 Early ingrowth of thalamocortical afferents to the neocortex of the prenatal rat. Proc Natl Acad Sci USA 88:2999–3003

Caviness VS Jr 1988 Architecture and development of the thalamocortical projection in the mouse. In: Bentivoglio M, Spreafico R (eds) Cellular thalamic mechanisms. Elsevier Science, Amsterdam, p 489–499

Caviness VS Jr, Frost DO 1980 Tangential organization of thalamic projections of the neocortex in the mouse. J Comp Neurol 194:355–367

Clarke E, O'Malley CD 1968 The human brain and spinal cord: a historical study illustrated by writings from antiquity to the twentieth century. University of California Press, Berkeley, CA

Daw NW, Reid SNM, Wang X-F, Flavin HJ 1995 Factors that are critical for plasticity in the visual cortex. In: Development of the cerebral cortex (Ciba Found Symp 193) p 258–276

Douglas RJ, Martin KAC 1991 Opening the grey box. Trends Neurosci 14:286–293

Driesch H 1914 The history and theory of vitalism. Macmillan, London

Felleman DJ, Van Essen DC 1991 Distributed hierarchical processing in the primate visual cortex. Cereb Cortex 1:1–47

Finlay BL, Pallas SL 1989 Control of cell number in the developing mammalian visual system. Prog Neurobiol 32:207–234

Fregnac Y, Schulz D, Thorpe S, Bienenstock E 1992 Cellular analogs of visual cortical epigenesis. I. Plasticity of orientation selectivity. J Neurosci 12:1280–1300

Friauf E, Shatz CJ 1991 Changing patterns of synaptic input to subplate and cortical plate. J Neurophysiol 66:2059–2071

Fox K, Daw NW 1993 Do NMDA receptors have a critical function in visual cortical plasticity? Trends Neurosci 16:116–122

Ghosh A 1995 Subplate neurons and the pattterning of thalamocortical connections. In: Development of the cerebral cortex (Ciba Found Symp 193) p 150–172

Glickstein M 1988 The discovery of the visual cortex. Sci Am 256:118–127

Herrmann K, Antonini A, Shatz CJ 1994 Ultrastructural evidence for synaptic interactions between thalamocortical axons and subplate neurons. Eur J Neurosci 6:1729–1742

Heumann D, Leuba G 1983 Neuronal death in the development and aging of the cerebral cortex of the mouse. Neuropathol Appl Neurobiol 9:297–311

Hubel DH, Livingstone MS 1987 Segregation of form, color, and stereopsis in primate area 18. J Neurosci 7:3378–3415

Hubel DH, Wiesel TN 1977 Functional architecture of macaque monkey visual cortex. Proc R Soc Lond B Biol Sci 198:1–59

Jhaveri S, Erzurumlu RS, Crossin K 1991 Barrel construction in rodent neocortex: role of thalamic afferents versus extracellular matrix molecules. Proc Natl Acad Sci USA 88:4489–4493

Kasper EM, Larkman AU, Lubke J, Blakemore C 1994a Pyramidal neurons in layer 5 of the rat visual cortex. I. Correlation among cell morphology, intrinsic electrophysiological properties and axon targets. J Comp Neurol 339:459–474

Kasper EM, Lubke J, Larkman AU, Blakemore C 1994b Pyramidal neurons in layer 5 of the rat visual cortex. III. Differential maturation of axon targeting, dendritic morphology and electrophysiological properties. J Comp Neurol 339:495–518

Kind P, Innocenti GM 1990 The development of cortical projections. In: Raymond PA, Easter SS Jr, Innocenti GM (eds) Systems approaches to developmental neurobiology. Plenum, New York, p 113–152

Komatsu Y, Toyama K 1988 Relevance of NMDA receptors to the long-term potentiation in kitten visual cortex. Biomed Res 9:39–41

LeVay S, Wiesel TN, Hubel DH 1980 The development of ocular dominance columns in normal and visually deprived monkeys. J Comp Neurol 191:1–51

Livingstone MS, Hubel DH 1984 Anatomy and physiology of a color system in the primate visual cortex. J Neurosci 4:309–356

Lund RD, Mustari MJ 1977 Development of the geniculocortical pathway in rats. J Comp Neurol 173:289–305

Luskin MB, Shatz CJ 1985a Neurogenesis of the cat's primary visual cortex. J Comp Neurol 242:611–631

Luskin MB, Shatz CJ 1985b Studies of the earliest generated cells of the cat's visual cortex: cogeneration of subplate and marginal zones. J Neurosci 5:1062–1075

McConnell SK 1989 The determination of neuronal fate in the cerebral cortex. Trends Neurosci 12:342–349

McConnell SK, Kaznowski CE 1991 Cell cycle dependence of laminar determination in developing neocortex. Science 254:282–285

Marin Padilla M 1971 Early prenatal ontogenesis of the cerebral cortex (neocortex) of the cat (*Felis domestica*): a Golgi study. I. The primordial neocortical organization. Z Anat Entwicklungsgesch 134:117–145

Molnár Z, Blakemore C 1995 Guidance of thalamocortical innervation. In: Development of the cerebral cortex (Ciba Found Symp 193) p 127–149

O'Leary DDM 1989 Do cortical areas emerge from a protocortex? Trends Neurosci 12:400–406

O'Leary DDM, Stanfield BB 1986 A transient pyramidal tract projection from the visual cortex in the hamster and its removal by selective collateral elimination. Dev Brain Res 27:87–99

O'Leary DDM, Stanfield BB 1989 Selective elimination of axons extended by developing cortical neurons is dependent on regional locale: experiments utilizing fetal cortical transplants. J Neurosci 9:2230–2246

O'Leary DDM, Borngasser DJ, Fox K, Schlaggar BL 1995 Plasticity in the development of neocortical areas. In: Development of the cerebral cortex (Ciba Found Symp 193) p 214–230

Oudega M, Varon S, Hagg T 1994 Distribution of corticospinal motor neurons in the postnatal rat—quantitative evidence for massive collateral elimination and modest cell death. J Comp Neurol 347:115–126

Parnavelas JG, Mione MC, Lavdas A 1995 The cell lineage of neuronal subtypes in the mammalian cerebral cortex. In: Development of the cerebral cortex (Ciba Found Symp 193) p 41–58

Peters A, Jones EG (eds) Cerebral cortex, vol 1: Cellular components of the cerebral cortex. Plenum, New York

Price DJ, Blakemore C 1985 Regressive events in the postnatal development of association projections in the visual cortex. Nature 316:721–724

Price J, Williams BP, Götz M 1995 The generation of cellular diversity in the cerebral cortex. In: Development of the cerebral cortex (Ciba Found Symp 193) p 71–84

Rakic P 1976 Prenatal genesis of connections subserving ocular dominance in the rhesus monkey. Nature 261:467–471

Rakic P 1977 Prenatal development of the visual system in the rhesus monkey. Philos Trans R Soc Lond B Biol Sci 278:245–260

Rakic P 1981 Developmental events leading to laminar and areal organization of the neocortex. In: Schmitt FO (ed) The organization of the cerebral cortex. MIT Press, Cambridge, MA, p 7–28

Rakic P 1988 Specification of cerebral cortical areas. Science 241:170–176

Reznikov KY, Fulop Z, Hajos F 1984 Mosaicism of the ventricular layer as the developmental basis of neocortical columnar organization. Anat Embryol 170:99–105

Rose FC 1994 The neurology of Ancient Greece—an overview. J Hist Neurosci 3: 237–260

Schlaggar BL, O'Leary DDM 1991 Potential of visual cortex to develop an array of functional units unique to somatosensory cortex. Science 252:1556–1560

Shatz CJ, Luskin MB 1986 Relationship between the geniculocortical afferents and their cortical target cells during development of the cat's primary visual cortex. J Neurosci 6:3655–3668

Shatz CG, Chun JJM, Luskin MB 1988 The role of the subplate in the development of the mammalian telencephalon. In: Peters A, Jones EG (eds) Cerebral cortex, vol 7: Development and maturation of cerebral cortex. Plenum, New York, p 35–58

Smart IHM, McSherry GM 1982 Growth patterns in the lateral wall of the mouse telencephalon. II. Histological changes during and subsequent to the period of isocortical neuron production. J Anat 134:415–442

Swindale NV, Vital-Durand F, Blakemore C 1981 Recovery from monocular deprivation in the monkey. 3. Reversal of anatomical effects in the visual cortex. Proc R Soc Lond B Biol Sci 213:435–450

Tan SS, Breen S 1993 Radial mosaicism and tangential cell dispersion both contribute to mouse neocortical development. Nature 362:638–640

Uylings HBM, van Eden CG, Parnavelas JG, Kalsbeek A 1990 The prenatal and postnatal development of rat cerebral cortex. In: Kolb B, Tees RC (eds) The cerebral cortex of the rat. MIT Press, Cambridge, MA, p 35–76

Valverde 1967 Apical dendritic spines of the visual cortex and light deprivation in the mouse. Exp Brain Res 5:274–292

Van der Loos H, Woolsey TA 1973 Somatosensory cortex: structural alterations following early injury to sense organs. Science 179:395–398

Walsh C, Cepko CL 1992 Widespread dispersion of neuronal clones across functional regions of the cerebral cortex. Science 255:434–440

Walsh C, Cepko CL 1993 Clonal dispersion in proliferative layers of developing cerebral cortex. Nature 362:632–635

Walsh C, Reid C 1995 Cell lineage and patterns of migration in the developing cortex. In: Development of the cerebral cortex (Ciba Found Symp 193) p 21–40

Welker E, Van der Loos H 1986 Quantitative correlation between barrel-field size and the sensory innervation of the whiskerpad: a comparative study in six strains of mice bred for different patterns of mystical vibrissae. J Neurosci 6:3355–3373

Young MP, Scannell JW, Burns GAPC, Blakemore C 1994 Analysis of connectivity: neural systems in the cerebral cortex. Rev Neurosci 5:227–249

Cell lineage and patterns of migration in the developing cortex

C. Walsh and C. Reid*

*Neurogenetics Laboratory, Beth Israel Hospital/Harvard Medical School, Alpert 347, 200 Longwood Avenue, Boston, MA 02115 and *Washington University School of Medicine, St Louis, MO 63110, USA*

Abstract. Knowledge of cell lineage in the cortex is important for understanding normal development as well as brain malformations. We studied cell lineage in rats by injecting a library of up to 3400 retroviruses, distinguishable by PCR analysis and encoding alkaline phosphatase, at E14–19. Histological analysis at P15 revealed normal cell morphology and allowed identification of about 80% of all labelled cells. PCR amplification of DNA tags allowed clonal analysis. Cortical cells labelled at E15 formed clustered or widespread clones with equal frequency. Clustered clones contained one to four cells within about 1 mm that had similar morphology and laminar location. However, 48% of cortical clones contained multiple cell types with widely different locations (2.1–6.7 mm; mean, 3.8 mm). Widespread clones contained two to four 'subunits' (one to five neurons each), spaced at apparent intervals of 2–3 mm, with each subunit morphologically indistinguishable from a clustered clone. Distinct subunits in the same clone usually differed in laminar location suggesting sequential formation. Clones labelled at E17 contained fewer neurons and up to two subunits. Clustered clones seem to be produced by stationary progenitors, whereas progenitors of clusters may themselves be produced by migratory, multipotential cells.

1995 Development of the cerebral cortex. Wiley, Chichester (Ciba Foundation Symposium 193) p 21–40

The cerebral cortex must overcome several problems during development. For example, the cortex consists of six layers of cells with distinctive morphologies. These cells must be formed in appropriate numbers and at appropriate times. In addition, distinct regions of the cortex develop distinct structural and functional characteristics. Many of the papers in this volume deal with the simultaneous acquisition of these two sets of properties. However, we will focus mainly on recent studies of cell lineage in the cortex.

Cell lineage analysis is fundamental to understanding the cellular development of many tissues. In the vertebrate haematopoietic system, a large body of evidence suggests that most or all cells arise ultimately from stem cells. These stem cells are defined by three criteria (Potten & Loeffler 1990): (1) they are multipotential and give rise to most or all cell types in the tissue; (2) they are self regenerating, i.e. they divide to both regenerate a stem cell and produce a second cell that generally has different properties; (3) they are immortal. The description of haematopoietic cell lineage has led to the isolation and characterization of many factors that direct specific lineage decisions.

The determination of a lineage tree for every cell in *Caenorhabditis elegans* was the first step in the identification of mutants that affected specific steps in the lineage. This led to the identification of genes that regulate lineage decisions. In *Drosophila* several aspects of the development of the nervous system have also become amenable to genetic study following cell lineage analysis. For example, the mechanoreceptors of the peripheral nervous system are formed as a series of determinate cell divisions. Mutations of *cut* and *numb* affect specific lineage decisions and cause particular cells to adopt cell fates that are different from their normal fates (Jan & Jan 1995).

The study of cell lineage is particularly important for the interpretation of pathological disorders of human brain development (Roberts et al 1995, Jones 1995). Human genetic disorders of the cerebral cortex are associated with specific pathological findings (Sarnat 1992). The pathogenesis of these malformations is not well understood, although understanding the pattern of cellular development will aid in the interpretation of these disorders.

Early events in cortical development

The cerebral cortex develops along two distinct pathways with different developmental patterns and roles. The oldest cells in the cortex form a precociously developed structure known as the preplate (Marin-Padilla 1978, Allendoerfer & Shatz 1994). The bulk of the cortex, however, is derived from a structure that is formed later (the cortical plate). Cortical plate cells are formed originally as a densely packed layer of cells that develops within the preplate and divides the preplate into an outer layer, called layer I or the marginal zone, and a deeper layer called the subplate. Later, the cortical plate becomes extremely dense, dwarfing the older cells in numbers. Finally, the cortical plate neurons differentiate to form the bulk of the neurons in the definitive, six-layered adult cortex (Marin-Padilla 1978). A remarkable feature of the cortical plate is that the neurons are formed in an inside out sequence, so that neurons destined ultimately for layer VI are formed first (Bayer & Altman 1991). Cell lineage analysis has focused on several questions, many of which remain unanswered. Do separate progenitor cells produce neurons destined for different layers of the cortex, or are there distinct progenitors for different cell types? Is the radial organization of the mature cortex into columns a consequence of constraints on the migration of cortical cell types?

Approaches to cell lineage

In *C. elegans*, lineage can be traced by simple observation of the developing embryo. However, this is impossible in vertebrates because of the smaller cell size. Direct injection of progenitor cells has been successful in some vertebrates, but individual cell marking *in vivo* is not feasible in mammals because the embryo develops deeply embedded in several layers of extraembryonic tissue. On the other hand, mammalian nervous tissue can be maintained in culture for days at a time, so that cellular marking experiments can be combined with time-lapse video imaging. Cells marked with lipophilic dyes such as DiI (1,1′dioctadecyl-3,3,3′,3′-tetramethylindocarbocyanine perchlorate) have been traced during their migrations (Fishell et al 1993, O'Rourke et al 1992), although these experiments have not provided direct information about cell lineage.

Genetic chimeras were one of the earliest tools for elucidating cell lineage in the cerebral cortex. Patterns of cell mixing can be observed by combining cells from embryos of different strains or species that can be distinguished histologically (Fishell et al 1990, Goldowitz 1987, Crandall & Herrup 1990). A recent variant on the chimera approach has been the development of a transgenic mouse that carries a β-galactosidase transgene integrated into the X chromosome (Tan & Breen 1993). All cells express the transgene in males, whereas in females only about half of the cells express the transgene because individual progenitor cells inactivate one of the X chromosomes randomly. These patterns of inactivation are inherited in a clonal fashion. The chimera and mosaic studies provide evidence for broad radial constraints among some cells, and also suggest that strict radial migration is not the exclusive pattern of development.

Retroviral studies

The retroviral method takes the chimera method a step further because it allows the analysis of single or small numbers of progenitor cells in isolation. Retroviruses label dividing cells by integrating their genome into the genome of the target cell, and they are useful markers of cell lineage because the retroviral genome is inherited clonally. However, it is impossible to control precisely which or how many progenitors are infected in a single experiment. Therefore, without direct means of distinguishing cells that are in different clones, the interpretation of retroviral experiments requires certain assumptions to be made about the patterns of cell migration.

Retroviral lineage experiments in the retina (Turner & Cepko 1987) and tectum (Gray et al 1988) indicated that sibling cells were arranged in a remarkable radial order. Similar patterns were uncommon in the cortex. Instead, labelled cells formed loose clusters scattered over as much as a few hundred micrometres (Luskin et al 1988, Walsh & Cepko 1988, Price & Thurlow 1988), or they occurred side-by-side as cells of similar morphology and laminar position (Parnavelas et al 1991). Analysis of clustered, retrovirally labelled cells by Price et al (1995, this volume) and Parnavelas (1995, this volume) suggests that cells within these clusters are often uniform in their morphological or neurochemical phenotype.

Retroviral libraries

Retroviral libraries simplify the interpretation of retroviral studies by marking clonal relationships directly (Walsh & Cepko 1992). These experiments utilize a mixture of retroviruses that are distinguishable because they contain distinct DNA tags (85–100 distinct tags in the earliest libraries and up to thousands of tags in the later experiments) that can be amplified (even in single, retrovirally labelled cells) using PCR analysis. Clonal relationships are marked directly by the DNA tag, so assumptions about migratory behaviour are not required and migratory behaviour can be analysed directly.

Retroviral library studies were first applied to the cerebral cortex in order to analyse cell dispersion between different functional regions (Walsh & Cepko 1992). Injections of the retroviral library were made at embryonic day (E) 15, and the brains were analysed at postnatal day (P) 9 or later. The cortex was usually trimmed and cut tangentially to the cortical layers, then stained for cytochrome oxidase to reveal the functional subdivisions of the cortex. The results of the first experiments were surprising. Some clones (more than half of all clones with more than one neuron) showed a wide dispersion, defined as greater than 1.5 mm, and occasionally covering much of the surface of the neocortex. This wide clonal dispersion across functional regions of the cortex suggested that cell lineage was not solely responsible for dividing the cortex into functional areas. However, retroviral library experiments also showed that earlier retroviral studies, which did not use a library, could produce misleading conclusions. Single marker studies are subject to two types of errors: lumping (erroneous interpretation of different clones together as a single clone, which occurs in 10–12% of clones) and splitting (erroneous interpretation of distinct parts of the same clone as different clones, which occurs in about 50% of clones or more) (Walsh & Cepko 1992). Consequently, single retroviral marker studies cannot demonstrate conclusively that a progenitor is committed to forming a single cell type—the same progenitor may give rise to other cells elsewhere in the cortex that fail to be interpreted as sibling cells.

Tangential preparations of the cortex are well suited to topographic analysis, but they do not illustrate the cortical layers clearly and are not optimal for interpretation of cellular morphology. Therefore, the combination of clonal analysis with determination of cell type was initially limited. Furthermore, preparation of tangential sections involved trimming away some of the neocortex, and most of the archicortex and hippocampus. Therefore, early preparations underestimated both the proportion of widespread clones and the amount of dispersion of sibling cells because some sibling cells were lost to clonal analysis. The early retroviral library experiments (Walsh & Cepko 1992) were also criticized because some of them were performed with larger viral inoculums. These might cause spurious clonal relationships due to coincidental infection of more than one progenitor cell with the same DNA tag (Kirkwood et al 1992). Consequently, subsequent retroviral library experiments used small viral inoculums in which the probability of spurious clonal relationships was usually less than 0.05 per experiment, i.e. fewer than one in 50 clones (Walsh et al 1992). Developmental time course studies of clonal dispersion (Walsh & Cepko 1993) using the retroviral method involved taking serial coronal sections and performing three-dimensional reconstructions after injection at E15 of a retroviral library encoding β-galactosidase. Brains were then examined for clonal patterns at E18, E21 or P3 by PCR analysis. Table 1 summarizes these results. The proportion of widespread clones is shown for each experiment. The remainder of clones consisted of either single cells, single clusters of neurons or glial cells. Widespread clones are defined at P15, P3 and E21 as showing greater than 1.5, 1.0 and 0.6 mm dispersion, respectively. The analysis at E18 was a key control because widespread dispersion was not seen in nine clones that were analysed. In contrast, analysis at E21 or P3 showed that the proportion of clones that were widespread was about 50%. This figure is higher than in earlier experiments (24%) because the entire telencephalon was analysed. Widespread clones accounted for about 70% of all labelled neurons. Analysis of the proliferative layers at E21 showed that 35% of widespread clones contained two or more dispersed sibling cells, suggesting that sibling cells disperse within the proliferative regions, presumably while still dividing, before postmitotic neurons begin their well-characterized radial ascent into the cortex.

Widespread dispersion in the proliferative layers beneath the cortex has been seen independently with a complementary method. Fishell et al (1993) developed an explant culture system for the cortex and labelled proliferative zone cells with DiI. Time-lapse video imaging indicated that the majority of cells in the proliferative zone were moving at up to 10–100 μm/h. These results indicate that dispersion of sibling cells within the proliferative layers may be a major cause of widespread dispersion. Another potential mechanism of widespread clonal dispersion was proposed by O'Rourke et al (1992) after they performed time-lapse video imaging of cortical slices. They noticed that

TABLE 1 Proportion of clones that are widespread as a function of experimental conditions

Age (injection–analysis)	Label	Widespread clones (% of total clones)	Reference
E15–E18	β-gal	0	Walsh & Cepko (1993)
E15–E21	β-gal	45[a]	Walsh & Cepko (1993)
E15–P3	β-gal	58[b]	Walsh & Cepko (1993)
E15–P15	AP	48[c]	Reid et al (1995)
E17–P15	AP	12.5[d]	Reid et al (1995)

[a]Different from E15–E18 at $P < 0.0005$
[b]Different from E15–E18 at $P < 0.003$
[c]Different from E15–E18 at $P < 0.00005$
[d]Different from E15–P15 at $P < 0.003$
AP, alkaline phosphatase; β-gal, β-galactosidase; E, embryonic day; P, postnatal day

the great majority of migrating cells in the intermediate zone took a radial course, but also that a minority (12%) left the radial course and migrated perpendicular to the glial fibres within the intermediate zone. Therefore, widespread dispersion may also arise from this non-radial migration in the intermediate zone.

Widespread clonal dispersion and formation of cortical cell types

A new retroviral library has allowed the analysis of clonal dispersion as well as the generation of the morphological cell types in the cerebral cortex. This new library encodes alkaline phosphatase, a histochemical marker that is unusually sensitive and, as a membrane-bound enzyme, tends to outline neurons particularly well (Fields-Berry et al 1992). The library was prepared from 3400 retroviral constructs, although its actual complexity is still being determined empirically. Analysis with the alkaline phosphatase-encoding library involved labelling rats at E15 or E17, and analysing the clonal patterns at P15 (Reid et al 1995). As in the β-galactosidase studies, close to half (48%) of clones labelled at E15 were widespread (i.e. more than 1.5 mm dispersion), accounting for 73% of all retrovirally labelled neurons (Table 1). This finding emphasizes that in experiments with single retroviruses, the majority of all cells analysed have sibling cells that are distributed widely.

The alkaline phosphatase results also suggest that sibling cells in widespread clones are not arranged randomly, but are instead spaced at periodic intervals of 1.5–3 mm (Reid et al 1995). The same spacing was noted in an earlier β-galactosidase study (Fig. 1), although it was not emphasized at that time. Within some widespread clones, single neurons or clusters of two to four

Lineage and migration

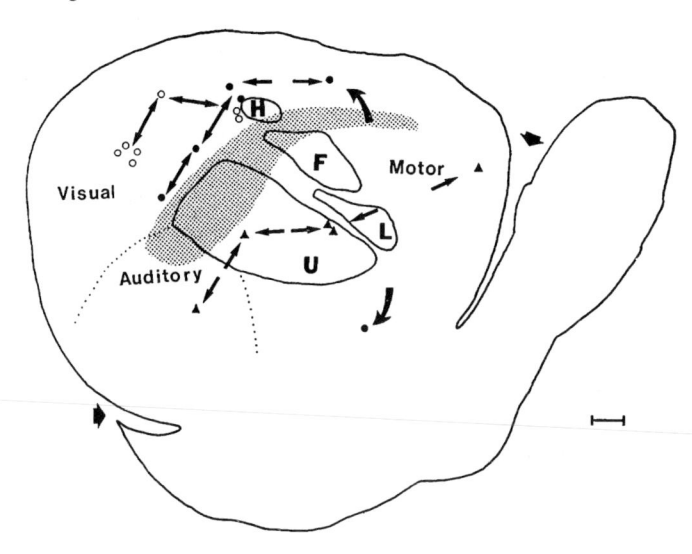

FIG. 1. Selected widespread clones labelled by retroviral library injection at embryonic day (E) 15 illustrating the periodic spacing of some sibling cells. Three widespread clones are shown, indicated by open circles, closed circles and triangles. Each clone is taken from a different experiment, but they are superimposed on a single outline of a flattened preparation of the cerebral cortex so that the patterns can be compared. Regions of the cerebral cortex are indicated: F, forepaw; H, hindpaw; L, lower lip; U, upper lip. Auditory, Motor and Visual indicate approximate locations of distinct functional regions of the cortex. The rhinal sulcus, the lateral border of neocortex, is indicated by two large arrows, with the entorhinal cortex below the arrows and the olfactory bulb rostral to the cortex. Widespread clones consist of single cells or clusters of cells that are arranged at apparent intervals of 1.5–3 mm. The clone indicated by the open circles (found in a brain analysed at postnatal day 9) contains a cluster of four cells (some or all of them are neurons), a single neuron at a distance of about 1.8 mm and another pair of cells at a distance of about a further 1.8 mm. The clone indicated by closed circles shows several single cells, and a single pair of cells, at intervals of about 2.0 mm. In addition, this clone shows a rostrolateral cell that is widely separated mediolaterally from the other cells (curved arrows). A third clone, located more laterally (triangles) shows similar intercell distances (about 2.2 mm), except for the distance between the rostral-most cells which is about 4 mm. The patterns of dispersion of these clones (indicated with arrows) parallel the orientation and shape of the lateral ventricle, which is shown schematically (by shading) based on three-dimensional reconstructions. Bar = 1.0 mm. (Modified from Fig. 6 of Walsh & Cepko (1992) and reproduced with permission.)

neurons occur in arc-shaped, rostral-caudal arrangements that parallel the outline of the lateral ventricle, although there are exceptions.

A relationship between the periodic spacing of cells and the morphological subtypes of cells within widespread clones is also observed in the AP library experiments (Reid et al 1995). The results confirm previous suggestions from single retroviral experiments that clustered sibling cells usually share

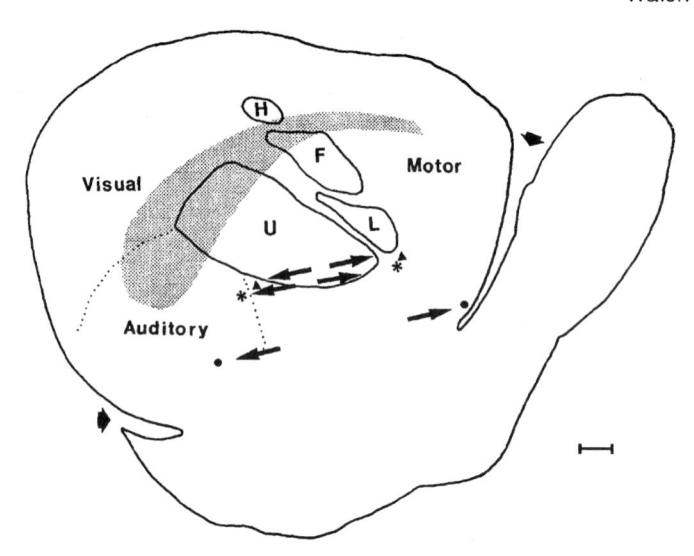

FIG. 2. Widespread clones labelled by retroviral library injection at embryonic day (E) 17 in the rat. Three clones are illustrated, which were taken from three different experiments and superimposed on one outline for comparison (using external landmarks such as the barrel field for alignment). Note that two clones (asterisk, triangle) show two cells at almost precisely corresponding locations in the cortex, and with the same spacing between cells (about 5.0 mm, which is about twice the spacing seen in Fig. 1). A third clone (circle) is located slightly further laterally, and contains two cells at similar relative locations to one another, but the spacing between cells is about 50% greater than that seen for the other two clones. The approximate shape and orientation of the lateral ventricle (obtained from three-dimensional computer reconstructions) is indicated by shading. Note that a line connecting each pair of cells runs roughly parallel to the lateral ventricle at this point, or to the rhinal sulcus (two large arrows). Regions of the cerebral cortex are indicated: F, forepaw; H, Hindpaw; L, lower lip; U, upper lip. Auditory, Motor and Visual indicate approximate locations of distinct functional regions of the cortex. Bar = 1.0 mm.

morphological features and laminar location (Parnavelas et al 1991, Mione et al 1994). However, in addition, sibling cells in widespread clones that are near to each other (less than 1.5 mm dispersion) have similar properties, whereas sibling cells that are far apart (more than 1.5 mm dispersion) usually have different morphology and/or laminar location. Therefore, 'subunits' are definable within widespread clones by two different criteria: intersibling spacing and cell morphology. In contrast to labelling at E15, labelling at E17 produced fewer widespread clones (12.5% compared with 48%, Table 1) that were also smaller (consisting of up to two widespread cells). Widespread clones labelled at E17 also had a consistent location (in the far lateral zone of the neocortex) and orientation (Fig. 2) that was roughly parallel to the rhinal sulcus or lateral ventricle.

A general model for cell lineage of cortical plate-derived cortical neurons

A model for cell lineage needs to account for the observation that the progenitors of widespread clones not only form different cell types in different places, but can also form multiple clusters of more than one (similar) cell each, with the clusters located in different cortical regions. Multiple clusters suggest that the progenitor of the widespread clone may give rise to secondary progenitors in some cases. A hypothetical model combines the notion of two types of progenitors with the observation of progenitor migration in the proliferative zone (Fig. 3). In this model, widespread clones result from multipotential, migratory progenitors that divide in a stem cell fashion, generating either single neurons or secondary progenitors at intervals across the cortex. The spacing between divisions of the migratory progenitor is determined by the rate and pathway of migration of the progenitor, and the length of the cell cycle of that progenitor. Secondary progenitors given off by the migratory progenitor are proposed to be non-migratory and give rise to cells of a similar phenotype (for at least one or two cell divisions). Postmitotic neurons take a primarily radial route up to the cortex, with occasional exceptions as described by O'Rourke et al (1992).

How does this model account for the patterns of retroviral labelling? Injections at E15 are postulated to infect migratory progenitors. Given what is known about retroviral infection, there is an equal probability that the virus integrates into either one or the other daughter cell. If the retrovirus integrated into the non-migratory daughter cell, a single cluster of cells would be labelled (a single-cell clone or a single cluster of similar cells). In contrast, if the retrovirus integrated into the migratory daughter, multiple single cells or clusters of cells would be labelled, producing a widespread clone. The equal probability of integration into either daughter cell might account for the observation that 50% of clones labelled at E15 are widespread and 50% are not. If, however, retroviral infection was performed two days later at E17, the migratory progenitor would have fewer remaining cell divisions before the end of neurogenesis. Therefore, regardless of which daughter cell was labelled, there would be a lower probability of labelling a widespread clone and labelled widespread clones would contain fewer subunits (Fig. 3). The proposed model is consistent with single retroviral results from many laboratories which suggest that sibling cells often form clusters with a uniform phenotype (this volume: Parnavelas 1995, Price et al 1995). The model is also consistent with *in vitro* observations of dispersion in the proliferative zone (Fishell et al 1993) and dispersion above the proliferative zone (O'Rourke et al 1992). By suggesting that most non-radial dispersion occurs in proliferative cells, the model also reconciles widespread clonal dispersion with the well-known radial migration of postmitotic neurons to the cortex (Rakic 1988).

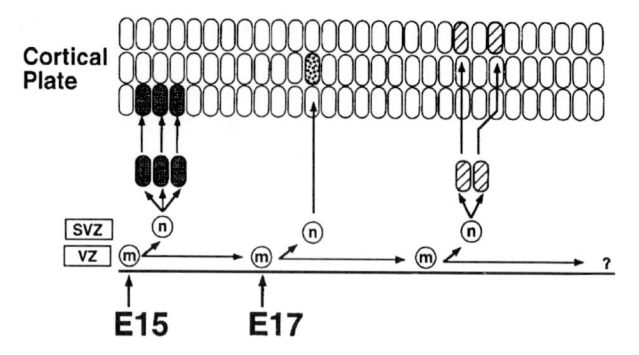

FIG. 3. A model for cell lineage in the cerebral cortex that is based on retroviral library injections at embryonic day (E) 15 and E17. A migratory, multipotential progenitor (m) divides at intervals, with the distance between cell divisions determined by the rate of cellular migration and the length of the cell cycle. The migratory cell generates a non-migratory cell (n) and another migratory cell in a stem cell fashion. The non-migratory cell has several potential fates: dividing multiple times, directly differentiating or dividing only once. It is not known whether divisions of the non-migratory progenitor cells are symmetrical or asymmetrical. Progeny of the non-migratory progenitors generally take a radial migratory course, although there are some exceptions. Retroviral infection of the migratory progenitor at E15 would result in the integration of the retrovirus into the migratory daughter or the non-migratory daughter with equal probability. Integration into the non-migratory daughter would label a cluster or single cell, whereas integration into the migratory daughter would label a widespread clone. Infection at E17 would label a migratory daughter with fewer remaining cell divisions. Therefore, regardless of which daughter cell retained the retrovirus, widespread clones would be labelled with lower frequency and would be smaller. SVZ, subventricular zone; VZ, ventricular zone.

Transplantation results from McConnell & Kaznowski (1991) have suggested that cortical progenitor cells are initially multipotential, but acquire commitment to generating specific cell fates at set points in the cell cycle. The proposed model postulates a migratory, multipotential cell and a non-migratory cell with different properties, and so appears to fit well with the transplantation results. Of course, lineage analysis alone cannot test whether progenitors are committed to particular fates, so there is as yet no retroviral evidence that the non-migratory progenitors are committed.

Other elegant transplantation studies from Levitt and co-workers (Barbe & Levitt 1991, Levitt et al 1995) have suggested that some aspects of regional identity in the cerebral cortex are sensitive to transplantation at early stages, but soon become resistant to transplantation. These transplantation results are most easily interpreted by suggesting that some progenitor cells, as well as postmitotic cells, already have areal specificity. Interestingly, the lineage analysis shows that some sibling cells disperse from the neocortex into the perirhinal and archicortex (Reid et al 1995). However, if it was the migratory (presumably non-specified) progenitor that was responsible for this widespread migration, the non-migratory progenitors would be capable of acquiring information about regional specificity as well as cell type specificity. Therefore, this model can also be reconciled with regional transplantation experiments. In contrast, cell fate within the neocortex may be substantially more plastic, as suggested by O'Leary et al (1995, this volume), even after neurons are postmitotic. The widespread dispersion of sibling cells across the entire neocortex in the most extreme cases suggests that areal fating within the neocortex is not contained within the progenitor cells of the cortical plate.

The proposed model also fits well with *in vitro* studies of cell lineage among cortical progenitors. Studies from several labs (this volume: Price et al 1995, Bartlett 1995, Reynolds & Weiss 1992) suggest the existence of a multipotential progenitor cell that can be isolated from the immature cortex. Recent studies suggest that some cortical progenitors have the properties of true stem cells (Davis & Temple 1994). The retroviral library model may represent the *in vivo* correlate of the *in vitro* results.

The model we propose and the results we present relate to retroviral infection at E15 or later (day of conception defined as E0). Retroviral experiments using earlier injections, e.g. E15 injections by Parnavelas (1995, this volume), might produce different patterns (conception defined as E1). O'Leary and co-workers (O'Leary & Borngasser 1992, O'Leary et al 1995, this volume) have found that progenitor cells do not migrate over significant distances in the proliferative zone at early stages of development, i.e. during formation of the preplate and subplate. They have suggested that this absence of early migration is important for the determination of positional information in the cortex. Our analyses of later stages suggest that positional information is not obviously encoded in cell lineage patterns when the cortical plate is developing. It is possible that the oldest preplate and subplate cells have mechanisms for imprinting regional specificity upon the cortical plate cells that are generated later.

Applications of cell lineage to genetic disorders and mechanisms

The availability of a model for cell lineage allows interpretations of genetic disorders of cortical development. For example, tuberous sclerosis is a disorder that shows clonal lesions with two distinct morphologies (Gomez 1988). The first are subependymal nodules, which are present in the proliferative zone and consist of small, undifferentiated cells that can undergo oncogenic transformation to produce tumours. These lesions might correspond to disorders of the migratory progenitor. Secondly, tuberous sclerosis also shows cortical tubers. These contain large numbers of dystrophic neurons, aberrant glia and intermediate cells of uncertain identity (Stefansson & Wollman 1980). Perhaps these lesions, which are characteristically limited to smaller cortical zones and sometimes originate from a point source in the subventricular zone, represent disorders of the non-migratory, clustered progenitor.

Other disorders of cortical development may represent disorders of the migratory, multipotential progenitor. For instance, periventricular heterotopia is inherited as a dominant trait (Huttenlocher et al 1994). The heterotopias consist of continuous bands of neural cells (probably neurons) that line the entire proliferative region of the cortex bilaterally. It is possible that these lesions represent a disorder of the migratory progenitor before it is constrained regionally. As our knowledge of cell lineage in the cortex becomes more refined, we will be able to develop clearer hypotheses about the many important neurological disorders that impinge upon the developing cerebral cortex.

Acknowledgments

Research from the authors' laboratory is supported by the NIH through grants GM14862 (to C. R.), KO8 NS01520 (to C. W.), and RO1 NS33769 (to C. W.). C. W. is also a fellow of the Rita Allen Foundation, and is supported by the Klingenstein Foundation Fund. We thank Benxiu Ji, Wenjiang Yu and Ivan Liang for assistance.

References

Allendoerfer KL, Shatz CJ 1994 The subplate, a transient neocortical structure: its role in development of connections between thalamus and cortex. Annu Rev Neurosci 17:185–218

Barbe MF, Levitt P 1991 The early commitment of fetal neurons to the limbic cortex. J Neurosci 11:519–533

Bartlett P, Richards LR, Kilpatrick TJ et al 1995 Factors regulating the differentiation of neural precursors in the forebrain. iN: Development of the cerebral cortex. Wiley, Chichester (Ciba Found Symp 193) p 85–99

Bayer SA, Altman J 1991 Neocortical development. Raven Press, New York.

Crandall J, Herrup K 1990 Patterns of cell lineage in the cerebral cortex reveal evidence for developmental boundaries. Exp Neurol 109:131–139

Davis AA, Temple S 1994 A self-renewing multipotential stem cell in embryonic rat cerebral cortex. Nature 372:263–266

Fields-Berry SC, Halliday AL, Cepko CL 1992 A recombinant retrovirus encoding alkaline phosphatase confirms clonal boundary assignments in lineage analysis of murine retina. Proc Natl Acad Sci USA 89:693–697

Fishell G, Mason CA, Hatten ME 1993 Dispersion of neural progenitors within the germinal zones of the forebrain. Nature 362:636–638

Fishell G, Rossant J, Vanderkooy D 1990 Neuronal lineages in chimeric mouse forebrain are segregated between compartments and in the rostrocaudal and radial planes. Dev Biol 141:70–83

Goldowitz D 1987 Cell partitioning and mixing in the formation of the CNS: analysis of the cortical somatosensory barrels in chimeric mice. Dev Brain Res 35:1–9

Gomez M 1988 Tuberous sclerosis. Raven Press, New York

Gray GE, Glover JC, Majors J, Sanes JR 1988 Radial arrangement of clonally related cells in the chicken optic tectum: lineage analysis with a recombinant retrovirus. Proc Natl Acad Sci USA 85:7356–7360

Huttenlocher PR, Taravath S, Mojtahedi S 1994 Periventricular heterotopias and epilepsy. Neurology 44:51–55

Jan YN, Jan LY 1995 Maggot's hair and bug's eye: role of cell interactions and intrinsic factors in cell fate specification. Neuron 14:1–5

Jones EG 1995 Cortical development and neuropathology in schizophrenia. In: Development of the cerebral cortex. Wiley, Chichester (Ciba Found Symp 193) p 277–295

Kirkwood TBL, Price J, Grove EA 1992 The dispersion of neuronal clones across the cerebral cortex. Science 258:317

Levitt P, Ferri R, Eagleson K 1995 Molecular contributions to cerebral cortical specification. In: Development of the cerebral cortex. Wiley, Chichester (Ciba Found Symp 193) p 200–213

Luskin MB, Pearlman AL, Sanes JR 1988 Cell lineage in the cerebral cortex of the mouse: studies *in vivo* and *in vitro* with a recombinant retrovirus. Neuron 1:635–647

Marin-Padilla M 1978 Dual origin of the mammalian neocortex and evolution of the cortical plate. Anat Embryol 152:109–126

McConnell SK, Kaznowski CE 1991 Cell cycle dependence of laminar determination in developing neocortex. Science 254:282–285

Mione MC, Danevic C, Boardman P, Harris B, Parnavelas JG 1994 Lineage analysis reveals neurotransmitter (GABA or glutamate) but not calcium-binding protein homogeneity in clonally related cortical neurons. J Neurosci 14:107–123

O'Leary DDM, Borngasser D 1992 Minimal dispersion of neuroepithelial cells and their progeny during generation of the cortical preplate. Soc Neurosci Abstr 18:925

O'Leary DDM, Borngasser DJ, Fox K, Schlaggar BL 1995 Plasticity in the development of neocortical areas. In: Development of the cerebral cortex. Wiley, Chichester (Ciba Found Symp 193) p 214–230

O'Rourke NA, Dailey ME, Smith SJ, McConnell SK 1992 Diverse migratory pathways in the developing cerebral cortex. Science 258:299–302

Parnavelas J, Mione MC, Lavdas A 1995 The cell lineage of neuronal subtypes in the mammalian cerebral cortex. In: Development of the cerebral cortex. Wiley, Chichester (Ciba Found Symp 193) p 41–58

Parnavelas JG, Barfield JA, Franke E, Luskin MB 1991 Separate progenitor cells give rise to pyramidal and nonpyramidal neurons in the rat telencephalon. Cereb Cortex 1:463–468

Potten CA, Loeffler M 1990 Stem cells: attributes, cycles, spirals, pitfalls and uncertainties: lessons for and from the crypt. Development 110:1001–1020

Price J, Williams BP, Götz M 1995 The generation of cellular diversity in the cerebral cortex. In: Development of the cerebral cortex. Wiley, Chichester (Ciba Found Symp 193) p 71–84

Price J, Thurlow L 1988 Cell lineage in the rat cerebral cortex: a study using retroviral-mediated gene transfer. Development 104:473–482

Rakic P 1988 Specification of cerebral cortical areas. Science 241:170–176

Reid CB, Liang I, Walsh C 1995 Systematic widespread clonal organization in cerebral cortex. Neuron 15:1–20

Reynolds BA, Weiss S 1992 Generation of neurons and astrocytes from isolated cells of the adult mammalian central nervous system. Science 255:1707–1710

Roberts GW, Royston MC, Goetz M 1995 Pathology of cortical development and neuropsychiatric disorders. In: Development of the cerebral cortex. Wiley, Chichester (Ciba Found Symp 193) p 296–321
Sarnat HB 1992 Cerebral dysgenesis: embryology and clinical expression. Oxford University Press, New York
Stefansson K, Wollman R 1980 Distribution of glial fibrillary acidic protein in central nervous system lesions of tuberous sclerosis. Acta Neuropathol 52:135–140
Tan SS, Breen S 1993 Radial mosaicism and tangential cell dispersion both contribute to mouse neocortical development. Nature 362:638–640
Turner D, Cepko CL 1987 Cell lineage in the rat retina: a common progenitor for neurons and glia persists late in development. Nature 328:131–136
Walsh C, Cepko CL 1988 Clonally related neurons show several patterns of migration in cerebral cortex. Science 255:1342–1345
Walsh C, Cepko CL 1992 Widespread dispersion of neuronal clones across functional regions of the cerebral cortex. Science 255:434–440
Walsh C, Cepko CL 1993 Clonal dispersion in proliferative layers of developing cerebral cortex. Nature 362:632–635
Walsh C, Cepko CL, Ryder EF, Church GM, Tabin C 1992 The dispersion of neuronal clones across the cerebral cortex: response. Science 258:317–320

DISCUSSION

O'Leary: Your model represents a useful framework for further investigations, but in your 1992 *Science* paper (Walsh & Cepko 1992) you suggested that areal specification occurs after neuronogenesis, i.e. after the cells are generated and have migrated to the cortical plate. Now you're suggesting that progenitor cells can stay in the epithelium (although they may migrate), become respecified and then give rise to additional cells. These cells then migrate with new specification instructions compared with earlier cells generated from the same progenitor cell.

Walsh: I'm suggesting that there are two types of progenitors: a migratory progenitor that is hypothetically in the ventricular zone; and a non-migratory progenitor that is possibly in the subventricular zone. The migratory progenitor cell gives rise to non-migratory cells, which divide in the same area creating a cluster of cells that in turn migrate more or less radially to the cortex. The migratory progenitor cell also regenerates another migratory cell that can migrate to somewhere else and create another non-migratory cell, and so on. What we described in the *Science* paper, and what we continue to see, is that there is a widespread dispersion of sibling cells across the cortex, so areal specification does not seem to arise exclusively from a lineage-based mechanism. Regional differences in the cortex may exist, possibly caused by the regional expression of markers or extracellular factors, and the migratory cells may move from one of these regional microenvironments to another relatively freely. Progenitor cells may receive a specification signal that is different in each area. Therefore, the wide dispersion can be reconciled with microenvironmental differences, but it doesn't require that there are any. I'm trying to explain the lineage results, but it requires other people to determine whether these microenvironmental differences exist.

O'Leary: This brings up the intriguing possibility that the progenitor cell itself may be respecified, as opposed to a model in which its progeny are pleuripotential when they leave the neuroepithelium and move out to the cortical plate, after which they take on a specific areal fate. Fishell et al (1993) described the movement of progenitor cells within the epithelium as random, rather than directed migration, which seems to be required by your new model. Can you explain this in terms of your model?

Walsh: The progenitor cells don't necessarily need to be respecified. They may exist within a developmental gradient. John Rubenstein mentioned that *Dlx2* and *Mash1* are expressed in specific strata within the wall of the striatal primordia (Porteus et al 1994). It is possible that the most undifferentiated progenitors do not know where they are or what they're doing. For example, they may not express neuronal or glial markers, but they may still become specified at a later stage. Therefore, they wouldn't need to be continuously respecified.

Secondly, you mentioned the work of Fishell et al (1993) and suggested that our model requires the cells to move rapidly in a directed fashion. However, the amount of dispersion that we see does not require migration over large distances within the ventricular zone. The apparent periodicity is 2–3 mm in the adult but there's approximately a fourfold increase in the length of the brain between postnatal day (P) 15, when we analyse the cells, and about embryonic day (E) 17 or E18 when these hypothetical movements would be occurring. The apparent spacing at E17/E18 would, therefore, only be about 500 μm. This is 500 μm in the cortex before radial migration has occurred, but the ventricular zone is smaller than the cortex, probably by another twofold, and so movement of only a couple of hundred micrometres would be required in the proliferative zone to explain the observed periodicity at P15. Fishell et al (1993) described movement at rates of 5–20 μm/h over an eight-hour period, which is more than enough to generate the dispersion that we observe. However, as you pointed out, the movement they observed did not appear to be directed primarily in a rostrocaudal direction. On the other hand, we did see rostrocaudal dispersion in retroviral experiments in which sibling cells were analysed after short post-injection survival times (Walsh & Cepko 1993). Fishell et al (1993) also described that cells in the cortical ventricular zone would contact the border between the cortex and the striatum, but they would not cross that border. Instead, cells migrated rostrally or caudally along the border. One way to reconcile the findings, therefore, is that as the migratory progenitor cells contact the border between cortical and striatal ventricular zone, random movement is translated into directed rostral or caudal movement.

LaMantia: The Fishell paper raises a different issue. They studied the specification of major functional regions of the telencephalon, but not of cortical areas. Different mechanisms may be operating in these different areas.

Chris Walsh, you showed how varying numbers of E15-labelled, restricted and non-restricted clones fitted your model. These results suggest that most progenitor cells stay in the same area and only a small population of them migrate. At a particular time, 48% of the clones are generated from the original progenitor cells. This suggests that many of their offspring are either eliminated or that directed movement of a particular class of cells occurs at a particular time. Is a subclass of progenitor cells specialized for mobility in the ventricular zone or is this the general behaviour of progenitor cells?

Walsh: I don't know the answer to that. However, this model alone accounts for all our results. If a retrovirus infects a particular cell at E15, it has a 50:50 chance of integrating into either of the daughter cells of the original infected cell. If it integrates into a non-migratory daughter cell, only a clustered clone will be labelled. On the other hand, if it integrates into a migratory daughter cell, then a widespread clone will be labelled. This may be why we see 50% widespread clones and 50% clustered clones. This is a consistent finding if we inject the retrovirus at E15, and then analyse at E21, P3 or P15. However, if we inject the retrovirus at E17, we see fewer and smaller widespread clones. This fits with the model because you would expect to miss some of the earlier migratory events if you injected later in development.

Rakic: I first want to congratulate Chris Walsh for his excellent presentation and for elaborating a complex and interesting model that reconciles results from his work on cell lineages in rodent cortex with the radial unit hypothesis (Rakic 1988). There has been a substantial evolution of the concepts that have emerged from the relatively new technique of retroviral labelling of cell lineages in the cerebral cortex. This evolution relates more to the interpretation of the results rather than to the results themselves. In 1988 perhaps we would have had a vigorous discussion on this subject, but now not much is left to discuss because Chris Walsh has developed a model that is compatible with the radial unit hypothesis of cortical development. David Kornack and I (unpublished results) have applied the retroviral gene transfer method to label clonally related cells in non-human primates using two distinct probes supplied to us by Joshua Sanes of Washington University in St Louis. We reasoned that the cortex of macaque monkeys, which is 100 times larger and develops 10 times slower, may provide a higher resolution of cellular developmental events and give us more confidence in identifying individual clones. Also, laminar and radial organization in the primate cortex is more prominent and precise than in rodents and, because of the similarity of the cortical neural organization among primates, it is likely that the results would be applicable to understanding corticogenesis in humans. The observed distribution patterns of retrovirally labelled progeny in this study is complex, but the results are compatible with Chris Walsh's results as well as with the radial migration results. My previous light and electron microscopy studies indicated that in primates radial migration is the predominant type of cell allocation in the neocortex (reviewed in Rakic 1988). However, I believe that this mode of postmitotic cell allocation is prevalent also in rodents, and in the first general discussion I will present some of Dr Seong-Seng Tan's results that support this statement.

Parnavelas: I differ with Chris Walsh's interpretation of the phenotypes of cells that make up the clones. With the exception of the cluster of glial cells, the cells did not look pyramidal. If you showed your results to individuals who have more experience in the cytology of the cerebral cortex, such as Ted Jones or Pasko Rakic, it would help to clarify whether progenitor cells produce only one cell phenotype or more than one, depending on which subunit they are in.

Walsh: I would be more than happy to show my results to anyone. However, many of the morphologies in Golgi-like alkaline phosphatase preparations are obvious. We do not classify cells if we are uncertain of their morphology.

Parnavelas: If cells in the ventricular zone move randomly according to the work of Fishell et al (1993), how can such a substantial proportion of labelled progenitor cells give rise to a number of subunits of cells spaced at regular intervals in the brain?

Walsh: The retroviral labelling experiments suggest that primarily there is rostrocaudal dispersion. This rostrocaudal movement is possible if an exception to random movement occurs when the cells encounter the border between the cortex and striatum. Movement then seems to be directed rostrocaudally. Basically, the periodic spacing that we observe requires a constant cell cycle length and directed migration at the border. That's one way of reconciling the results but there may be others.

Levitt: It may not be possible to compare these two sets of results. Chris Walsh looked at a broad zone, whereas Fishell et al (1993) looked at a discrete area in the lateral region of the ventricular zone, next to the lateral ganglionic eminence. We cannot assume that the movements in the ventricular zone are random.

Another layer of complexity in your model may be that, irrespective of movement within a zone, what matters to a particular cell (i.e. what it's going to become and where it is located) is where it exits the zone. I propose that we need to identify transcription factors that may regulate the environment in which cells exit the cell cycle, rather than factors that specify the progenitor cells themselves. Then it would not be necessary to talk about respecification of the progenitor cells. A similar situation occurs in the *Drosophila* retina, where the fate of a progenitor cell is determined by its nearest neighbour when it exits the cycle. My chapter presents evidence for environmental differences in the early ventricular zone (Levitt et al 1995, this volume).

Walsh: Our model is also compatible with your transplantation experiments (Barbe & Levitt 1991, 1992, 1995, Levitt et al 1995, this volume).

Bartlett: One of the predictions that can be made from this model is that the precursor cell is a multipotential stem cell that is dividing asymmetrically, self-renewing and giving rise to a single committed progenitor. How strong are the results which suggest that the progenitor is restricted to the glial or neuronal cell lineages?

Walsh: We only have a relatively small data set and our PCR analysis only works for about 40–70% of the cells. The clusters have been accurately defined by Jack Price and John Parnavelas.

Bartlett: One of the problems with retroviral labelling is that it cannot be done early enough, so we're labelling at a stage when neurogenesis has already started and some commitment may have already occurred. It may be more helpful to label before neurogenesis has commenced when one might predict that all the clones would be geographically dispersed.

Walsh: One might predict the opposite if one took Dennis O'Leary's results into account (O'Leary et al 1995, this volume). We've been frustrated by the inability to make retroviral injections at an earlier stage, which is why we have given our retrovirus a different coat that allows it to infect any species. We have given this retrovirus to Pasko Rakic, and we will give it to anyone else who is interested in using it in another species. We have started to do these experiments in ferrets where we have labelled subplate cells reasonably efficiently by injecting the retrovirus as early as E29. We expected to see more radially oriented clones at that stage, and we also thought that labelled subplate cells might be accompanied by at least some labelled sibling cells in the overlying cortical plate. We don't have many results on this so far, but we have observed that some of the cortical plate cells are dispersed widely in a similar manner. The labelled subplate cells do not seem to be accompanied by labelled neurons in the overlying cortical plate, but we haven't done any PCR analysis yet (Reid & Walsh 1994). E29 is still relatively late in terms of subplate formation in the ferret. I'm hoping that we can label several days earlier so that more preplate cells are labelled. The clonal patterns may then change again.

Bartlett: One would predict that the precursors which eventually give rise to the preplate will migrate rostrocaudally, and that they will give rise to progenitors of cortical clusters.

Walsh: Many things are possible. It may not be obligatory for multipotential cells to be migratory at all stages or at all times—migration may stop, start and stop again. It may just have been a fortunate accident that the cells are migratory at this stage because it seems to allow widespread clones to be broken into and analysed as several subunits.

Levitt: In the rodent the amount of rostrocaudal growth is much less between E12 and E14 than between E14 and E18. Therefore, one of the reasons why you might not see as dramatic a dispersion if you label earlier is that there's less rostrocaudal extension of the wall itself.

There is a rostrocaudal and a lateral to medial gradient of neurogenesis in the rodent, but it's not nearly as sharp as it is in the primate. However, because of this gradient, you may expect to see more dispersion in certain areas of the cortical wall than you actually do. An injection caudally and medially at E17 might, therefore, be equivalent to an injection rostrally and laterally at E15.

Walsh: Yes, that's an excellent point. Widespread clones observed after an injection at E17 have a distinctive shape and location. This does not necessarily agree with the gradients of neurogenesis revealed with [^{3}H]thymidine (Bayer & Altman 1991). We now have seven clones, either with the β-galactosidase or alkaline phosphatase library, that are oriented in the same way, i.e. parallel to the rhinal sulcus, in the far lateral neocortex. I would have predicted from the [^{3}H]thymidine gradients that widespread clones labelled late would be in the medial part of the cortex, but they always seem to be in the lateral part of the cortex.

Levitt: There are two populations of cells: moving and stationary. The stationary population is not necessarily committed to produce only cells of one type. If one moved them physically to another environment they might change and become something else.

Walsh: I agree. One thing that intrigues me about this model is that the adhesive properties of the progenitor cells may be regulated coordinately with their level of commitment, so that when a cell stops moving, it generates different sorts of progeny. Also, when cells stop moving, their stickiness changes and they may become more attached to radial glia.

O'Leary: One may also predict from your model that, if the caudal cortex was infected with a retrovirus, one would observe a higher frequency of isolated clones that didn't have clonally related siblings elsewhere in the cortex. Bearing in mind the rostral to caudal flow of the progenitor cells, do you see this?

Walsh: We can't determine where the cell was when it was labelled.

O'Leary: It is not necessary to determine that. You only have to determine that isolated clones, which do not have siblings elsewhere, are most commonly found in the caudal cortex.

Walsh: That's an interesting point, but one that we haven't looked at.

References

Barbe MF, Levitt P 1991 The early commitment of fetal neurons to limbic cortex. J Neurosci 11:519–533

Barbe MF, Levitt P 1992 Attraction of specific thalamic input by cerebral grafts depends on the molecular identity of the implant. Proc Natl Acad Sci USA 89: 3706–3710

Barbe MF, Levitt P 1995 Age-dependent specification of the cortico-cortical connections of cerebral grafts. J Neurosci 15:1819–1834

Bayer SA, Altman J 1991 Neocortical development. Raven Press, New York

Fishell G, Mason CA, Hatten ME 1993 Dispersion of neural progenitors within the germinal zones of the forebrain. Nature 362:636–638

Levitt P, Ferri R, Eagleson K 1995 Molecular contributions to cerebral cortex specification. In: Development of the cerebral cortex. Wiley, Chichester (Ciba Found Symp 193) p 200–213

O'Leary DDM, Borngasser DJ, Fox K, Schlaggar BL 1995 Plasticity in the development of neocortical areas. In: Development of the cerebral cortex. Wiley, Chichester (Ciba Found Symp 193) p 214–230

Porteus MH, Bulfone A, Lin JK, Puelles L, Lo LC, Rubenstein JLR 1994 *Dlx-2*, *Mash-1* and *Map-5* expression and bromodeoxyuridine incorporation define molecularly distinct cell populations in the embryonic mouse forebrain. J Neurosci 14:6370–6383

Rakic P 1988 Specification of cerebral cortical areas. Science 241:170–176

Reid C, Walsh C 1994 Lineage of subplate and cortical plate neurons in ferret studied with an alkaline phosphatase (AP) retroviral library. Soc Neurosci Abstr 24:1487

Walsh C, Cepko CL 1992 Widespread dispersion of neuronal clones across functional regions of the cerebral cortex. Science 255:434–440

Walsh C, Cepko CL 1993 Widespread clonal dispersion in proliferative layers of cerebral cortex. Nature 362:632–635

The cel lineage of neuronal subtypes in the mammalian cerebral cortex

J. G. Parnavelas, M. C. Mione and A. Lavdas

Department of Anatomy and Developmental Biology, University College London, Gower Street, London WC1E 6BT, UK

Abstract. We have studied the lineage relationships of pyramidal and non-pyramidal neurons, the principal neuronal types in the cerebral cortex, using a recombinant retrovirus that carries the gene encoding *Escherichia coli* β-galactosidase as a lineage marker. The phenotype of every cell of clones of β-galactosidase-labelled neurons generated by intraventricular injection of recombinant retrovirus in rat embryos at different stages of cortical neurogenesis was identified using light and electron microscopy as well as immunohistochemistry for known markers of neuronal subtypes. We found that clonally related neurons in adult rats showed the same morphological and neurotransmitter phenotypes, suggesting that lineages of pyramidal and non-pyramidal neurons are specified as early as E14, the time of onset of neurogenesis. However, when we followed the development of cortical cell lineages, we noted that a significant number of neuronal clones showed a mixed pyramidal/non-pyramidal cell composition during the first three weeks of life. We suggest that the change in the composition of neuronal clones between the third week of postnatal life and adulthood may either be due to changes in the phenotype of some developing neurons or, more likely, to selective cell death.

1995 Development of the cerebral cortex. Wiley, Chichester (Ciba Foundation Symposium 193) p 41–58

The cerebral cortex is, in many respects, a unique structure. It is similar to other parts of the central and peripheral nervous systems in that it is composed of an array of heterogeneous and highly specialized cells. However, in contrast to regions where different cell types are generated at different times (e.g. the spinal cord, Altman & Bayer 1984) or where heterogeneity is the result of the assemblage of nuclei of different origins (e.g. the thalamus, Jones 1985), the different cell types in the cortex are generated at the same time from a seemingly homogeneous population of neuroepithelial cells lining the embryonic telencephalic ventricles (Berry & Rogers 1965, Rakic 1974).

Neurons in the cerebral cortex are organized into functionally different areas that show distinctive cytoarchitectonic features. Cytoarchitectonic areas were initially defined solely on the basis of the size, shape and arrangement of the constituent neurons, but each area is also unique with regard to its afferent and efferent connections and the physiological properties of its neurons. All the areas of the neocortex have a common basic structure, with neurons arranged in six layers or laminae oriented parallel to the surface of the cortex. Each layer contains a complement of pyramidal and non-pyramidal neurons, the principal neuronal cell types of the cortex. The question of how this fine organization is achieved has been the subject of numerous studies during the last two centuries (see Ramón y Cajal 1960 for references).

It is now known that neurons and glia in the cerebral cortex acquire their phenotypes both from their genetic material and their environment (McConnell 1991). We examined the importance of inherited factors in determining the phenotype of cortical neurons by using a recombinant retrovirus as a lineage marker. The findings support the hypothesis that lineage plays an important role in determining the phenotype of cortical neurons and the transmitter(s) they contain. Work in the central nervous system of invertebrates also supports the notion that lineage influences greatly the acquisition of neuronal phenotype (Huff et al 1989, Sternberg et al 1992, see McConnell 1991 for review). However, neurons in other parts of the vertebrate central nervous system most likely acquire their phenotypes as a result of factors intervening after they have become postmitotic (retina, Turner et al 1990; spinal cord, Leber et al 1990; tectum, Gray & Sanes 1991).

One question that arose from our lineage studies was whether other aspects of a neuron's phenotype, such as those related with the acquisition of physiological properties, are also dictated by lineage. Calcium-binding proteins are used as markers of subpopulations of non-pyramidal neurons according to their connections and functional properties (Demeulemeester et al 1989, Baimbridge et al 1992, Hornung & Celio 1992). Therefore, we investigated whether clonally related non-pyramidal neurons expressed the same calcium-binding protein. This analysis showed that clonally related neurons do not contain the same calcium-binding protein(s), suggesting that the expression of these proteins is not genetically programmed and is likely to be induced by external factors. Knowledge of the roles of both the genetic and epigenetic influences and their relative importance in cell phenotype determination is essential if we are to understand what determines the choice of differentiation pathway taken by individual epithelial cells (progenitors) and when these choices are made during development.

Lineage of pyramidal and non-pyramidal neurons

All areas of the cerebral cortex contain two broad classes of neurons, the pyramidal and the non-pyramidal cells, in approximately the same proportions (Rockel et al 1980). These neurons show characteristic morphological

(Szentágothai 1973), neurochemical (Parnavelas et al 1989) and functional (Gilbert 1983) properties. Pyramidal neurons, the projection cells of the cerebral cortex, utilize the excitatory neurotransmitters L-glutamate (Glu) or L-aspartate at their sites of projection in the cortex and in a number of subcortical areas. The non-pyramidal neurons are the cortical interneurons that exert an inhibitory effect on cells of the cortex by releasing γ-aminobutyric acid (GABA). The pyramidal and non-pyramidal neurons in each of the six cortical layers show a characteristic size, shape and packing density. They also share physiological properties and axonal connections (Gilbert 1983). Several subtypes have been recognized on the basis of their location, as well as unique morphological, neurochemical or physiological features (Parnavelas et al 1989). A fundamental, yet unresolved, question is whether the different cell types that make up the cerebral cortex are generated by intrinsically different lineages. We studied the lineage relationships of pyramidal and non-pyramidal neurons in the cortex by using a recombinant retrovirus as a lineage marker (Sanes et al 1986, Price et al 1987). We examined clusters of clonally related neurons carrying the reporter gene *lacZ*, which encodes *Escherichia coli* β-galactosidase, in adult rats that had previously received one intraventricular injection of BAG (β-gal-at-gag) retrovirus (Price et al 1987) during cortical neurogenesis, i.e. between embryonic day (E) 14 and E21 (the first day of pregnancy was designated E1). The clonal relationship of β-galactosidase-labelled neurons in lightly infected brains (fewer than seven clones per hemisphere) was inferred by their close proximity within the cortex (maximum spread of 500 μm both in the anteroposterior and mediolateral dimensions) (see Mione et al 1994 for details). Our emphasis in this analysis was the reliable morphological identification of clonally related cells. We decided to use electron microscopy because the enzymatic reaction used to detect β-galactosidase produces an electron-dense precipitate, and it allows labelled pyramidal and non-pyramidal neurons to be unambiguously classified on the basis of well-established ultrastructural features (see Parnavelas et al 1991 for references). We also examined the neurochemical properties of clonally related cells using a postembedding immunohistochemical technique on resin-embedded, 1 μm thick sections (Mione et al 1994) (Fig. 1). We found that clonally related cells were immunoreactive for Glu or GABA. Glu and GABA are amino acid neurotransmitters contained in cortical pyramidal and non-pyramidal neurons, respectively. By applying both electron microscopy and immunohistochemistry for Glu and GABA, we were able to classify several hundred cells reliably. These cells made up nearly 100 clones from 32 lightly infected brains. We found that all but three clones were homogeneously composed of all pyramidal or all non-pyramidal neurons (Figs. 1,2A). The three mixed clones included pyramidal neurons and astrocytes. We concluded from these observations that separate progenitor cells for pyramidal and non-pyramidal neurons are present in the ventricular zone of the rat throughout the

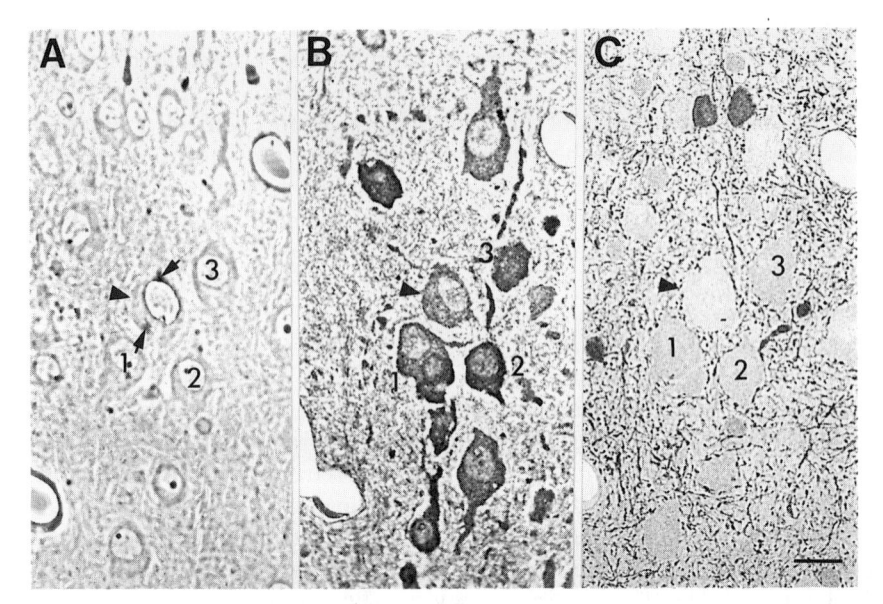

FIG. 1. Light micrographs of unstained and immunostained 1 μm thick sections taken through β-galactosidase+ (β-gal+) neurons in the cerebral cortex of rats injected with retrovirus at embryonic day (E) 16, showing a pyramidal neuron (arrowhead). This is part of a cluster of six clonally related pyramidal neurons (not shown). The pattern of β-gal staining in this cell was typical of many pyramidal neurons (A). It consisted of a ring of blue reaction product outlining the nucleus and spots of reaction product at the origin of two of the dendrites (arrows). After immunostaining, this neuron revealed immunoreactivity for L-glutamate (B), but not for γ-aminobutyric acid (C). The same neighbouring neurons (1, 2 and 3) are shown in these three consecutive sections. Bar = 20 μm.

period of neurogenesis. These findings show clearly that the lineages for pyramidal and non-pyramidal neurons diverge by the onset of neurogenesis (E14).

We performed a further immunohistochemical analysis in GABA-immunoreactive, clonally related neurons. This consisted of the localization of three calcium-binding proteins, parvalbumin, calbindin and calretinin, that are thought to characterize selective subpopulations of cortical non-pyramidal neurons (Demeulemeester et al 1989, Celio 1990). We used the postembedding immunohistochemical method on resin-embedded, 1 μm thick sections through clonally related neurons. The presence of these proteins has been correlated with specific afferent systems (Hornung & Celio 1992) and with common physiological properties (Baimbridge et al 1992). We were interested in whether these features could be dictated by lineage. In this case our study failed to demonstrate the same calcium-binding protein(s) in clonally related non-pyramidal neurons (Fig. 3), suggesting that the expression of these proteins

Lineage of neuronal subtypes

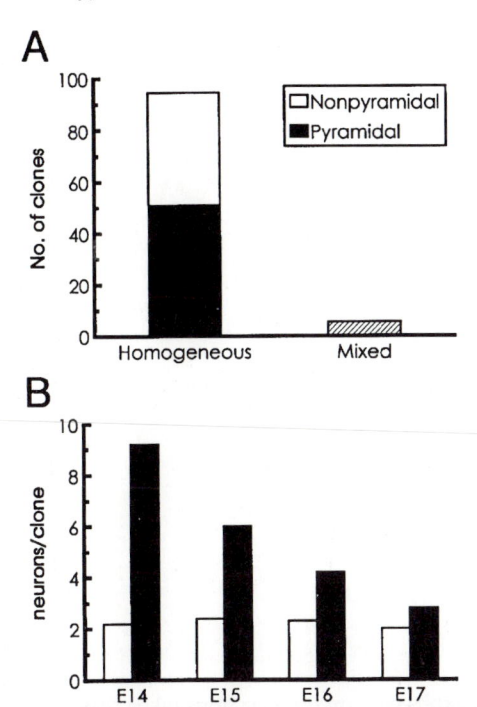

FIG. 2. (A) Histogram illustrating that the vast majority of neuronal clones examined in the cerebral cortex of the adult rat after injection of retrovirus at different embryonic stages were composed of either pyramidal or non-pyramidal neurons. Only a small proportion of clones contained cells of different phenotypes. (B) Histogram illustrating the size of pyramidal and non-pyramidal clones after injections of retrovirus at four embryonic (E) stages. Clones of pyramidal neurons were large in size after earlier injections, diminishing after later injections. Clones of non-pyramidal neurons were typically composed of only two or three neurons irrespective of the time of injection.

may be induced by environmental cues that reach the neurons after they attain their final positions in the cortex. This notion is in accordance with the late appearance of immunohistochemically detectable calcium-binding proteins in cortical interneurons (Alcántara et al 1993, Soriano et al 1992). Taken together, these results support the hypothesis that the expression of neuronal phenotypic traits in the cortex involves both genetic determinants and environmental cues.

We examined a large number of clonally related neurons from brains injected at different ages, and we observed remarkable differences in the number and laminar distribution of pyramidal and non-pyramidal neurons marked with retrovirus. Clones of non-pyramidal neurons were typically composed of two or three cells, and they resided in the cortical layers that were just being generated at the time of injection. Clones of pyramidal neurons were

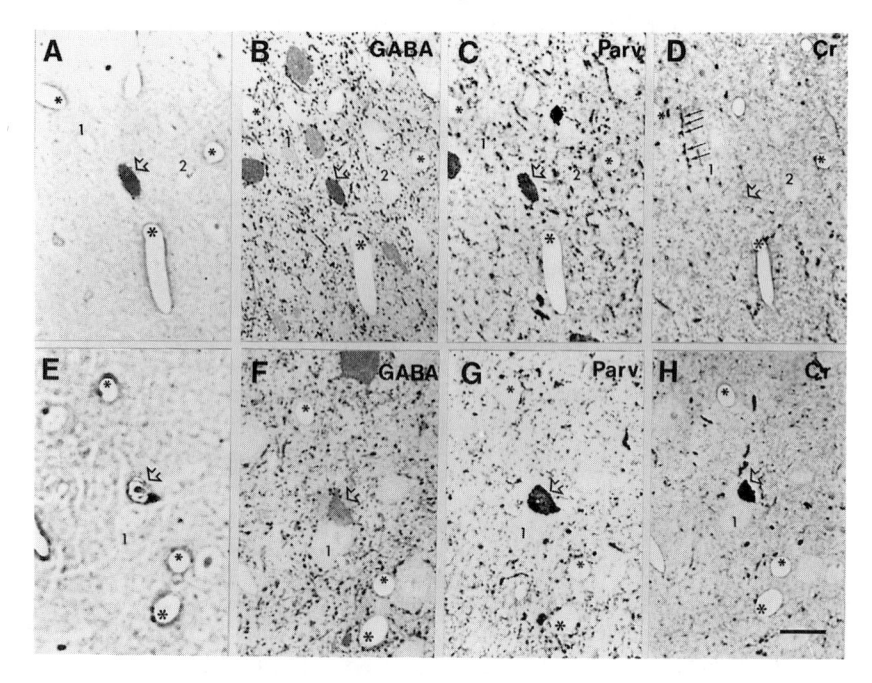

FIG. 3. Light microgaphs of 1 μm thick sections taken through two clonally related β-galactosidase+ (β-gal+) non-pyramidal neurons in the somatosensory cortex of a rat injected with retrovirus at embryonic day (E) 14. Figures A–D show one of the two cells (open arrow) in an unstained section, (A); after γ-aminobutyric acid (GABA) immunostaining (positive), (B); after immunostaining for parvalbumin (parv, positive), (C); after immunostaining for calretinin (cr, negative), (D). The same neighbouring neurons (1, 2) are shown in the four sections. Asterisks mark the same blood vessels, and small arrows in (D) point to calretinin-immunoreactive boutons. Figures E–H show the other cell (arrow) in an unstained section, (E); after immunostaining for GABA (positive), (F); after immunostaining for parvalbumin (positive), (G); after immunostaining for calretinin (positive), (H). The same neighbouring neuron (1) is shown in the four sections. Asterisks mark the same blood vessels. Note that the two cells of this cluster were homogeneously immunoreactive for GABA and parvalbumin, but not calretinin. Bar = 20 μm.

larger and were dispersed in several layers following earlier injections, and their size and laminar distribution were progressively reduced following later injections (Fig. 2B). This difference may explain the preponderance of pyramidal neurons in the mature cerebral cortex (Rockel et al 1980). This finding also suggests the existence of different mechanisms that generate the two main neuronal subtypes of the cerebral cortex. Thus, progenitors of non-pyramidal neurons are able to divide only once or twice to give rise to two or three postmitotic neurons; whereas progenitors of pyramidal neurons produce several postmitotic cells as well as several mitotically active progeny, which results in the appearance of larger clones (up to 23 cells after injection at E14). Therefore, we may speculate that the cells which give rise to non-pyramidal neurons arise from multipotential progenitors in the ventricular zone. These progenitors, which can generate progenitors of other cell types, including those that end up in the subventricular zone (Mione & Parnavelas 1994), coexist in the ventricular zone with cells committed to producing pyramidal neurons or glia (astrocytes or oligodendrocytes) (Luskin et al 1993) until at least E17. We consider it unlikely that non-pyramidal neurons arise from an ensemble of quiescent progenitors in the ventricular zone because earlier studies (Waechter & Jaensch 1972, Takahashi et al 1993) have shown that nearly all cells in the proliferative zone are mitotically active during the earlier stages of corticogenesis.

Development of neuronal lineages

A number of developmental events take place in the rat cerebral cortex during the first few weeks of postnatal life. These include the continuous production of glial cells, naturally occurring cell death, neuronal migration and the arrival of cortical afferents. The last two events are likely to expose late generated neurons to different external signals (Berry 1974, Ferrer et al 1992, Hatten 1993). What effect do these processes have on the composition of neuronal clones in the cortex? We investigated whether the apparent morphological and neurochemical homogeneity observed in clonally related neurons in the adult cortex is maintained during development.

We followed the development of cortical cell lineages labelled with retroviral injections at E16 in rats at postnatal day (P) 7, P14 and P21. We used GABA and Glu immunohistochemistry in the P7 rats to identify cell phenotype because earlier studies (Parnavelas & Lieberman 1979) suggested that cortical neurons at this age have not yet developed their characteristic morphological features. However, in both the P14 and P21 rats we used both electron microscopy and transmitter immunohistochemistry to identify cell phenotype. Analysis of a total of 58 clones from the three postnatal ages showed that a significant number at P7 and P14, and fewer at P21, showed mixed pyramidal/non-pyramidal composition (Figs. 4,5). These clones were typically composed of two or three cells, with one non-pyramidal neuron and one or two pyramidal cells.

We sometimes observed that these mixed neuronal clones at these ages contained cells immunoreactive for both Glu and GABA. In the overall population of cortical neurons these bireactive cells represented about 4% of all neurons at P7, 2% at P14 and less than 0.5% in adult rats. This suggests that the change in the composition of neuronal clones between the third and fourth week of postnatal life, when neuronal clones appear to be homogeneous, may be due to a change in the phenotype of some developing neurons. In the light of available evidence, it is possible that cells undergo a change in

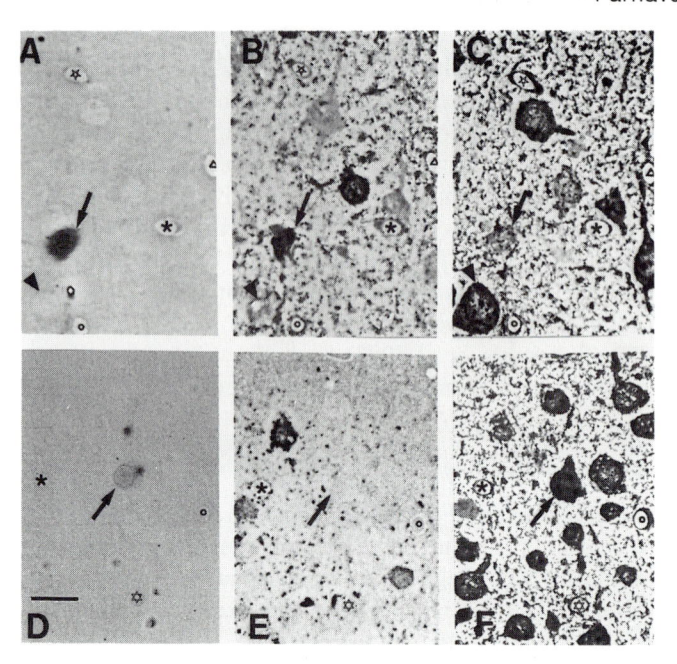

FIG. 4. An example of a two-cell mixed clone in layer V of a rat at postnatal day (P) 14. The first cell (arrows in A–C) is shown in a section stained histochemically for β-galactosidase (β-gal) (A) and in two consecutive sections stained immunohistochemically for γ-aminobutyric acid (GABA) (B) and glutamate (C). In A, the cell is labelled heavily, a feature characteristic of non-pyramidal neurons. This identity was confirmed in the subsequent sections because the cell was GABA positive (B) and glutamate negative (C). The second cell of the clone (arrows in D–F) shows features typical of many pyramidal neurons, and it was GABA negative (E) and glutamate positive (F). To aid identification of the cell in question, we have used symbols to show the position of the same blood vessel in each of the three serial selections. Nearby cells are indicated by arrowheads in A–C. Bar = 20 μm.

transmitter phenotype. For some neurons, maturation of a particular phenotype may involve a trial expression of one or the other (or both) neurotransmitters, and the bireactive neurons observed during the first three weeks of life may represent cells in the process of changing their phenotype. It is more difficult to envisage that cortical neurons undergo morphological transformation after the third postnatal week because this would involve a change of a full complement of nuclear, cytoplasmic, dendritic and synaptic features (Parnavelas & Lieberman 1979, Miller 1988). It is also possible that cell death could turn a number of mixed clones into homogeneous clones or single cells, depending on the original size of the clone and the number of cells eliminated. Although we do not have direct evidence to support this

Lineage of neuronal subtypes

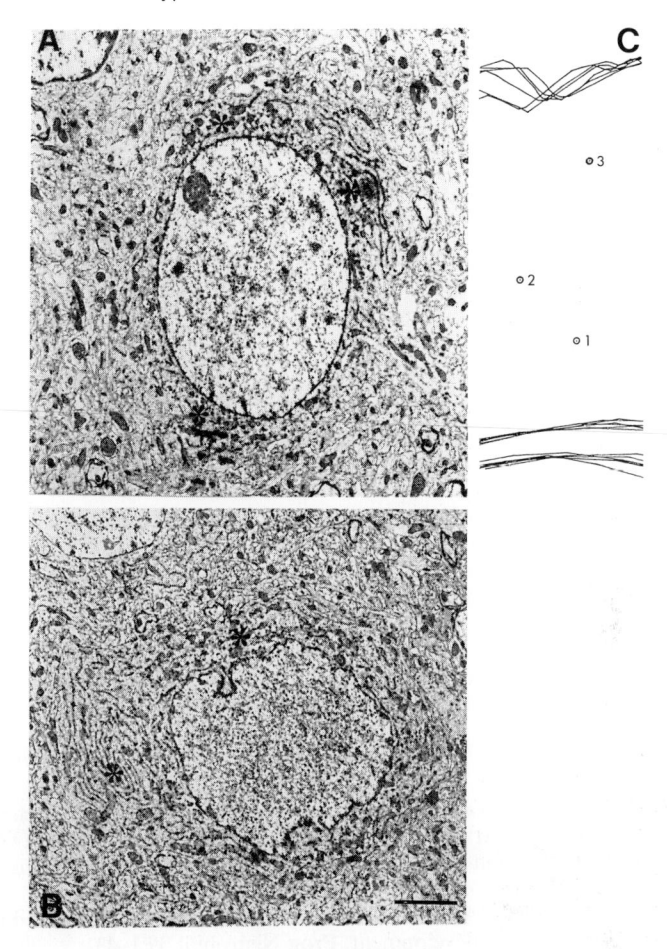

FIG. 5. An example of a three-cell mixed clone in the cerebral cortex of a rat at postnatal day (P) 14. Reconstruction of this clone from a number of consecutive sections is shown in C (pia towards the top). The first cell (A) shows nuclear, cytoplasmic and synaptic features typical of pyramidal neurons, whereas the second cell (B) shows features characteristic of non-pyramidal cells. In both electron micrographs, the β-galactosidase reaction product is associated with a number of organelles (asterisks). The phenotype of these cells was confirmed with γ-aminobutyric acid and glutamate immunocytochemistry. The third cell (not shown) was also identified as a non-pyramidal neuron. Bar = 2.5 μm.

interpretation, naturally occurring cell death is at its peak during the first three weeks of postnatal life in the rat (Ferrer et al 1992). During this time we have observed clones containing both pyramidal and non-pyramidal neurons. Non-pyramidal cells seem to be selectively affected in the process of naturally occurring cell death in the cortex (Parnavelas & Cavanagh 1988, Miller 1994). The finding of a significant increase in the number of single β-galactosidase-labelled cells between P7 and adulthood supports this interpretation (Mione et al 1994).

Acknowledgements

We are grateful for the excellent technical help provided by Peter Boardman, Brett Harris and John Cavanagh. Our work was supported by the Medical Research Council and the Wellcome Trust.

References

Alcántara S, Ferrer I, Soriano E 1993 Postnatal development of parvalbumin and calbindin D28K immunoreactivities in the cerebral cortex of the rat. Anat Embryol 188:63–73

Altman J, Bayer SA 1984 The development of the rat spinal cord. Adv Anat Embryol Cell Biol 85:1–164

Baimbridge KG, Celio MR, Rogers JH 1992 Calcium-binding proteins in the nervous system. Trends Neurosci 15:303–308

Berry M 1974 Development of the rat cerebral cortex. In: Gottlieb G (ed) Studies on the development of behavior and the nervous system, vol. 2: Aspects of neurogenesis. Academic Press, New York, p 7–67

Berry M, Rogers AW 1965 The migration of neuroblasts in the developing cerebral cortex. J Anat 99:691–709

Celio MR 1990 Calbindin D-28K and parvalbumin in the rat nervous system. Neuroscience 35:375–475

Demeulemeester H, Vandesande F, Orban GA, Heizmann CW, Pochet R 1989 Calbindin D-28K and parvalbumin immunoreactivity is confined to two separate neuronal subpopulations in the cat visual cortex, whereas partial coexistence is shown in the dorsal lateral geniculate nucleus. Neurosci Lett 99:6–11

Ferrer I, Soriano E, DelRio JA, Alcántara S, Auladell C 1992 Cell death and removal in the cerebral cortex during development. Prog Neurobiol 39:1–43

Gilbert CD 1983 Microcircuitry of the visual cortex. Annu Rev Neurosci 6:217–247

Gray GE, Sanes JR 1991 Migratory patterns and phenotypic choices of clonally related cells in the avian optic tectum. Neuron 6:211–225

Hatten ME 1993 The role of migration in central nervous system neuronal development. Curr Opin Neurobiol 3:38–44

Hornung J-P, Celio MR 1992 The selective innervation by serotinergic axons of calbindin-containing interneurons in the neocortex and hippocampus of the marmoset. J Comp Neurol 320:457–467

Huff R, Furst A, Mahowald AP 1989 *Drosophila* embryonic neuroblasts in culture: autonomous differentiation of specific neurotransmitters. Dev Biol 134:146–157

Jones EG 1985 The thalamus. Plenum Press, New York

Leber SM, Breedlove SM, Sanes JR 1990 Lineage, arrangement, and death of clonally related motoneurons in chick spinal cord. J Neurosci 10:2451–2462

Luskin MB, Parnavelas JG, Barfield JA 1993 Neurons, astrocytes and oligodendrocytes of the rat cerebral cortex originate from separate progenitor cells: an ultrastructural analysis of clonally related cells. J Neurosci 13:1730–1750

McConnell SK 1991 The generation of neuronal diversity in the central nervous system. Annu Rev Neurosci 14:269–300
Miller MW 1988 Development of projection and local circuit neurons in neocortex. In: Peters A, Jones EG (eds) Cerebral cortex, vol. 7: Development and maturation of cerebral cortex. Plenum Press, New York, p 133–175
Miller MW 1994 Relationship of the time of origin and death of neurons in rat somatosensory cortex: barrel versus septal cortex and projection versus local circuit neurons. Soc Neurosci Abstr 20:689
Mione MC, Parnavelas JG 1994 Origin of subventricular zone cells. Soc Neurosci Abstr 20:1275
Mione MC, Danevic C, Boardman P, Harris B, Parnavelas JG 1994 Lineage analysis reveals neurotransmitter (GABA or glutamate) but not calcium-binding protein homogeneity in clonally related cortical neurons. J Neurosci 14:107–123
Parnavelas JG, Cavanagh ME 1988 Transient expression of neurotransmitters in the developing neocortex. Trends Neurosci 11:92–93
Parnavelas JG, Lieberman AR 1979 An ultrastructural study of the maturation of neuronal somata in the visual cortex of the rat. Anat Embryol 157:311–328
Parnavelas JG, Dinopoulos A, Davies SW 1989 The central visual pathways. In: Björklund A, Hökfelt T, Swanson SW (eds) Handbook of chemical neuroanatomy, vol. 7: Integrated systems of the CNS, part II, Elsevier, Amsterdam, p 1–164
Parnavelas JG, Barfield JA, Franke E, Luskin MB 1991 Separate progenitor cells give rise to pyramidal and nonpyramidal neurons in the rat telencephalon. Cereb Cortex 1:463–468
Price J, Turner D, Cepko C 1987 Lineage analysis in the vertebrate nervous system by retrovirus-mediated gene transfer. Proc Natl Acad Sci USA 84:156–160
Rakic P 1974 Neurons in rhesus monkey visual cortex: systematic relation between time of origin and eventual disposition. Science 183:425–427
Ramón y Cajal S 1960 Studies on vertebrate neurogenesis (translated from French by L. Guth). Charles C. Thomas, Springfield, Il, p 325–335
Rockel AJ, Hiorns RW, Powell TPS 1980 The basic uniformity in structure of the neocortex. Brain 103:221–244
Sanes JR, Rubenstein JLR, Nicolas J-F 1986 Use of a recombinant retrovirus to study post-implantation cell lineage in mouse embryos. EMBO J 5:3133–3142
Soriano E, Del Rio JA, Ferrer I, Auladell C, De Lecea L, Alcántara S 1992 Late appearance of parvalbumin-immunoreactive neurons in the rodent cerebral cortex does not follow an 'inside-out' sequence. Neurosci Lett 142:147–150
Sternberg PW, Liu K, Chamberlin HM 1992 Specification of neuronal identity in Caenorhabditis elegans. In: Shankland M, Macagno ER (eds) Determinants of neuronal identify. Academic Press, San Diego, p 1–45
Szentágothai J 1973 Synaptology of the visual cortex. In: Jung R (ed) Handbook of Sensory Physiology, vol VII/3: Central processing of visual information. Springer-Verlag, Berlin, part B, p 269–324
Takahashi T, Nowakowski RS, Caviness VS Jr 1993 Cell cycle parameters and patterns of nuclear movement in the neocortical proliferative zone of the fetal mouse. J Neurosci 13:820–833
Turner DL, Snyder EY, Cepko CL 1990 Lineage-independent determination of cell type in the embryonic mouse retina. Neuron 4:833–845
Waechter RV, Jaensch B 1972 Generation times of the matrix cells during embryonic brain development: an autoradiographic study in rats. Brain Res 46:235–250

DISCUSSION

Blakemore: Some of the β-galactosidase labelling in your embedded sections looks surprisingly weak. Is it possible that expression of the *lacZ* construct may be selectively down-regulated so that, for example, the expression of the gene may be down-regulated more quickly in non-pyramidal than in pyramidal cells? Selective down-regulation might account for the different number of cells per local clone.

Parnavelas: I am puzzled by your comment. The labelling in my sections was robust. The sections you are referring to were not tampered with in any way, and they were photographed before they were postfixed and processed for electron microscopy.

J. Price: Weak labelling is probably not due to alterations in gene regulation. There is a technical problem in that β-galactosidase staining fades quickly, so one often sees a faded picture of a cell that was originally clearly positive. A high level of expression is generally observed with the retrovirus that John Parnavelas is using, and detecting positive cells isn't a problem. It is possible that the way John Parnavelas makes these thinner sections and mounts them may exacerbate the faded appearance.

Parnavelas: Fading of stained cells is not a problem in our experiments. Labelled cells do not fade after postfixation with osmium tetroxide and subsequent processing for electron microscopy. We are able to detect product with the light microscope in sections that are 0.5–1.0 μm thick.

Kennedy: The presence of strongly labelled and weakly labelled cells within a single clone may determine whether different degrees of retroviral expression occurs.

Parnavelas: The degree of labelling does not differ appreciably among cells of a given clone. There are a few examples in our material where some cells in a clone show extensive dendrite staining and others don't, but in the majority of clones, cells stain remarkably similarly. Small differences in the intensity of staining may depend on factors such as the metabolic activity of individual cells.

Kennedy: Is the consensus among people using this technique that the degree of expression within the cells is equal?

J. Price: It is not necessarily the degree of expression that may change, rather it may be the intensity of the histochemical labelling. The intensity of this blue stain definitely varies between cells in culture, but we do not know what this represents. From my experience, the reaction can continue for a prolonged period with little background, which generates a situation where distinguishing positive cells is not a problem.

Walsh: I was also frustrated by the variable intensity of the blue stain. That's one of the reasons why we switched to another marker which is better at labelling different cell types. We have also performed controls to address

Lineage of neuronal subtypes

whether unlabelled cells are infected with the retrovirus. Using PCR analysis, we removed the labelled neurons and then looked in the remaining unlabelled tissue to see if the same viral tag is present in the cells that are not expressing β-galactosidase. We've performed this control experiment a couple of times with cortical neurons, and we have not found any cells hiding in the tissue in the cortex, although cells can be found in the olfactory bulb, subcortical white matter or striatum (Walsh & Cepko 1992). Another control we've done is to look at the laminar location of the different cortical cell types labelled by injections at different ages and compare it with the tritiated thymidine results on the formation of cell types (Bayer & Altman 1991). This is easier to do in the ferret than in the rat. We found that in the ferret, when we injected early, cells in all the different layers were labelled; but, if we inject later on, the label is limited to the superficial layers. However, if we take the results from an embryonic day (E) 29 retroviral label and make a correction for the differences in cell density between the different cortical layers (Beaulieu & Colonnier 1989), we find that although cells that are postmitotic at the time of injection label inefficiently, cells formed after the time of the injection label remarkably equally (Reid & Walsh 1994). This suggests that the progenitors are behaving the same as unlabelled progenitors.

Jones: In my opinion, the intensity of staining in all of those cells was equal, but the extent of staining in the processes was clearly not equal. John Parnavelas, do you agree with that?

Parnavelas: Yes.

Jones: The extent of staining depends on the size of the cell and the size of its processes—a large cell with thick processes will usually be stained more heavily. It will not necessarily have a more intense stain, but the stain will extend further throughout the processes. With any kind of stain, it's usually easier to identify pyramidal cells because they are large and have thicker dendrites than non-pyramidal cells. It may, therefore, be erroneous to say that levels of expression are different in these cells. Instead, what is different is the capacity of the levels of expression to stain the cell in a way which enables us to categorize it.

I compliment you on the meticulous way in which you have tried to identify the derivatives of each precursor cell. You have characterized the cells on the basis of being pyramidal or non-pyramidal, but is it possible to take this analysis further? A more fundamental classification of cells in the cortex is whether they are glutamatergic or γ-aminobutyric acid (GABA)ergic. Your classification does not take into account the non-pyramidal cells that are glutamatergic, namely the spiny stellate cells. Clones that are primarily glutamatergic would include both pyramidal and spiny stellate cells, whereas clones that are GABAergic would include all other non-pyramidal cells. This is a fundamental distinction. Then you could go back further and ask what is the elementary classification, the elementary delineation, of a cell before this fundamental glutamatergic/GABAergic divergence occurs?

There are two things that I observed in your results. Firstly, pyramidal cells are present throughout the thickness of the cortex in your preparations. They may be distinguished into an elementary pyramidal subclass, and then a further delineation of these cells may be made on the basis of their axonal projections and their connectivity, which is a lamina-specific classification. I realize that there are subtle distinctions, but there must be something which is superimposed upon that fundamental delineation into a pyramidal subclass. This appeals to me personally because the pyramidal cell is an archetype that is then modulated and determined by other epigenetic influences. I'm not sure that I can make a similar archetypal plea for the GABAergic non-pyramidal cells because these may fall into six or seven different classes which are fundamentally different in their morphology and in their target cells. In the two categories of clones that you have, one could ask whether you could delineate clones which contain only one class of non-pyramidal cell, or whether there are multiple classes. In the latter case, the modifications that have to occur in terms of cell type are larger than those required to modify pyramidal cells.

Parnavelas: We are currently doing experiments to address this question. We are trying to distinguish between the subpopulations of non-pyramidal cells on the basis of morphology and peptide content.

You also mentioned spiny stellate cells in the cerebral cortex. We have analysed some of these cells, using both immunocytochemistry and electron microscopy, and we have shown that they are glutamatergic and that they possess some of the features of pyramidal neurons. However, it is difficult to analyse them extensively in the rat cerebral cortex because they are not abundant.

Jones: Perhaps the fundamental difference is that there is a modulation of the spiny glutamatergic cells which leads to most cells acquiring a pyramidal shape in the rat, whereas in primates and cats, a more extensive remodelling occurs so that some of these cells acquire a stellate shape.

Innocenti: Spiny stellate cells have a number of properties in common with pyramidal cells; for example, their axons go to other areas of the cortex. They are a relatively recent acquisition of the cerebral cortex because they are not present in the dolphin or hedgehog, for example. Recently, we found that at least one type of these neurons, those with callosal axons, are initially indistinguishable from pyramidal cells (Vercelli et al 1992). They develop an apical dendrite which subsequently atrophies, transforming them from a pyramidal to a spiny stellate morphology. This may also have happened during evolution. L. Katz has made some similar observations in the somatosensory cortex of the rat, although these have not yet been published.

Bolz: There are pyramidal cells and stellate cells in layer IV of cat visual cortex, but there is a fundamental difference in the way in which they respond to thalamic afferents (Bolz et al 1995, this volume, Kossel et al 1995). I was surprised to see that about 25% of the cells in layer IV are pyramidal cells, 50%

of the cells are spiny stellate cells and 10% of the cells have a morphology which is in-between. Lorente de Nó (1938) might have called these cells 'star pyramids'. Spiny stellate cells close to the upper and lower margin of geniculocortical afferents and near the borders between ocular dominance columns have highly asymmetric dendritic fields. Pyramidal cells in layer IV, however, do not exhibit dendritic asymmetries near laminar or columnar borders. Thus, spiny stellate cells and pyramidal cells represent two distinct cell types, rather than a continuum of cell morphologies.

LaMantia: John Parnavelas has described a single binary choice, i.e. whether a cell becomes pyramidal or non-pyramidal. Within that, there is a matrix of other properties which some cells acquire. There are other places in the nervous system where dichotomies of cell types are present that don't necessarily address all the phenotypic characteristics. For example, in the spinal cord, there is a motor versus sensory dichotomy based on position. How is a dichotomy of precursor cells set up in the ventricular zone where the cells are all squeezed in next to each other? How can the binary choice that is clearly observed in your results be achieved mechanistically?

Parnavelas: We would like to address that question with molecular biology.

LaMantia: At some point you would expect to have more mixed clones than you actually observe.

J. Price: It is difficult to argue precisely what you would expect.

LaMantia: Yes, but there is still the question of whether or not at some point these two precursor populations ever merge into one pre-precursor.

Blakemore: Can anyone comment on clonal relationships between preplate cells and true cortical plate cells? Has anyone injected early enough?

Walsh: We've labelled the subplate cells, but we haven't done any PCR analysis yet. I hesitate to talk about the results without the PCR analysis, but the retrovirally labelled subplate cells do tend to be isolated (Reid & Walsh 1994).

Molnár: There is also an interesting difference during X-chromosome inactivation—both labelled and unlabelled columns are present in the cortical plate but not in the subplate (Tan & Breen 1993).

Rakic: One reason why the subplate is labelled uniformly may be because X-chromosome inactivation occurs at midgestation (between E7.5 and E9 or E10), so subplate cells are generated and express β-galactosidase before X-chromosome inactivation occurs. If they were generated after inactivation, only half of them would express β-galactosidase.

Molnár: But the subplate is generated between E11–13 in the mouse.

Rakic: X-chromosome inactivation may not occur in all cells and all animals at the same time.

Molnár: Yes, but there is still a discrepancy in the timing. Even if we make the assumption that in the mouse all the subplate neurons are generated by E10, X-chromosome inactivation still occurs at E8.5.

Rakic: It may not be exactly E8.5. However, you may be correct, it is possible that there is another explanation.

Boncinelli: Which cell types are found in layer I?

Rakic: Layer I neurons in rodents consist of cells that are generated early. However, in primates layer I contains other cells which are presumably generated later.

O'Leary: One of the major differences in the cortex in terms of projection is the distinction between callosal neurons and the layer V subcortically projecting neurons (Koester & O'Leary 1994). These two cell types could have different lineages. Some evidence for this possibility may already be available. For instance, the Walsh & Cepko (1988) paper illustrated some isolated layer V clones, in which several cells were scattered tangentially over about 500 µm. This may suggest that layer V subcortically projecting cells might be derived from a different lineage than callosally projecting cells, which are found throughout all layers. Do you have any evidence that pyramidal cell clones which are localized to a particular layer are typically found in layer V?

Parnavelas: We rarely observe this to be the case. Pyramidal clones that result from early injections are invariably distributed over a number of layers. If the retrovirus is injected later, e.g. at E16 or E17, the clones tend to be confined to fewer layers, usually one or two adjacent layers. I don't recall an example where a cluster of pyramidal cells was localized in layer V.

Rakic: Dennis O'Leary made a thought-provoking comment on the possible relationship between cell lineages and types of prospective connectivity. I would like to give a concrete example of a possible test for this. A study using double-labelling methods has indicated that a subclass of cells which migrate to the superficial layers of the monkey frontal cortex extend their axons to the contralateral hemisphere (Schwartz et al 1991). [^{3}H]Thymidine labelling indicates that these cells are not subplate neurons, which are generated earlier, but are callosal neurons that are generated later and are migrating to layers II and III. Other cells that migrate at the same time across the same terrain do not have axons that extend to the contralateral side. Presumably, many of these cells form local circuit neurons in layers II and III. Thus, there is a distinction between cells as to the type of their prospective connectivity before they reach the same layer of the cortical plate. Combining similar anatomical approaches with the retroviral gene transfer labelling method could, perhaps, provide a more definitive answer regarding whether these two basic cell classes come from separate clones.

Ghosh: Is it possible to address the issue of callosal versus subcortical projection directly by repeating your experiment, but also labelling the cells retrogradely from the other hemisphere or from subcortical targets? This may address the question of whether pyramidal cells in the same layer that project to one target or another are related by lineage.

Parnavelas: That was one of our original ideas. But, when we saw that clonally related cells, especially after early injections, were spread over most of the cortical thickness, we decided that there was no point in doing that experiment. Our results suggested that clonally related pyramidal neurons project to different targets.

Ghosh: Some models of laminar cell fate determination depend on the presence of a certain molecule in the ventricular zone at a particular developmental time. However, the co-generation of cells in layer V, which project to distinct targets, places additional constraints on the interactions within the ventricular zone that may specify cell fate.

J. Price: I would like to add a cautionary note in relation to this discussion. It is a mistake to think that cell lineage has got to follow anatomy. The distinctions that one sees between different cell types are obviously genuine differences, but the factors that mould cell lineage will be driven by developmental and evolutionary events. These events will not necessarily lead to the most related cells having the most similar anatomy. Let us take the retina as an example. People originally thought that retinal cells which looked similar were related in terms of their lineage. But, when Turner et al (Turner & Cepko 1987, Turner et al 1990) looked at cell lineage in the retina, they discovered that all the cells came from one population of precursor cells, and that these precursors could make any or all of the retinal cell types. This demonstrates that it is a mistake to argue backwards from anatomy towards lineage.

Innocenti: But the pyramidal cell that maintains an apical dendrite and the pyramidal cell that loses an apical dendrite must be different in some way.

J. Price: That may be the case, but I'm cautioning against believing that two cells which appear to be similar in a variety of different ways are more related than two cells which are more dissimilar. It's not something that we can justify a priori, even though it might turn out to be correct.

Let me give you another example. Nerve cells and muscle cells appear all over the *Caenorhabditis elegans* lineage tree. In other words, two entirely different lineages can give rise to two cells that are morphologically identical. That's an extreme case, and I'm not saying that it also occurs in mammals. If two pyramidal neuronal types shared the same developmental pattern until late in development, at which point their behaviour diverged slightly, I would be surprised if they did not share a common lineage. Nonetheless, I'm just making a general point that you can't assume that anatomical similarity equals common lineage.

Innocenti: This is the same situation that you may have experienced when you are in another continent, and you meet somebody who looks exactly like a friend of yours who lives 3000 km away. They may not be clonally related, but they must have some common genes.

J. Price: That's right, your friend is bald and the chap you're meeting now is bald, but this doesn't mean that they're related.

Ghosh: I would like to make another point about the relationship between cell morphology and cell lineage. In *Drosophila* neurogenesis many genes, such as *notch*, that influence cell–cell interactions can regulate neuronal differentiation (Jan & Jan 1995). Although neurons are distributed widely, each neuron is generated independently from a small cluster of four precursor cells—one cell becomes a neuron and other three cells become supporting cells. In this case, the interpretation that two neurons are related clonally because of their similar morphology would be misleading.

References

Bayer SA, Altman J 1991 Neocortical Development. Raven Press, New York

Beaulieu C, Colonnier M 1989 Number of neurons in individual laminae of area-3B, area-4-γ, and area-6A-alpha of the cat cerebral cortex: a comparison with major visual areas. J Comp Neurol 279:228–234

Bolz J, Kossel A, Bagnard D 1995 The specificity of interactions between cortex and thalamus. In: Development of the cerebral cortex. Wiley, Chichester (Ciba Found Symp 193) p 173–191

Jan YN, Jan LY 1995 Maggot's hair and bug's eye: role of cell interactions and intrinsic factors in cell fate specification. Neuron 14:1–5

Koester SE, O'Leary DDM 1994 Development of projection neurons in mammalian cerebral cortex. Prog Brain Res 102:207–215

Kossel A, Lowel S, Bolz J 1995 Relationships between dendritic fields and functional architecture in striate cortex of normal and visually deprived cats. J Neurosci 15:3913–3926

Lorente de Nó R 1938 Cerebral cortex: architecture, intracortical connections, motor projections. In: Fulton JF (ed) The physiology of the nervous system. Oxford University Press, London, p 291–329

Reid C, Walsh C 1994 Lineage of subplate and cortical plate neurons in ferret studied with an alkaline phosphatase (AP) retroviral library. Soc Neurosci Abstr 24:1487

Schwartz ML, Rakic P, Goldman-Rakic PS 1991 Early phenotype expression of cortical neurons: evidence that a subclass of migrating neurons has callosal axons. Proc Natl Acad Sci USA 88:1354–1358

Tan SS, Breen S 1993 Radial mosaicism and tangential cell dispersion both contribute to mouse neocortical development. Nature 362:638–640

Turner D, Cepko C 1987 Cell lineage in the rat retina: a common progenitor for neurons and glia persists late in development. Nature 328:131–136

Turner DL, Snyder EY, Cepko CL 1990 Lineage-independent determination of cell type in the embryonic mouse retina. Neuron 4:833–845

Vercelli A, Assal F, Innocenti GM 1992 Emergence of callosally projecting neurons with stellate morphology in the visual cortex of the kitten. Exp Brain Res 90:346–358

Walsh C, Cepko CL 1988 Clonally related cortical cells show several migration patterns. Science 241:1342–1345

Walsh C, Cepko CL 1992 Widespread dispersion of neuronal clones across functional regions of the cerebral cortex. Science 255:434–440

General discussion I

Dendritic differentiation

Blakemore: It's clear that some of the diversity seen within individual clones could be attributed either to preprogrammed divergence of phenotype, or to epigenetic influences on individual cells during or after migration. Consider the two basic classes of layer V pyramidal cells—those projecting to the brainstem and those projecting to other cortical regions. The first obvious event that distinguishes the two types is the choice of axon pathway. This happens very early on, some days before birth in the rat, perhaps even before the cells have completed their migration. The cells remain morphologically indistinguishable until several days after birth and their distinctive electrophysical differences do not start to emerge until some two weeks after birth. It seems likely, therefore, that inherent properties of these neurons expressed differentially soon after mitosis determine the selection of pathways by their axons. However, it is conceivable that some later stages of differentiation are influenced by extrinsic events; for example, the transport of factors from target structures that might subsequently influence the phenotype.

LaMantia: Dendritic differentiation in the peripheral ganglia and the spinal cord is affected by targets, innervation and the establishment of subsequent connections. Consequently, this is a situation where differentiation may not depend so much on lineages, but on targets or sensitivity to a particular trophic factor or activity-dependent mechanism. It first seems that there are two precursor populations that make either one kind of dendrite or another, but what might be occurring instead is that there are just two precursor populations that are specified positionally in a particular region of the cortex, and that the other characteristics are acquired subsequently.

Levitt: Dendritic modelling involves detailed changes, but changes that are laid down on a basic plan. One can manipulate dendritic and axonal form quite dramatically, but I wonder if one is really changing the basic phenotypic properties of the cell. I would suggest that it is not possible to take a non-pyramidal stellate neuron and make it into a pyramidal cell.

LaMantia: But several people (Peinzdo & Katz 1990, Koester & O'Leary 1992a) have shown that it is possible to do that.

Bolz: I would like to comment on the phenotypic specification of two morphologically distinct types of pyramidal cells that project to different targets. In the primary visual cortex, corticotectal cells in layer V always have a prominent apical dendrite that forms a large tuft in layer I. Corticocortical cells in the same layer, either projecting contralaterally via the corpus callosum (Hübner & Bolz 1988) or ipsilaterally to higher visual areas (Hübner et al 1990), have a different morphology—their apical dendrite is short and they have fewer basal dendrites than corticotectal cells. Dennis O'Leary (Koester & O'Leary 1992a) and ourselves (Bolz et al 1991) looked at these cells very early in development, and we both found a common morphological prototype for cortical and subcortical projecting pyramidal cells in layer V. This suggests that the association between cell morphology and efferent target occurs at a later stage in cortical development. We took the immature cells in organotypic slice cultures before they reached their target, and we found that they developed their characteristic morphological features *in vitro*, i.e. in the absence of their projection target (Bolz et al 1991). Therefore, there is a significant degree of intrinsic specification in the development of cortical projections; that is, cortical cells know which target to innervate and which shape to develop. The difference in their axonal projections (callosal axons first grow towards the midline, whereas corticotectal axons first grow towards the internal capsule) is extremely sharp, and their axonal trajectories are already different once they first reach the white matter. This suggests that the decision where to grow is made at this early stage.

Blakemore: But the differences in soma-dendritic morphology do not appear until some time after birth, whereas the axons unerringly select their routes for growth at a much earlier stage.

Bolz: The axons have to make the decision whether to turn left or right. There must be cues present which tell the axons where to go. The fact that they respond differentially to these cues, even though they are still migrating, suggests that these cells already have a different identity. If you take these cells, not in organotypic slice cultures, but dissociated *in vitro*, this morphological distinction does not occur (Huettner & Baughman 1986). This suggests that cell identity and local interactions are involved.

Innocenti: In relation to this possibility that cortical neurons develop specialized dendritic phenotypes in the absence of their target, I should mention that we have identified dendritic differences between neurons that maintain their callosal axon in development and neurons that eliminate this axon (Weisskopf & Innocenti 1991). However, these differences remained when the contralateral target site was ablated, indicating that the target was not required for the particular dendritic phenotype that we were looking at. On the other hand, during normal development *in vivo* (and also presumably in culture), cortical axons have numerous initial collaterals. These collaterals could provide the retrograde signal required for establishing and maintaining dendrites in the absence of contact with other targets.

O'Leary: I agree with Jürgen Bolz. The decision to be a callosally or subcortically projecting neuron seems to be made extremely early, at least in rodents (Koester & O'Leary 1993). However, during the first four days of cortical axogenesis in the rodent, axons are highly polarized in their growth and they are directed laterally towards the internal capsule, i.e. along the subcortical trajectory (Koester & O'Leary 1992b, 1994). Few, if any, axons are directed medially towards the corpus callosum, even though many callosal cells have been generated and have already migrated into place. Therefore, their axogenesis is either delayed, that is the cells may wait several days before they begin to elaborate an axon, or alternatively, their axon grows radially out of the cortical plate but temporarily stalls upon reaching the point in its trajectory at which it makes a decision to go medially or laterally.

Cell lineage

Parnavelas: I would like to change the topic of this general discussion and return to Chris Walsh's model of cell lineage. When we look at sections through the ventricular zone with an electron microscope, we find that many epithelial cells form numerous gap junctions with neighbouring cells. One would predict from Chris Walsh's model that progenitor cells couple and uncouple continuously with epithelial cells in the ventricular zone.

Rubenstein: In that respect, can you infect all the cells in the ventricular zone, including the ones that are coupled to each other via gap junctions, or can you only infect those subclasses of cells within the ventricular zone that are more mobile?

Walsh: It is impossible to infect all the cells in the ventricular zone.

Rubenstein: What percentage do you infect?

Walsh: Only a small percentage, but we design these experiments to limit infection. It is possible that retrovirally labelled progenitors do not behave in the same manner as normal progenitors. However, one way to answer this is to determine whether the retrovirally labelled progenitor cells make the same ratios of the same types of neurons at the same times during development as unlabelled progenitors of the cortex. The simplest way to do that is to look at the pattern of retroviral labelling at different ages and see whether similar proportions of cells are labelled in the different layers of the mature cortex. One can infect ferrets with retrovirus at embryonic day (E) 29, E33 or E40 (Reid & Walsh 1994). At E40, only cells in layer II are labelled because it is the only layer that is still being generated to any significant extent. If one infects at an earlier stage, the superficial neurons and relatively smaller proportions of the deeper layers are predominantly labelled, as one would predict. This is also due to the fact that there are simply more neurons in the upper cortical layers than there are in the deeper cortical layers. One can calculate a corrected labelling index that controls for the counts of labelled cells in terms of the laminar differences in cell density. One then finds that infecting at E33 produces equal proportions of labelled cells in layers II/III and IV, whereas labelling indexes in layer V, layer VI and the subplate are lower. If one infects at E29, there is an equal amount of labelling for all of the different layers, with the exception of layer VI and the subplate. This suggests that retrovirus-labelled progenitors generate approximately the same numbers of neurons in layer II, III, IV and V as unlabelled progenitors in the same brain.

Rakic: This does not answer the question of whether retroviral vectors label all subtypes of cell precursors equally.

J. Price: I agree and the same is true for multiple cell types. One can be reasonably sure that it is possible to label all the precursors because it is possible to label all the cell types, but one can't assume that all the precursors are labelled in the appropriate proportions that are represented in the ventricular zone. This is because other unknown parameters may be involved. For example, the density of receptors for the retrovirus might vary between precursor cell types.

Kennedy: John Parnavelas, did you not find that late injections give rise to clusters of non-pyramidal cells restricted to the upper layers, and that early injections give rise to clusters of non-pyramidal cells restricted to the lower layers? Pyramidal clusters following early injections are the only type of cluster that span the full width of the cortex. If you don't believe that non-proliferating cells exist in the ventricular zone, where are the precursors of the cells that give rise to the upper layer non-pyramidal clusters?

Parnavelas: Early injections tend to label non-pyramidal cells that reside in the lower layers of the cerebral cortex. The results of our lineage studies agree with earlier analyses using [^{3}H]thymidine, which showed that non-pyramidal cells are generated in an inside out sequence but are more widespread than projection neurons born at the same time.

Blakemore: Henry Kennedy, are you asking what the progenitors of the upper layer non-pyramidal cells are doing early on in neurogenesis?

Kennedy: Yes.

Blakemore: Presumably, some of them might not have committed themselves to asymmetric division. Perhaps they undergo symmetrical proliferative division at an early stage and produce other precursors but not neurons. Many of the progeny of early divisions are new progenitors that may stay in the ventricular or subventricular zone. If they also tend to migrate laterally, their daughter neurons would be scattered and they wouldn't have been seen as clonally related by John Parnavelas.

Parnavelas: That's possible. However, it is unlikely that non-pyramidal neurons arise from dormant progenitors in the ventricular zone because earlier studies (Waechter & Jaensch 1972) have shown that nearly all cells in the proliferative zone were mitotically active during the earlier stages of corticogenesis. We speculate that the cells giving rise to non-pyramidal cells arise from multipotential progenitors in the ventricular zone. These progenitors must coexist in the ventricular zone with the population of committed progenitors producing pyramidal neurons or glia.

O'Leary: There's a disagreement between these two sets of results. Chris Walsh finds mixed clones and John Parnavelas finds either homogeneous pyramidal clones or homogeneous non-pyramidal clones. Chris Walsh's analysis has tended to focus on migratory progenitor cells, i.e. he's described examples of mixed clones within the migratory progenitor cell population. In contrast, John Parnavelas' analysis may favour the static progenitor cell population. Is it possible that migratory and static progenitors play to different rules?

Walsh: A single subunit of a widespread clone is not clearly distinguishable from a clustered clone, so one would not know just by looking at a cluster of neurons that the cluster also had sibling cells 2–3 mm or 8–10 mm away. John Parnavelas' working definition of a 'clone' as cells that are within about 500 µm and that are separated from other labelled cells is a limited definition. He's looking at what we would call a single subunit, without regard to whether that same clone has other subunits elsewhere. The model that I presented is intended as a general model. John Parnavelas is studying the terminal branches of clones that the retroviral library shows to be more widespread lineage trees. Our results and his results are consistent if you look at it in that way.

O'Leary: So clonal restrictions are basically artefacts of the technique?

Walsh: Artefact isn't the word I'd choose. However, the presence of multiple clusters with different fates in the same clone are important and suggests that many progenitors are multipotential, even if clusters show uniform phenotypes.

O'Leary: Are the clusters that derive from non-migratory progenitors homogeneous?

Walsh: In general, yes. They can be in a single layer or in adjacent layers. For example, two cells can both be in layer V, IV, III or II; or multiple cells can be in adjacent layers—II and III, or III and IV. Some clusters have non-pyramidal cells in layers II and IV, which represents an exception.

Bolz: Are they of the same morphological type?

Walsh: Yes, and so far we've never seen an E15 cluster that crosses layer IV. When we see pyramidal cells in layers III and V, even if they're close to each other in the same radially defined cortical region, they have so far been different clones. This is when labelling is performed at E15 or later.

I would also like to emphasize, however, that John Parnavelas' earliest injections are earlier than our earliest injections. We define the day of conception as E0, whereas he calls it E1. When we inject at E14, we see larger clustered subunits that cross layer IV (C. Reid & C. Walsh, unpublished results). The clusters we see after labelling at E15 tend to be more restricted and smaller. If we look at all our retroviral labelled neurons in the cortex and we ask what proportion of retrovirally labelled neurons consist of single clusters of clones and what proportion are subunits of widespread clones, we find that almost three quarters of all the labelled neurons are subunits of widespread clones. This suggests that the majority of the neurons that John Parvavelas is studying are subunits of widespread clones.

LaMantia: There are two issues concerning cell lineage: regional identity of populations of cortical neurons; and actual cellular identity based on morphology, projection and target-dependent or activity-dependent effects. In terms of terminal lineage, what conclusions can be drawn about the general rules by which cells acquire their regional or cellular fate at this stage? Cell lineage doesn't seem to play a predictive role, but there are other factors that can influence a cells' identity at this late stage, such as trophic factors, activity, other classes of signalling molecules and cell–cell interactions. Can we say that cell lineage is an indeterminate mechanism, based on what you've observed?

Walsh: I favour the view that there are two types of progenitors: one that knows nothing and one that knows a lot. The latter type may have some level of commitment towards a particular cell type; for example, pyramidal versus non-pyramidal. It may also have a certain level of commitment in terms of its regional identity. Pat Levitt's studies (Levitt et al 1995, this volume) suggest that there are some progenitors that can divide and still express a regionally specific phenotype. Therefore, there may be a sequence of progenitors originating from a non-specified progenitor. It is possible that a non-specified progenitor is mobile and moves from one cortical zone to another that has a level of regional identity, but the lineage studies are not going to address the possible existence of diverse regional microenvironments within the proliferative zone.

Rakic: I would like to present a figure, relevant to our present discussion, that was kindly given to me by Seong-Seng Tan of the University of Melbourne (Fig. 1). He developed a remarkable line of transgenic mice that can shed some light on the problem of cell allocation. He has observed distinct radial columns of blue (transgene active) and clear (transgene negative) cells in the cortex of transgenic mice in which *lacZ*, the gene encoding β-galactosidase, was introduced serendipitously into the X chromosome. Therefore, in female mice, after X-chromosome inactivation during mid-gestation, only about one half of the progenitors and their descendants are labelled. Seong-Seng Tan mentioned to me that the dramatic columnar organization observed in Fig. 1 is not encountered in every transgenic mouse. In fact, in many cases, although the radial borders are clearly visible, a larger number of blue cells are present within the unlabelled columns and vice versa than Fig. 1 would suggest. He has interpreted this 'mixing' of the cell types as evidence of progenitor dispersion within the ventricular zone or lateral migration of neurons beyond radial boundaries (Tan et al 1993).

Cell lineage 65

FIG. 1. Parasagittal section of a transgenic mouse cerebrum showing β-galactosidase-positive (dark grey bands) and β-galactosidase-negative (clear bands) radial columns running from the white matter to the pial surface. Differential marking was accomplished by expression of the β-galactosidase transgene in approximately half of the cells in females. This was achieved by inserting the transgene into one of the two X chromosomes and taking advantage of the natural X-inactivation of the transgene in female descendants during the middle of gestation (courtesy of Dr S. S. Tan, University of Melbourne, with permission).

I believe that the pattern, variability and significance of β-galactosidase labelling in the transgenic mice can be explained in a different way, as illustrated schematically in Fig. 2. Several years ago, I suggested, on the basis of the pattern of [^{3}H]thymidine labelling, that ontogenetic columns in the cortex are polyclones, i.e. that they are formed by the contribution from several founder cells composing a proliferative unit in the ventricular zone (Rakic 1988). Thus, in Seong-Seng Tan's transgenic mice, before the X-chromosome inactivation occurs, all the progenitor cells will express β-galactosidase, whereas when X-chromosome inactivation is complete, about one half of cells will not express β-galactosidase. Therefore, one can expect that each radial column is formed by a mixture of β-galactosidase-positive and β-galactosidase-negative cells. My interpretation is that random mixtures of both cell types in each transgenic mouse, and in each column in their cortex, would follow a Gaussian curve distribution. Suppose, for the sake of argument, that 10 founder cells contribute to each column (Fig. 2). The chances that in a single mouse all columns are either blue or unstained are negligible, and in most columns the expected ratio of the two cell types would be

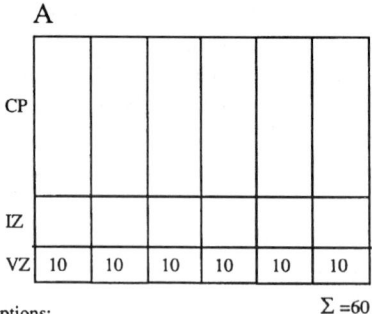

Assumptions:

1. Each polyclone unit in the VZ is composed of 10 founder cells
2. Probability of each founder cell to express β-gal actosidase is 0.5
3. Radical unit are preserved in the IZ and CP

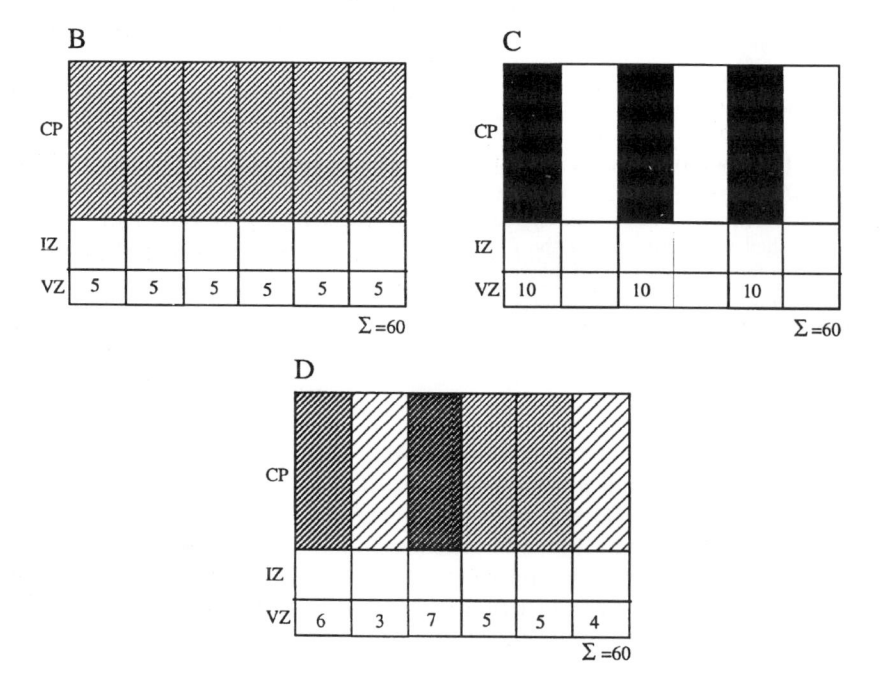

FIG. 2. A model for the interpretation of the results obtained from transgenic mice. The assumptions 1 and 2 listed in A are reasonable according to our present knowledge. Assumption 3 makes the point that, even if there is no lateral dispersion, one can interpret the results as being fully compatible with the radial unit hypothesis. CP, cortical plate; IZ, intermediate zone; VZ, ventricular zone; Σ, total number of founder cells for six presumptive columns.

Cell lineage

between 4:6 and 6:4 (Fig. 2D). Therefore, when one encounters 30% blue cells in the otherwise unstained column, this does not necessarily mean that lateral migration has occurred, but instead that the ratio of blue to unstained founder cells contributing to this particular column is 3:7. When two adjacent columns have the same ratio, there would be no contrast in the staining between them and one would obtain a double column. If the ratio in all columns of the single animal is 5:5, one would not see any indication of radial columns (Fig. 2B), but there is an infinitesimally small chance of obtaining such a combination. Equally unlikely is the possibility that all columns would have a ratio of 10:0 or 0:10, both of which would obscure any radial boundaries. However, the borders are visible when adjacent columns have different ratios, and the bigger the difference in ratio between adjacent columns, the more obvious is the border between them (as in Fig. 1). According to the Gaussian distribution, the ratios of 4:6, 5:5 or 6:4 should be most frequent. A ratio of unstained to blue of 9:1 or 10:0 would occur only occasionally, and the chance that a mostly blue column (e.g. 8:2) is situated next to a mostly unstained column (2:8) is even smaller. Nevertheless, cases where such juxtaposition occurs are the most informative because they set up the lowest limit of the possible tangential dispersion. So, if the mostly unstained column encountered in any of the transgenic mice contains only 15% blue cells, I suggest that no more than 15% of the cells in the mouse neocortex can be dispersed laterally. Even this may be a conservative estimate because the histochemical stain does not allow a clear distinction between neurons and glia and, therefore, it is likely that a substantial fraction of the blue cells in each column are actually glial. Thus, dispersion of clonally related cells is probably smaller than presently assumed and the radial allocation is, according to this interpretation, the predominate mode of distribution. However, irrespective of their number, cells that obey radial constraints may be functionally significant for the development of cortical architecture.

In conclusion, blue cells in the predominantly unstained column do not necessarily signify lateral dispersion, rather they represent the proportion of β-galactosidase-positive cells that contribute to this particular column after X-chromosome inactivation. I believe that the proposed model explains the results in transgenic mice most parsimoniously and provides the most direct evidence for the radial unit hypothesis (Rakic 1988).

Parnavelas: This section (Fig. 1) was taken close to the midline. If one looks at a more lateral section, the boundaries are not as clear. Is this because neuronal migration is not as radial?

Rakic: It is more difficult to interpret sections that are cut more laterally. However, even in these cases Seong-Seng Tan sometimes sees distinct radial columns.

Walsh: I believe that the blue and white partially segregated regions are only seen in parasagittal sections. If you cut the material in the frontal plane, would you still see them?

Rakic: Yes, you would see them. Seong-Seng Tan has some coronal sections of transgenic mice (unpublished results) that have radial columns that are well-defined. Further, Nakatsuji and his colleagues in Japan have developed mouse chimeras that also reveal distinct radial columns in the coronally cut cortex (Nakatsuji et al 1991). It is difficult to explain these columns in any other way but by some clonal restrictions.

Kennedy: There is a problem in that the mosaics produced by Tan & Breen (1993) suggest that there is perhaps more radial migration than what is observed in the retroviral studies. We need to consider that the retrovirus in the ventricles will only infect those cells lining the ventricles and that they represent only a minute fraction of the total number of precursor cells. Precursors in the subventricular zone may behave differently from those in the ventricular zone. Is it possible that the less widespread clones following late injections are due to the ventricular zone proper shutting down at these stages, and that we are observing the consequence of infection of the subventricular zone?

Walsh: Our retroviral results are not inconsistent with the mosaic findings of Tan & Breen. The model being developed is that there is a clustered or relatively radially constrained element during development and a stage of development that is not constrained radially. These two stages might form some sort of developmental sequence. The mosaic studies show that there is a certain amount of non-radial intermingling and that there are also some broad radial constraints acting on at least some cells.

Kennedy: Yes, but Pasko Rakic is arguing that there is a lot of radial migration.

Ghosh: Is the spacing of zones in the X-chromosome inactivation experiments related to the spacing of the cells in your widespread clones?

Walsh: We are comparing the rat and mouse, but it is possible that there is some structural element that defines the spacing because the migration does not appear to be random.

One cautionary note is that the chimeras do not enable us to distinguish easily between neurons and glial cells. We all agree that some glia show an extreme radial constraint in their clonal properties, particularly in the somatosensory cortex where most of Tan & Breen's sections are taken, because glial migration is constrained by the barrels. The β-galactosidase staining in the chimeras doesn't allow the easy separation of neuronal and glial patterns, which may behave differently. Glia make up roughly two thirds of the cells in the cortex, so they will tend to produce a columnar background in the chimeras.

Bartlett: One potential problem with this interpretation is that the expression of *lacZ* is controlled by a 3-hydroxy-3-methylglutaryl coenzyme (HMGCoA) reductase promoter, which can lead to different levels of expression of *lacZ* both in different cell types (low expression in astrocytes compared with neurons) and potentially in different regions.

J. Price: Yes, but if the transgene inserts anywhere other than the X chromosome, expression is observed in all cells, so there's no evidence for regulation of the transcriptional unit.

Bartlett: There is evidence that cholesterol levels can influence the activity of the HMGCoA reductase promoter and thus influence *lacZ* expression (O'Meara et al 1991).

References

Bolz J, Hubener M, Kehrer I, Novak N 1991 Structural organization and development of identified projection neurons in primary visual cortex. In: Bagnoli P, Hodos W (eds) The changing visual system: maturation and aging in the central nervous system, Plenum Press, New York, p 233–246
Hübener M, Bolz J 1988 Morphology of identified projection neurons in layer 5 of rat visual cortex. Neurosci Lett 94:76–81
Hübener M, Schwarz C, Bolz J 1990 Morphological types of projection neurons in layer 5 of cat visual cortex. J Comp Neurol 301:655–674
Huettner JE, Baughman RW 1986 Primary culture of identified neurons from the visual cortex of postnatal rats. J Neurosci 6:3044–3060
Koester SE, O'Leary DDM 1992a Functional classes of cortical projection neurons develop dendritic distinctions by class-specific sculpting of an early common pattern. J Neurosci 12:1382–1393
Koester SE, O'Leary DDM 1992b Regional polarization in early axon extension by cortical neurons. Soc Neurosci Abstr 18:223
Koester SE, O'Leary DDM 1993 Connectional distinction between callosal and subcortically projecting cortical neurons is determined before axon extension. Dev Biol 160:1–14
Koester SE, O'Leary DDM 1994 Development of projection neurones in mammalian cerebral cortex. Prog Brain Res 102:207–215
Levitt P, Ferri R, Eagleson K 1995 Molecular contributions to cerebral cortical specification. In: Development of the cerebral cortex. Wiley, Chichester (Ciba Found Symp 193) p 200–213
Nakatsuji M, Kadokawa Y, Suemori H 1991 Radial columnar patches in the chimeric cerebral cortex visualized by use of mouse embryonic stem cells expressing β-galactosidase. Dev Growth & Differ 33:571–578
O'Meara NMG, Devery RAM, Owens D, Collins PB, Johnson AN, Tomkin GH 1991 Serum lipoproteins and cholesterol metabolism in two hypercholesterolaemic rabbit models. Diabetologia 34:139–143
Peinzdo A, Katz LC 1990 Development of cortical spiny stellate cells: retraction of a transient apical dendrite. Soc Neurosci Abstr 16:1127
Rakic P 1988 Specification of cerebral cortical areas. Science 241:170–176
Reid C, Walsh C 1994 Lineage of subplate and cortical plate neurons in ferret studied with an alkaline phosphatase (AP) retroviral library. Soc Neurosci Abstr 24:1487
Tan SS, Breen S 1993 Radial mosaicism and tangential cell dispersion both contribute to mouse neocortical development. Nature 362:638–640

Waechter RV, Jaensch B 1972 Generation times of the matrix cells during embryonic brain development: an autoradiographic study in rats. Brain Res 40:235–250

Weisskopf M, Innocenti GM 1991 Neurons with callosal projections in visual areas of newborn kittens: an analysis of their dendritic phenotype with respect to the fate of the callosal axon and of its target. Exp Brain Res 86:151–158

The generation of cellular diversity in the cerebral cortex

Jack Price, Brenda P. Williams* and Magdalena Götz

*SmithKline Beecham Pharmaceuticals, New Frontiers Science Park, Harlow, Essex CM19 5AW, UK and *Boston Biomedical Research Institute, Boston, Mass, USA*

> *Abstract.* We have used retroviral vectors to study cell lineage in the embryonic rat cerebral cortex both *in vivo* and in dissociated cell culture. We provide evidence that during the late phase of corticogenesis, most precursor cells of the ventricular zone are specified for the production of a single cell type, either neurons or one of the glial cell types. Although specified, the precursor cells that generate neurons can apparently generate both pyramidal and non-pyramidal cells. Earlier stages of development are dominated by a different type of precursor cell with a number of properties that lead us to believe that it is the founding, multipotential precursor cell of the cerebral cortex. We discuss a possible model of cell lineage which unifies these various observations.
>
> *1995 Development of the cerebral cortex. Wiley, Chichester (Ciba Foundation Symposium 193) p 71–84*

We have known for some time that all the neurons and macroglial cells that comprise the cerebrum arise from the precursor cells of the ventricular zone (VZ). These cell types are generated in a progressive fashion. The first cells to arise are cells of the preplate that are destined to be lamina I neurons and the cells of the subplate. This is followed by the true cortical neurons, first the deeper layers, then the more superficial layers. At some point during the process of neuronal generation, gliogenesis begins, first with the production of astrocytes, then finally with the production of oligodendrocytes.

Although we know that these cell types arise from the VZ, an issue that is only now being resolved is whether individual precursor cells have the same developmental potential as the population as a whole. If each VZ cell were multipotential, the principal problem in generating cellular diversity would be: how do cells switch between the production of different cell types as corticogenesis proceeds? This is a feasible model of development, and the available evidence on a number of regions of the central nervous system), for example, the retina, suggest that this is how cellular diversity is controlled (Turner et al 1990, Turner & Cepko 1987).

There is, however, no reason a priori to assume that each precursor cell is multipotential. It is possible that early in development, subpopulations of cells are generated, each of which has a different developmental potential. The VZ would be a mosaic of diverse precursor cell types, and corticogenesis would be controlled, not by switching of precursor cells between fates, but by the sequential activation of precursor cell populations. Clearly, these alternative models involve quite different control mechanisms. If the second model were correct, for example, there would be no point in asking why precursor cells were switching to the production of astrocytes at the time when astrocytes were beginning to appear, because cells would not be making that decision at that time.

Evidence for specified precursor cells

We have performed a series of cell lineage experiments in an attempt to address this issue, using retroviral vectors to label individual VZ cells *in situ* (Price & Thurlow 1988, Grove et al 1993). By repeating this experiment many times, we have built up a complete picture of the potential expressed by different VZ cells. We labelled VZ cells at embryonic day (E) 16 and analysed the resultant marked clones at postnatal stages. We found discrete clusters of virally labelled cells by following the expression of the *lacZ* gene encoded by the virus, and we discovered that in between 80 and 90% of cases, each cluster was composed of a single cell type. Using cell-type specific markers and other anatomical techniques, we were able to distinguish five such types of clusters. These were composed of neurons, grey matter astrocytes, white matter astrocytes, oligodendrocytes or a cell type that resembled an astrocyte, but did not stain with an antibody against glial fibrillary acidic protein. In addition, we discovered clusters composed of more than one cell type, the best identified of which being those composed of neurons and oligodendrocytes.

These results constitute prima-facie evidence for the mosaic model of cortical development. Each cluster was composed of a single cell type, so it appeared that each precursor cell had generated a single cell type, and that each had, therefore, a different restricted potential. There are, however, two problems with this interpretation: one technical and the other conceptual. The technical problem was that we had no independent evidence that each observed cluster was the entire progeny of a single infected precursor cell. One of the limitations of the retroviral approach to cell lineage is that neither the number nor the position of infected precursor cells is known precisely. These have to be inferred from the distribution of the labelled progeny. Two types of error are possible, and Walsh & Cepko (1992) have termed these the 'lumping' and the 'splitting' errors. The former is when two clones are mistaken for one because the two infected founder cells happened to be close neighbours, causing the two clones to have overlapping distributions. The latter is when the progeny of one precursor cell disperse sufficiently widely to be mistaken for separate clones. The lumping error was unlikely to be a problem in our experiments given the nature of the result—if clones overlapped, mixed clonal types would tend to prevail, yet the overwhelming majority of our clones was composed of a single cell type. The splitting error, however, was conceivable, especially because we now know from a variety of experiments that there is considerable dispersion of cells during cortical development (Price & Thurlow 1988, Walsh & Cepko 1992, Fishell et al 1993, O'Rourke et al 1992, Tan & Breen 1993).

The conceptual problem relates to the type of conclusion that can be made from these results. Cell lineage studies can show what cells do, i.e. their presumptive fate. They cannot indicate the state of determination of precursor cells because the fate taken up by cells during normal development may not constitute their full developmental potential. Thus, it is not valid to conclude from our results that the precursor cells are determined or even that they are specified.

This problem, applicable to all cell lineage studies, takes a particular twist when a retrovirus is used as the lineage marker. Retroviruses, such as Moloney murine leukaemia viruses, can only integrate into dividing cells, probably because the virus only gains access to the chromosomal DNA of the infected cell when the nuclear membrane breaks down during mitosis (Roe et al 1993). As we have demonstrated (Hajihosseini et al 1993), one consequence of this is that when a precursor cell is infected, only one of its daughter cells actually inherits the virus (Fig. 1A). We have discussed the consequences of this for cell lineage studies in detail elsewhere (Price 1992), but the main point is that when precursor cells divide symmetrically to give one daughter that is a stem cell and another that is specified to a restricted fate, the daughter with the restricted fate will inherit the label in 50% of cases (Fig. 1B). Thus, although the originally infected precursor cell may have been multipotential, the labelled progeny will be restricted. The ultimate case occurs in the neuroepithelium, where the labelled progeny may comprise a single postmitotic neuron. Indeed, this observation probably accounts for the high incidence of one-cell clones observed in lineage experiments of this type (Price & Thurlow 1988, Walsh & Cepko 1992, Grove et al 1993).

Tissue culture experiments

To some extent, we have been able to deal with both the technical and the conceptual problems by repeating the experiment in dissociated cell culture (Williams et al 1991). Embryonic cortex was dissociated, labelled with virus,

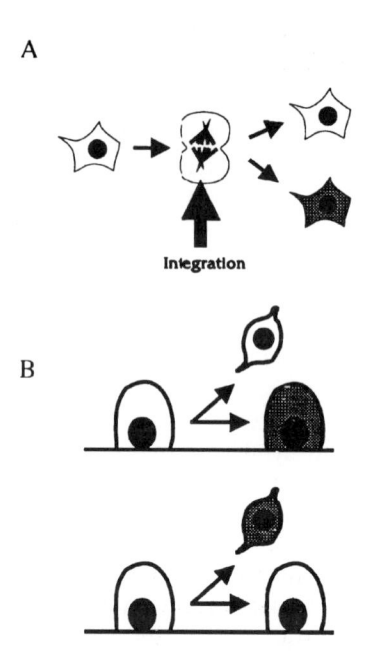

FIG. 1. (A) Virus integration occurs only into dividing cells, and only when the nuclear membrane breaks down during mitosis. The cell's DNA has already been duplicated by this stage of the cycle, so integration can only occur into the DNA inherited by one daughter (the shaded cell). If both of the daughters have a similar fate, i.e. if the division is symmetrical, the consequence of this late integration is principally quantitative—the lineage has effectively been labelled one cell cycle later than might have been thought. (B) The consequences can be more profound if the infected cell divides asymmetrically. If a ventricular zone (VZ) cell gives rise to another VZ cell and a specified cell (for example, a postmitotic neuron), the daughter remaining in the VZ will inherit the virus in 50% of cases, and again the lineage will be labelled just one cell cycle later than might have been thought. However, in the other 50% of cases, the specified cell will inherit the virus and give rise to a different outcome.

and grown on an astrocytic monolayer. The clones were identified using an anti-β-galactosidase serum, and a variety of monoclonal antibodies were used to identify cell type. These results confirmed the *in vivo* finding that the majority of precursor cells generated a single cell type and, as *in vivo*, the principal exceptions were clones composed of neurons and oligodendrocytes. We know from time-lapse micrography that the movement of cells in the cultures is modest, so we can discount splitting as a problem. Also, the behaviour of precursor cells in culture is indistinguishable from that *in vivo*, even though their environment is substantially altered, so it is reasonable to conclude that they are specified in terms of the fate they adopt. In fact, we have since repeated these experiments under a variety of culture conditions, with similar results. We think, therefore, that the precursor cells are largely determined to their restricted fates.

Neuronal precursor cells

The above results suggest that by E16, between 80 and 90% of VZ cells are specified, or possibly determined, to generate a single cell type. We will deal with two specific questions that such a conclusion raises. How is the diversity of neuronal types generated and where do the specified precursor cells themselves come from?

There is a great diversity of neurons in the cerebral cortex, but for simplicity we can reduce them to two broad types: pyramidal and non-pyramidal. The pyramidal cells are the glutamatergic projection neurons of the cortex; the non-pyramidal cells are diverse in form and function, but can be considered to be all γ-aminobutyric acid (GABA)ergic interneurons. We asked above whether individual precursor cells have the potential to generate multiple cell types. Similarly, we can ask whether specified neuronal precursor cells have the potential to generate multiple neuronal types or whether there are separate subpopulations among the neuronal precursor cells for each neuronal type. Others have addressed this question *in vivo*, and they have found that neuronal clones were composed of either pyramidal or non-pyramidal cells (Mione et al 1994); however, we wanted to know what the outcome would be with the cells in tissue culture (Götz et al 1995). We used antisera against glutamate and GABA to identify the two neuronal subtypes in neuronal clones from E16 cultures. We discovered that the majority of neuronal clones were composed of either GABAergic or glutamatergic cells, but that a minority of clones comprised both types.

This result indicates that at least a proportion of precursor cells can generate both types of neuron, but can it be explained by the existence of bipotential neuronal precursor cells alone? About 80% of cortical neurons are glutamatergic and few neuronal clones are composed of more than three cells, so clones composed entirely of pyramidal cells would be expected to arise by chance even if all the precursor cells were bipotential. Pure GABAergic clones would also be expected to arise albeit at a lower frequency. We asked, therefore, whether the proportions of mixed and pure clones that we observed were the same as those predicted by chance if all the precursor cells were bipotential. We found that our results were statistically indistinguishable from the results predicted from such a model. The most parsimonious explanation, therefore, is that there is a single population of neuronal precursor cells which are neuronally specified but can generate either pyramidal or non-pyramidal cells.

The apparent discrepancy of these results from the *in vivo* results of Mione et al (1994) is informative. The neuronal precursor cells appear to express a broader developmental potential in culture than they do *in vivo*. This is quite a common finding in developmental biology and suggests either that influences exist *in vivo*, which tend to restrict the ability of precursor cells to switch between the production of different neuronal types, or that we have introduced influences into the cultures, which are not found *in vivo*, that induce the precursor cells to express the broader potential. We cannot currently distinguish between these two alternatives.

Multipotential precursor cells

The second question arising from our initial results that we will address is the origin of the specified precursor cells. The results we have presented so far suggest that by mid-corticogenesis, the VZ is a mosaic of precursor cells of different developmental potentials. These populations must have arisen at some stage of development from cells with a broader potential. What were those cells, and when did the transition occur?

The obvious way of addressing this question is to label cells with virus at earlier stages. We have been unsuccessful so far in our attempts to label cells *in vivo* during early corticogenesis, but we have addressed the question in culture (Williams & Price 1995). When we cultured cells from embryos between E12 and E18, we discovered that under our standard culture conditions, the results varied only slightly between the youngest and oldest embryos. In each case, the majority of precursor cells gave a single cell type. We now know, however, that this result is misleading. There is an early population of cells with different properties from those of the specified precursor cells, but these cells are highly sensitive to culture density and to the period of culture. In order to find appreciable numbers of these early cells, we had to culture at high density and for no more than seven days. Under these optimal conditions, about 50% of the labelled precursor cells generate clones of what we have called neuroepithelial cells. These cells are clearly neural since they label with antibodies raised against microtubule-associated protein 5 (MAP-5) and nestin. They are precursor cells because they divide rapidly and symmetrically giving rise to large clones that include neurons, astrocytes and oligodendrocytes, although some cell types appear in low numbers and not all appear under the same conditions. They are an early neural precursor type because their numbers decrease rapidly with developmental time, even under optimal conditions, so that by E16 they comprise less than 5% of all the labelled clones. The NE cells have, therefore, most of the properties that would be expected of the founding cortical precursor cells, and our preliminary conclusion is that this is what they represent.

Unfortunately, the neuroepithelial cells demonstrate a number of confounding features under all the culture conditions that we have investigated so far. The majority of these cells fail to differentiate in culture; rather, most of them die

Generation of cellular diversity 77

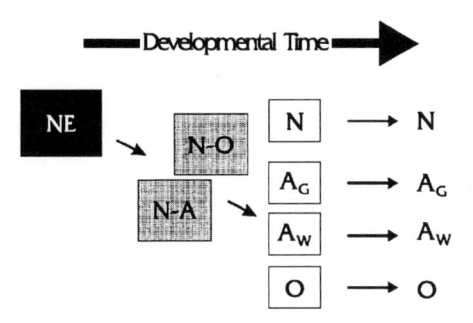

FIG. 2. A model for cortical cell lineage. The earliest cells are the neuroepithelial (NE) cells, the putative multipotential founder cells of the cerebral cortex (black box). The ultimate precursor cells are the different specified populations (white boxes), each generating a single cell type. N, neurons; A_G, grey matter astrocytes; A_W, white matter astrocytes; O, oligodendrocytes. N-O and N-A precursor cells are assumed to lie in the middle of the developmental profile, but in fact the relationships between each of the precursor types is unknown.

during the second week in culture without differentiating. We presume that this represents apoptosis due to an absence of the appropriate survival and differentiation factors, but we have yet to identify the missing factors. In addition, the staining of the neuroepithelial cell population with MAP-5 and nestin (and also with a monoclonal antibody we have generated against an as yet unknown determinant) is heterogeneous. The reason why some cells stain with these markers and some do not is not clear. Nonetheless, the evidence we have leads us to believe that the neuroepithelial cells are the founding, cortical multipotential precursor cells.

A working model

Figure 2 shows a model which summarizes our results, and condenses our view of how cell diversity is generated in the cerebral cortex. Early cortical development is dominated by the neuroepithelial precursor cell population, which is likely to be the founding precursor cell population of the cerebral cortex. By E16, this population has almost disappeared, and it has been replaced by specified precursor populations (shown in white boxes), each of which generates a single cell type. What is still unclear is the path between the two phases of development, i.e. the process of specification. We know that some cells can generate neurons and oligodendrocytes (N-O cells). We have also recently discovered that under different conditions, some cells generate neurons and astrocytes (N-A cells). Either one or the other of these two populations tends to appear under any particular culture condition suggesting that these represent different outcomes for the same population of cells. However, these clones do not represent more than 5% of the total, so we have been unable to address this point experimentally. In terms of our model, it seems likely that the N-O/N-A precursor cells fall somewhere between the neuroepithelial cells and the specified populations, but we do not know how this might happen. We can summarize by saying that we know that each of the populations exist because we can label them, but we cannot tell from cell lineage results alone how the populations link together. Clearly, this mechanistic question is the next major issue that needs to be addressed.

Acknowledgements

We would like to thank Andrew Matus, Tony Frankfurter and Ron McKay for kindly providing us with antibodies. We thank the Medical Research Council, the Multiple Sclerosis Society of Great Britain and Northern Ireland, the Commission of the European Communities and the British Council for financial support.

References

Fishell G, Mason CA, Hatten ME 1993 Dispersion of neural progenitors within the germinal zones of the forebrain. Nature 362:636–638

Götz M, Williams BP, Bolz J, Price J 1995 The specification of neuronal fate: a common precursor for neurotransmitter subtypes in the rat cerebral cortex *in vitro*. Eur J Neurosci 7:889–898

Grove EA, Williams BP, Li DQ, Hajihosseini M, Friedrich A, Price J 1993 Multiple restricted lineages in the embryonic cerebral cortex. Development 117:553–561

Hajihosseini M, Iavachev L, Price J 1993 Evidence that retroviruses integrate into post-replication host DNA. EMBO J 12:4969–4974

Mione MC, Danevic C, Boardman P, Harris B, Parnavelas JG 1994 Lineage analysis reveals neurotransmitter (GABA or glutamate) but not calcium-binding protein homogeneity in clonally related cortical neurons. J Neurosci 14:107–123

O'Rourke NA, Dailey ME, Smith SJ, McConnell SK 1992 Diverse migratory pathways in the developing cerebral cortex. Science 258:299–302

Price J 1992 Making sense of cell lineage. Perspect Dev Neurol 1:139–148

Price J, Thurlow L 1988 Cell lineage in the rat cerebral cortex: a study using retroviral-mediated gene transfer. Development 104:473–482

Roe TY, Reynolds TC, Yu G, Brown PO 1993 Integration of murine leukemia virus DNA depends on mitosis. EMBO J 12:2099–2108

Tan SS, Breen S 1993 Radial mosaicism and tangential cell dispersion both contribute to mouse neocortical development. Nature 362:638–640

Turner D, Cepko C 1987 Cell lineage in the rat retina: a common progenitor for neurons and glia persists late in development. Nature 328:131–136

Turner DL, Snyder EY, Cepko CL 1990 Lineage-independent determination of cell type in the embryonic mouse retina. Neuron 4:833–845

Walsh C, Cepko CL 1992 Widespread dispersion of neuronal clones across functional regions of the cerebral cortex. Science 255:434–440

Williams BP, Price J 1995 Evidence for multiple precursor cell types in the embryonic rat cerebral cortex. Neuron 8:1181–1188

Williams BP, Read J, Price J 1991 The generation of neurons and oligodendrocytes in the cerebral cortex from a common precursor cell. Neuron 7:685–693

DISCUSSION

Rubenstein: Not much is known about the subventricular zone during the early stages of development; whereas, postnatally, in the mouse the subventricular zone is important for gliogenesis. Can you speculate on whether any of the cells in the ventricular zone become sequestered into the subventricular zone?

J. Price: Our *in vivo* results suggest that we are only labelling ventricular zone cells. This is probably due to poor penetration by the retrovirus—we inject the retrovirus into the ventricle and we assume that only cells that are in contact with the ventricle are infected. I don't have any results to prove this, it's purely an assumption. We can't rule out the possibility that some subventricular zone cells could be labelled.

Rubenstein: I understand that the ventricular zone cells are relatively recalcitrant towards infection. Are the cells that are being labelled the ones that have stopped proliferating and are about to migrate?

J. Price: They are not recalcitrant, it's probably just a question of retroviral titre. About $0.5\,\mu m$ of retrovirus is injected, which has a titre of about 10^6 particles/ml, so infection occurs with only a few hundred particles, whereas there are probably millions of cells in the ventricular zone.

LaMantia: The distinction between glutamatergic and γ-aminobutyric acid (GABA)ergic cells is apparent not only in the cortex but also in the basal ganglia, the thalamus, the olfactory bulb and the hippocampus. Consequently, how specific are your results to the cortex? If you take pieces of neuroepithelium from each of those areas and do the same experiments (i.e. disperse the progenitor cells and assay the transmitter phenotype *in vitro*), do you get the same numbers?

J. Price: We are doing some of these experiments now, but we don't have any results just yet.

Levitt: Your results are interesting because Reh & Tully (1986) found the same effect in the frog retina. When they took away the central dopaminergic amacrine cells with 6-hydroxydopamine, more dopaminergic amacrine cells were produced in the retina. You are showing that GABA addition drives more cells to become glutaminergic.

D. Price: How do you envisage that this feedback mechanism operates *in vivo*?

J. Price: I could not answer that question until quite recently, but several people (van Eden et al 1989, Meinecke & Schwarz 1992) have now shown that there are cells that stain with anti-GABA antibodies in and around the ventricular zone. These cells are, therefore, a potential source of GABA, although I don't know what kind of cells they are.

D. Price: Is the hypothesis that they then regulate the fate of their neighbours?

J. Price: Yes.

LaMantia: Do these cells have GABA receptors on them?

J. Price: Yes. We've checked this and Kriegstein et al (1994) have too. We are now looking at the types of GABA receptors in relation to their physiological response.

Parnavelas: Kriegstein et al (1994) showed that cells in the ventricular zone have GABA and glutamate receptors, and also that the activation of these receptors negatively regulates DNA synthesis in proliferating cells, and therefore influences the cell cycle and neurogenesis.

J. Price: They may also influence migration.

Jones: Using *in situ* hybridization and looking specifically at the mRNAs for the different subtypes of GABA receptor, we have found that the $\alpha 5$ receptor is the subtype that is present on most of these cells (M. Huntsman & E. Jones, unpublished results).

I would like to raise a technical point concerning glutamate. A few years ago, if someone had tried to characterize the cells as glutamatergic on the basis of glutamate immunoreactivity, half of this audience would probably have shouted them down. The reason is that every cell in the brain, and probably every cell in the body, should theoretically show immunoreactivity towards glutamate, simply because of its presence in the metabolic pool. As glutamate immunostaining has been applied more and more to the nervous system in normal animals, it's remarkable that it is specific for those cells which we know by other criteria are actually glutamatergic. The metabolic pool is probably washed out during the preparative procedures or it may not be fixed in a way which enables it to be detected by conventional immunocytochemical techniques. On the other hand, there are some examples in normal animals where the largest GABAergic cells, namely the basket cells in the cerebral cortex, do show some immunoreactivity for glutamate. In this case, the assumption is that this is part of the metabolic pool which in a large cell is obviously much larger. So is glutamate a truly appropriate marker for identifying pyramidal cells or excitatory neurons in the cortex? Should we use another marker? This is important because in your experiments, in which the application of GABA immunostaining led to an apparent absence of GABA cells, the interpretation could be that GABA is simply causing cells to produce more glutamate because they have to produce more GABA.

J. Price: I hadn't thought of that. We are interpreting these results cautiously because there are numerous explanations other than that the clones switch from being GABAergic to glutamatergic.

Molnár: Is there a variation in the GABA to glutamatergic neuron ratio in different areas of the cerebral cortex?

Generation of cellular diversity 81

J. Price: I don't know.

Jones: I can't speak for rodents, but we've done extensive studies in monkeys where we based the percentage of GABAergic cells on either immunocytochemistry for GABA or on *in situ* hybridization for the 67 kDa form of glutamic acid decarboxylase. We found that this percentage ranges between 25 and 30%, so there's little areal variation (Hendry et al 1987, Jones et al 1994).

LaMantia: Does this suggest that, in terms of regional cellular specification, the distinction between GABAergic and glutamatergic cells is not informative, but only a general indication?

J. Price: Yes. We also observed unexpectedly that if we stained the neurons in the clones, 100% of them are either GABA or glutamate positive (Götz et al 1995), but if we stained the neurons in culture as a whole, we saw the occasional cell that was not labelled (Götz & Bolz 1994). It turns out, and this could be a fortuitous finding or a red herring, that if the cells are treated with dibromodeoxyuridine (dBrdU) just before they are dissociated, cells which label with dBrdU are the same cells that fail to make either GABA or glutamate in culture. This observation and the results of McConnell & Kaznowski (1991) suggest that these cells are making decisions around the time that they are passing through their final division. The implication is that if they are dissociated just as they're passing through the final S phase or shortly afterwards, the decision is disrupted because they never make either glutamate or GABA. This also fits with our hypothesis and accentuates that this point in time is an important moment in the life history of a neuron, i.e. when it has to make certain decisions.

Bolz: Evidence for signals from the local cortical environment that act early in the cell cycle for the specification of transmitter phenotypes of cortical neurons is presented by Götz & Bolz (1994).

Ghosh: Sue McConnell has suggested that part of the way you could specify laminar fate is by a developmentally regulated signal present in the ventricular zone (McConnell & Kaznowski 1991). As the cells are differentiating during a particular developmental stage, they might encounter a signal that specifies their laminar identity. The GABAergic and glutamatergic cells are distributed throughout the entire cortex, so how do you imagine that the signals which determine neurotransmitter phenotype affect particular cells throughout the period of neurogenesis?

J. Price: The results suggest that the two decisions occur at the same time in the ventricular zone; however, the signals may not be the same.

Ghosh: If the cells in the ventricular zone are plastic with regard to neurotransmitter phenotype, why do clonally related cells *in vivo* typically give rise to cells of the same transmitter phenotype?

J. Price: There is a difference between the *in vivo* results and the *in vitro* results, namely that in culture the cells aren't under the same constraint. There may be a residue of this constraint, in that each precursor cell does tend to continue to make a particular neuronal type once it has started to do so, although we have shown that this is not statistically significant. All that is necessary to translate that to the *in vivo* situation, is to say that once a cell starts to make neurons of a particular type *in vivo* it continues to do so, not because it doesn't have the potential to make other types of neurons, but because it does not receive instructions to do anything else. *In vivo* one would predict that switches would be relatively infrequent, whereas in culture the cells are bombarded with aberrant signals; therefore, I would argue, that it is more likely that they will be induced to switch and demonstrate their full developmental potential *in vitro*. I can't prove that hypothesis but it fits with the results.

LaMantia: Is there an age when the neuroepithelium from the cortical rudiment no longer gives rise to GABAergic cells versus glutamatergic cells?

J. Price: We haven't studied this, we have only looked at E16 embryos.

Bartlett: Jack Price, your results on the effect of cell density fit well with our previous results (Drago et al 1991) which show that precursor cells require fibroblast growth factor (FGF) at low cell density, but as you increase the density of neuroepithelial cells that dependency is progressively lost. If you block the action of endogenous FGF with inositol hexatrisphosphate, the precursors don't grow (Kilpatrick & Bartlett 1995). Have you looked to see if FGF is the endogenous factor that is stimulating your culture?

J. Price: FGF doesn't help neuroepithelial clones survive. It only helps neuronal clones.

Bartlett: With or without serum?

J. Price: Either. FGF is a survival factor and a mitogenic factor. It's impossible to say how it affects the neuronal precursors, but it doesn't affect the glial or neuroepithelial precursors as far as we can tell.

Bartlett: It's surprising that these clones die because when clones are generated from E10 mouse epithelium, they grow well and survive for up to three to four weeks *in vitro*. Therefore, it appears that neuroepithelial clones generated under your conditions receive a signal(s) from the surrounding cells to undergo apoptosis.

J. Price: Are the cells that give rise to neurons and astrocytes similar to neuroepithelial cells? In other words, are they multipotential cells or neuron–astrocyte (N–A) precursors? Our evidence suggests that N–A precursors are more sturdy—they behave like specified precursor cells which may be abused and continue to do what they were originally doing. This suggests that most of your multipotential cells, which come from cultures that are initially plated at high density and then seeded at low density, are N–A precursors. The neuroepithelial cells are probably disappearing from this culture.

Bartlett: Although we initially established clones this way, we now plate single neuroepithelial cells into wells, and we find that they generate morphologically identical clones. The clones consist of neuroepithelial cells that don't differentiate into neurons unless they are stimulated specifically with endogenous growth factors (Bartlett et al 1995, this volume).

Maffei: Is anything known about the electrical properties of the mature GABAergic and glutamatergic cells?

J. Price: We haven't looked at mature neurons. We're concentrating on precursor cells because this is an area that people have neglected in the past.

Rakic: The age of the cell may be important. For instance, Farrant et al (1994) showed that active NMDA receptors have different combinations of subunits during granule cell migration. When migration is completed this combination changes.

Parnavelas: Is this in the cerebellum?

Rakic: Yes, but similar experiments can also be performed in the telencephalon.

References

Bartlett PF, Richards LR, Kilpatrick TJ 1995 Factors regulating the differentiation of neural precursors in the forebrain. In: Development of the cerebral cortex. Wiley, Chichester (Ciba Found Symp 193) p 85–99

Drago J, Murphy M, Carroll SM, Harvey RP, Bartlett PF 1991 Fibroblast growth factor-2-mediated proliferation of insulin-like growth factors. Proc Natl Acad Sci USA 88:2199–2203

Farrant M, Feldmeyer D, Takahashi T, Cull-Candy SG 1994 NMDA-receptor channel diversity in the developing cerebellum. Nature 368:335–339

Götz M, Bolz J 1994 Differentiation of transmitter phenotypes in rat cerebral cortex. Eur J Neurosci 6:18–32

Götz M, Williams BP, Bolz J, Price J 1995 The specification of neuronal fate: a common precursor for neurotransmitter sub-types in the rat cerebral cortex *in vitro*. Eur J Neurosci 7:889–898

Hendry SHC, Schwark HD, Jones EG, Yan J 1987 Numbers and proportions of GABA-immunoreactive neurons in different areas of monkey cerebral cortex. J Neurosci 7:1503–1519

Jones EG, Huntley GW, Benson DL 1994 Alpha calcium/calmodulin-dependent protein kinase II selectively expressed in a subpopulation of excitatory neurons in monkey sensory-motor cortex: comparison with GAD-67 expression. J Neurosci 14:611–629

Kilpatrick TJ, Bartlett PF 1995 Cloned multipotential precursors from the mouse cerebrum require FGF-2, whereas glial precursors were stimulated with either FGF-2 or EGF. J Neurosci 15:3653–3661

Kriegstein A, Davis MBE, LoTurco J 1994 GABA and glutamate depolarise cerebral cortical progenitor cells and inhibit DNA synthesis. Soc Neurosci Abstr 20:458

McConnell SK, Kaznowski CE 1991 Cell cycle dependence of laminar determination in developing neocortex. Science 254:282–285

Meinecke DL, Schwarz ML 1992 Comparative development of glutamate and GABA-immunoreactive neuronal populations in the cerebral cortex of rat, ferret and monkey. Cerebral cortex 2:825–832

Reh T, Tully T 1986 Regulation of tyrosine hydroxylase-containing amacrine cell number in larval frog retina. Dev Biol 114:463–469

van Eden CG, Mrzljak L, Voorn P, Uylings HBM 1989 Prenatal development of GABAergic neurons in the neocortex of the rat. J Comp Neurol 289:213–227

Factors regulating the differentiation of neural precursors in the forebrain

P. F. Bartlett, L. R. Richards, T. J. Kilpatrick, P. T. Talman, K. A. Bailey*, G. J. Brooker, R. Dutton, S. A. Koblar, V. Nurcombe†, M. O. Ford†, S. S. Cheema, V. Likiardopoulos, G. Barrett and M. Murphy

*The Walter and Eliza Hall Institute of Medical Research, Royal Melbourne Hospital and Departments of *Pathology, and †Anatomy and Cell Biology, University of Melbourne, Parkville, Victoria 3050, Australia*

Abstract. Precursors from the neuroepithelium of the developing cortex and the adult subventricular zone can be cloned *in vitro* after stimulation with fibroblast growth factor 2 (FGF-2), and they have the potential to give rise to both neurons and glia. The generation of neurons from these clones can be stimulated by either a factor derived from an astrocyte precursor line, Ast-1, or FGF-1. We have shown that neuronal differentiation stimulated by FGF-1 can be inhibited by diacylglycerol lipase inhibitor and mimicked by arachidonic acid, suggesting that the neuronal differentiation is signalled through the phospholipase Cγ pathway. The sequential expression of FGF-2, followed by FGF within the developing forebrain neuroepithelium, fits with the different functions that the two FGFs play in precursor regulation. We have shown that the precursor response to FGF-1 is regulated by a heparan sulphate proteoglycan expressed within the developing neuroepithelium. Precursors restricted to the astrocyte cell lineage can be stimulated by epidermal growth factor or FGF-2; however, the differentiation into glial fibrillary acidic protein-positive astrocytes appears to require a cytokine acting through the leukaemia inhibitory factor-β receptor.

1995 Development of the cerebral cortex. Wiley, Chichester (Ciba Foundation Symposium 193) p 85–99

The cerebral cortex is the product of precursor cells contained within the neuroepithelium of the dorsal aspect of the telencephalon. The precursor cells proliferate rapidly and generate the first neurons which migrate to the pial surface to form the preplate, which later becomes the subplate. The preplate neurons have recently assumed increased importance because they may retain the positional information contained within their neuroepithelial precursors that is required for targeting the appropriate thalamic input (Ghosh et al 1990). In contrast to these early neurons, some of the later neuronal progeny of the neuroepithelium (also called the ventricular zone), which migrate primarily in association with radial glia (Rakic 1978) and penetrate the subplate to form the cortical plate, have been shown to be geographically unaligned with their parent precursor, and they probably lack positional information (for review see O'Leary et al 1994). Once the neuronal populations have been generated, the majority of cell proliferation occurs within the subventricular zone, the progeny of which appear to be predominantly glial cells.

In order to understand the foundation for the structural and cellular development of the cerebral cortex, we need to define the genetic and epigenetic events that regulate neurogenesis and gliogenesis, and characterize the lineage potential of precursors within the germinal zone. Previous studies using retroviral markers have suggested that the majority of precursors within the developing cerebral cortex are restricted to either the glial cell or the neuronal lineage (for review see Luskin 1994). This interpretation, however, is hampered by an inability to label cell precursors at the onset of neurogenesis. This is due to inadequate phenotyping of all the progeny of a single precursor (clone) and not being able to monitor progeny of a single clone over an extended period of time. Our studies have centred on sampling the precursor population at various times during neurogenesis. We have been able to examine the lineage repertoire of individual precursors, and we have gained some insight into the epigenetic factors that influence lineage determination of precursor cells by using defined growth conditions and clonal analysis *in vitro*.

Lineage potential of forebrain precursors

Cloning of precursors from the earliest neuroepithelium

In the mouse, embryonic day (E) 9–E10, is the earliest stage at which one can successfully isolate neuroepithelial cells from the developing telencephalon. At this stage less than 1% of the cells have neuronal markers (Drago et al 1991). Previous studies have shown that proliferation of neuroepithelial cells isolated at this stage can be stimulated by fibroblast growth factor 2 (FGF-2) (Murphy et al 1990), a factor that is expressed by the neuroepithelium during this developmental period (Drago et al 1991, Nurcombe et al 1993). This factor was, therefore, chosen to stimulate precursors plated out at one cell per well. Clones arose at the frequency of about one in 20 cells plated (Kilpatrick & Bartlett 1993) and they consisted of flat epithelioid cells. The clones continued to proliferate over two to three weeks, but contained few if any neurofilament-positive neurons. However, after 10 days many of the clones contained significant numbers of astrocytes, i.e. were glial fibrillary acidic protein (GFAP)-positive. The observation that many of the clones contained a low number of neurons as well as glia confirmed our previous studies using immortalized cell lines from E10, which suggested that some of the precursors were in fact multipotential (Bartlett et al 1988). However, the major finding was that the majority of clonal progeny failed to differentiate, an observation confirmed by the high recloning rate (up to 85%) of the clonal progeny (Kilpatrick & Bartlett 1993). At this time we had identified Ast-1 as a factor, present in medium conditioned by an astrocyte precursor cell line, that stimulated neuron differentiation in E10 neuroepithelium (Kilpatrick et al 1993). We were able to induce significant neuronal production within one to two days in a significant proportion of clones by application of this conditioned medium (Kilpatrick & Bartlett 1993), providing the first evidence that a discrete second signal was required for neuronal differentiation. Recent evidence suggests that Ast-1 is a transforming growth factor β-like molecule (M. Murphy, unpublished work 1994); however, its precise nature and expression pattern *in vivo* is not yet known. It now appears that a more potent signal for initiating neuronal differentiation is likely to be FGF-1 (see below).

Evidence for early commitment of neuronal precursors within the developing neuroepithelium

Although *in vitro* clonal analysis indicates that many precursors with significant proliferative potential are also bipotential, several studies using retroviral markers have indicated that some cells have restricted lineage capabilities. Recently, we have identified a cell surface marker that is associated with commitment of precursor cells to the neuronal lineage and we have demonstrated that as early as E10, many of the neuroepithelial cells are committed to the neuronal lineage. Unlike glia and virtually all other cell types in the body, neurons have an inability to express major histocompatibility complex (MHC) class I molecules on their surface in response to γ-interferon (Bartlett et al 1989). We, therefore, decided to examine the stage at which this unresponsiveness occurs. At E9, all the neuroepithelial cells can express MHC molecules, but as development of the cortex proceeds, an increasing percentage of the precursor cells lose their ability to be induced (Fig. 1). We have selected the non-inducible cells by cell sorting and we have demonstrated that when they are replated under a variety of conditions, they have the potential to give rise exclusively to neurons (Bailey et al 1994). Thus, it appears as if these precursors are committed to the neuronal lineage. Although the neuronal precursors have the ability to proliferate, this is more restricted than the uncommitted precursors. This may explain the limited size of neuron-only clones labelled by retroviral markers *in situ* in the forebrain (Walsh & Cepko 1992).

Evidence for lineage-committed precursors later in cortical development

To examine the potential of precursors later in development, we have cloned cells from the forebrain at E17, and we have found that FGF-2 stimulates three types of precursors: one that is restricted to producing astrocytes; one that

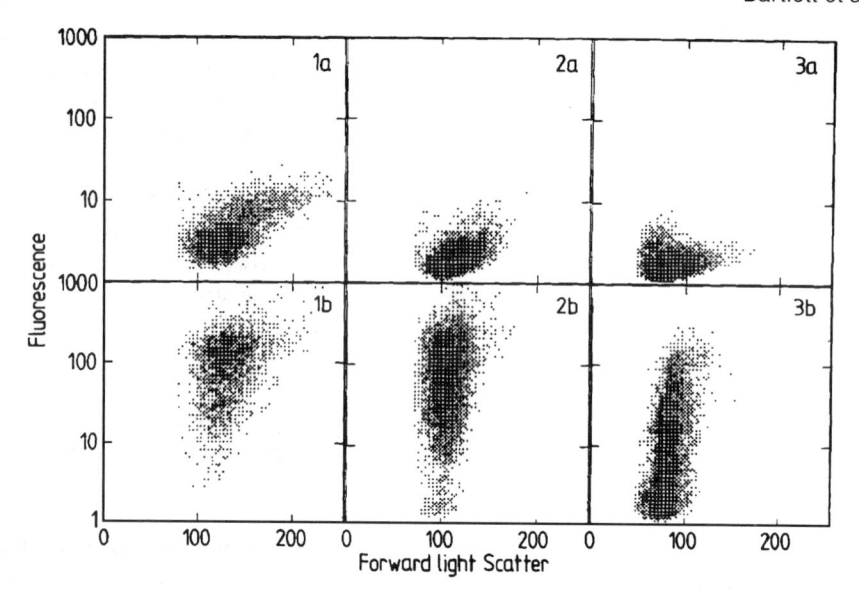

FIG. 1. Neuroepithelial cells taken from the telencephalon at embryonic day (E) 10 (1) E12 (2) and E14 (3) were incubated with γ-interferon for 24 h before being stained by immunofluorescence for major histocompatibility complex class I expression with anti-H-2^k antibodies (b) or with isotype-matched control serum (a). The flow cytometry profiles show an increasing percentage of cells which fail to express the class I antigen as development proceeds. The fluorescence axis is in arbitary units and forward light scatter is a measure of cell size (arbitrary units).

produces predominantly neurons; and one that is similar to the E10 precursors that generate both astrocytes and neurons (Kilpatrick & Bartlett 1995). In contrast to the E10 clones, the differentiation of these clones does not require an exogenous second signal. This indicates that these precursors have already received the appropriate signals *in situ*, and in the case of the multipotential clone, it may reflect the presence of factors produced by other cells within the clone, such as astrocytes. This result confirms the earlier studies with retroviral tracers, but it also indicates that numerous precursors cells capable of generating neurons are present towards the end of neuronal production in the mouse cortex. It was also found that the glial-committed precursor in the forebrain at E17 could be stimulated selectively by epidermal growth factor (EGF) (Kilpatrick & Bartlett 1995).

Presence of multipotential precursors in the adult mouse forebrain

We and another group have described the presence of neuronal precursors within the adult brain (Richards et al 1992, Reynolds & Weiss 1992). We have now localized these precursors to the subventricular zone of the lateral ventricle, the remnant of the germinative layer that gives rise to the cells of the cerebral cortex. Using the same procedures as for the E10 cells, we have generated epithelioid clones from cells of the subventricular zone by stimulating them with FGF-2 (Fig. 2a). Like the E10 clones, the subventricular zone clones generated with FGF-2 contain GFAP-positive cells but not neurofilament-positive cells (i.e. neurons). Unlike E10 precursors, however, about 50–60% of the subventricular zone clones, when stimulated with FGF-1, contained more than 50 neurons (Fig. 2b) that stained positively for the neuronal marker microtubule-associated protein 2 (MAP-2).

FGF-1 stimulates neuronal differentiation

FGF-1 and FGF-2 stimulate the same precursor

The observation that FGF-1 stimulated subventricular zone clones to produce neurons, whereas FGF-2-stimulated clones contained only glia, can be explained in two ways: either they stimulated different precursors; or FGF-1 induced neuronal differentiation in addition to proliferation. To test this, we first generated clones with FGF-2 and after five days exposed them to FGF-1. We found that about 55% of the switched clones produced neurons of an identical phenotype to that observed in clones generated with FGF-1, indicating that the precursor population stimulated with both factors was probably identical.

Requirement for heparan sulphate proteoglycan for the action of FGF-1

The concept that FGF-1 is the primary stimulator of neuronal differentiation was supported by our previous observations that FGF-1 expression in the neuroepithelium of the telencephalon coincided with the time neurogenesis commenced, i.e. at about E11 (Nurcombe et al 1993). FGF-1, however, failed to stimulate neuron production in E10 clones. This discrepancy may be explained in part by our observation that the action of FGF-1 required the presence of a brain-derived heparan sulfate proteoglycan (HSPG). We found that at the onset of FGF-1 expression in the neuroepithelium, there was concomitant expression of an HSPG (Ford et al 1994) which preferentially bound FGF-1 and was required for neuroepithelial stimulation (Nurcombe et al 1993). When E10 clones were generated in the presence of FGF-1 and the appropriate HSPG, MAP-2-positive neurons were generated in about 60% of clones. The requirement for HSPG is due to its ability to bind and present FGF-1 to its receptor. In addition, direct binding of the HSPG to the FGF receptor may be involved in signal transduction. The corollary of this result is that the adult precursors must express the appropriate HSPG on its cell surface constitutively. Although we have not shown this definitively, neuronal

(a)

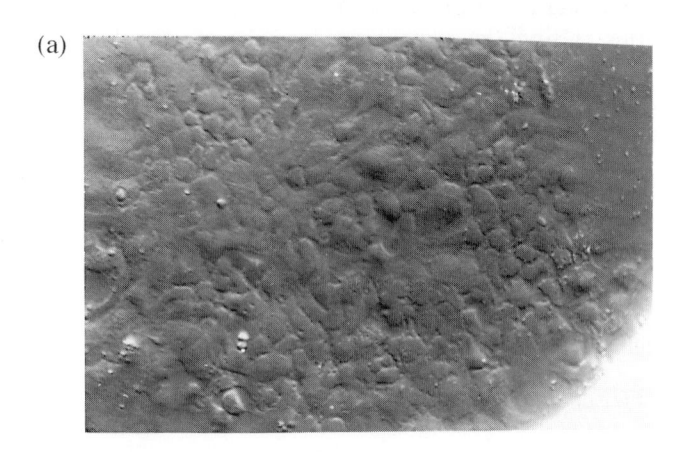

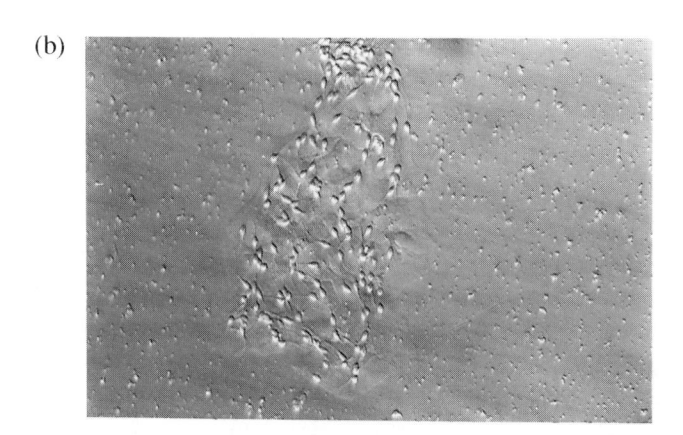

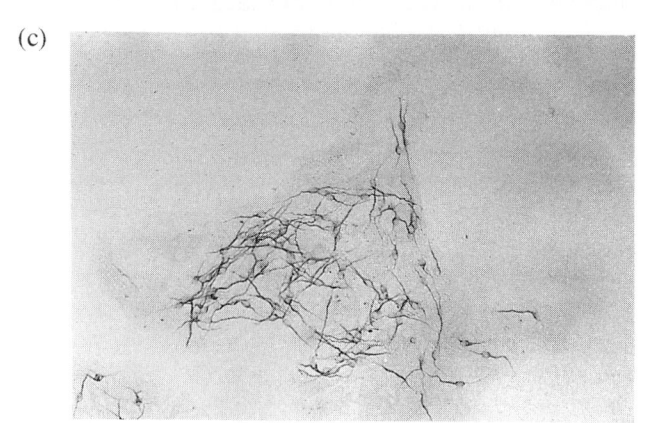

(b)

(c)

differentiation of adult precursors can be prevented by the addition of heparin, which is another FGF-1-presenting molecule. This demonstrates that it is the context in which FGF-1 is presented to its receptor that determines the signalling pathway.

FGF-1 stimulates neuronal differentiation through a second messenger pathway distinct from the proliferative pathway

The binding of FGF to its receptor (of which there are four well characterized types) leads to receptor dimerization and *trans*-autophosphorylation on certain tyrosine residues. This in turn creates binding sites for molecules with SH2 domains, including Src and phospholipase Cγ (PLCγ), and activates several signalling cascades which may be independent or synergistic (Schlessinger & Ullrich 1992). The proliferative signal initiated by FGF is not dependent on PLCγ activation (Peters et al 1992), but recent evidence suggests that activation of this molecule is required for neuronal functions, such as neurite outgrowth (Williams et al 1994). We have investigated whether this pathway (from PLCγ activation to diacylglycerol (DAG) production, the conversion of DAG to arachidonic acid by DAG lipase then finally to Ca^{2+} influx) was involved in stimulating neuronal differentiation by examining the effect of two components of this pathway: DAG lipase inhibitor and arachidonic acid. The addition of DAG lipase inhibitor in the dose range of 5–10 μg/ml to FGF-1-stimulated clones resulted in total inhibition of neuron formation, whereas the total number of clones was not substantially reduced (Fig. 3). This supports the notion that the neuronal differentiation pathway required the production of arachidonic acid. In addition, up to 90% of the clones which could be generated with arachidonic acid at 100 μg/ml were of the neuronal type, again supporting the idea that signalling through this pathway promotes predominantly neuron differentiation. The addition of arachidonic acid to FGF-2-stimulated clones also induced neuronal differentiation similar to that induced with FGF-1.

EGF stimulates non-neuronal precursors

As mentioned previously, EGF-stimulated precursors appeared to give rise only to clones containing astrocytes. This was also true for the adult precursors. In order to examine whether this reflected a different precursor

FIG. 2. Photomicrographs of clones generated from the adult subventricular zone stimulated with fibroblast growth factor 2 (FGF-2) (a) or FGF-1 (b). Note the phase bright neurons on top of the monolayers (b) which stain positively for microtubule-associated protein 2 expression using immunoperoxidase staining (c).

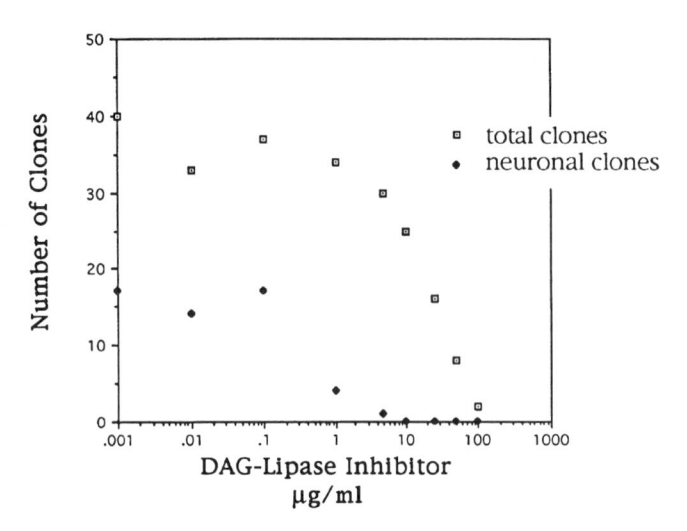

FIG. 3. Effect of diaclyglycerol (DAG) lipase inhibitor on the number of clones generated in the presence of fibroblast growth factor 1.

population or merely a failure to produce neurons, we added both arachidonic acid and FGF-1 to these cultures. However, none of the clones produced neurons. Clones generated with EGF under serum-free conditions are called neurospheres. These also failed to generate neurons under these conditions, which was in contrast to results reported previously by Reynolds & Weiss (1992). A possible explanation for this is that the previous study was not carried out at the clonal level. EGF supports the survival of mature neurons *in vitro* suggesting that these surviving neurons have been incorporated into the neurospheres. It is also possible that the conditions we have used are not conducive to the neuronal differentiation of EGF-stimulated precursors.

Astrocyte differentiation requires a factor acting through the leukaemic inhibitory factor β receptor

The above studies suggest that the production of astrocytes from their precursors does not require discrete epigenetic signalling apart from that required to stimulate cell division. Another explanation, however, could be that astrocyte production, as measured by GFAP expression, which does not occur until between seven and eight days after plating in an E10 clone (previous studies by Abney et al [1981] have shown the temporal expression of GFAP *in vitro* matches that *in vivo*), is due to the production of endogenous factors within the clone. Recent experiments tend to support this latter contention because E10 neuroepithelial cells cultured *in vitro* can be prevented from differentiating into astrocytes by antibodies which bind to, and block ligand binding to, the leukaemic inhibitory factor β (LIF-β) receptor (L. R. Richards, M. Murphy & P. F. Bartlett, unpublished results 1994). In addition, mice bearing a mutation in this receptor appear to have an astrocyte deficiency (Ware et al 1995) and we have recently shown that E10 neuroepithelial cells obtained from these mice fail to differentiate into astrocytes when plated *in vitro* (S. Koblar, C. B. Ware, R. Dutton, M. Murphy & P. F. Bartlett, unpublished results 1994). LIF and ciliary neurotrophic factor (CNTF), which both act through this receptor, have been shown to influence astrocyte production *in vitro*. However, mice with mutations in either of the genes coding for these ligands appear to have normal astrocyte production. Thus, other members of the family, like oncostatin M or an as yet unidentified ligand, may be responsible for regulating astrocyte differentiation because recent evidence suggests another CNTF-like molecule may be present during embryogenesis (G. Yancopoulos, personal communication).

Conclusions

The results of *in vitro* clonal analysis have demonstrated that the precursor cells which give rise to the cerebral cortex require a number of growth factors to regulate their proliferation and differentiation. It now appears that FGF-1, in association with a specific HSPG, stimulates neuronal differentiation through the PLCγ pathway, and a molecule acting through the LIF-β receptor regulates glial differentiation. A multipotential precursor persists throughout development, presumably by self renewal, and is also present in the adult subventricular zone, in addition to precursors that are committed to the neuronal or glial pathway. EGF is involved predominantly in stimulating precursors that are committed to the glial pathway. It should now be possible to examine the mechanisms regulating precursor differentiation into the different subtypes of neurons found in the cortex by clonal analysis and the use of defined growth factors. Ultimately, it should also facilitate an understanding of how epigenetic factors and genetic factors interact to regulate early cortical development.

Acknowledgements

This work was supported by the National Health and Medical Research Council of Australia, and the Motor Neurone Disease Research Institute.

References

Abney ER, Bartlett PF, Raff MC 1981 Astrocytes, ependymal cells, and oligodendrocytes develop on schedule in dissociated cell cultures of embryonic rat brain. Dev Biol 83:301–310

Bailey KA, Drago JD, Bartlett PF 1994 Neuronal progenitors identified by their inability to express class I histocompatibility antigens in response to interferon γ. J Neurosci Res 39:166–177

Bartlett PF, Reid HH, Bailey KA, Bernard O 1988 Immortalization of mouse neural precursor cells by the c-*myc* oncogene. Proc Natl Acad Sci USA 85:3255–3259

Bartlett PF, Kerr RSC, Bailey KA 1989 Expression of MHC antigens in the central nervous system. Transplant Proc 21:3163–3165

Drago J, Murphy M, Carroll SM, Harvey RP, Bartlett PF 1991 Fibroblast growth factor-mediated proliferation of central nervous system precursors depends on endogenous production of insulin-like growth factor-3. Proc Natl Acad Sci USA 88:2199–2203

Ford MD, Bartlett PF, Nurcombe V 1994 Co-localization of FGF-2 and a novel heparan sulphate proteoglycan in embryonic mouse brain. Neuroreport 5:565–568

Ghosh A, Antonini A, McConnell SK, Shatz CJ 1990 Requirement for subplate neurons in the formation of thalamocortical connections. Nature 347:179–181

Kilpatrick TJ, Bartlett PF 1993 Cloning and growth of multipotential precursors: requirements for proliferation and differentiation. Neuron 10:255–265

Kilpatrick TJ, Bartlett PF 1995 Cloned multipotential precursors from the mouse cerebrum require FGF-2, whereas glial precursors were stimulated with either FGF-2 or EGF. J Neurosci 15:3653–3661

Kilpatrick TJ, Talman PT, Bartlett PF 1993 The differentiation and survival of murine neurons *in vitro* is promoted by soluble factors produced by an astrocytic cell line. J Neurosci Res 35:147–161

Luskin MB 1994 Neuronal cell lineage in the vertebrate central nervous system. FASEB J 8:722–730

Murphy M, Drago J, Bartlett PF 1990 Fibroblast growth factor stimulates the proliferation and differentiation of neural precursor cells *in vitro*. J Neurosci Res 25:463–475

Nurcombe V, Ford M, Bartlett PF 1993 Developmental regulation of neural response to FGF-1 and FGF-2 by heparan sulfate proteoglycan. Science 260:103–106

O'Leary DDM, Schlagger BL, Tuttle R 1994 Specification of neocortical areas and thalamocortical connections. Annu Rev Neurosci 17:419–439

Peters KG, Marie J, Wilson E et al 1992 Point mutation of an FGF receptor abolishes phosphotidylinositol turnover and Ca^{2+} flux but not mitogenesis. Nature 358:678–681

Rakic P 1978 Neuronal migration and contact guidance in primate telencephalon. Postgrad Med J (suppl 1) 54:25–40

Reynolds BA, Weiss S 1992 Generation of neurons and astrocytes from isolated cells of the adult mammalian central nervous system. Science 255:1707–1710

Richards LJ, Kilpatrick TJ, Bartlett PF 1992 *De novo* generation of neuronal cells from the adult mouse brain. Proc Natl Acad Sci USA 89:8591–8595

Schlessinger J, Ullrich A 1992 Growth factor signaling by receptor tyrosine kinases. Neuron 9:383–391

Walsh C, Cepko CL 1992 Widespread dispersion of neuronal clones across functional regions of the cerebral cortex. Science 255:434–440

Ware CB, Horowitz MC, Renshaw BR et al 1995 Targeted disruption of the low-affinity leukemia inhibitor factor receptor gene causes placental, skeletal, neural and metabolic defects, and results in perinatal death. Development 121:1283–1299

Williams EJ, Furness J, Walsh FS, Doherty P 1994 Characterisation of the second messenger pathway underlying neurite outgrowth stimulated by FGF. Development 120:1685–1693

DISCUSSION

Walsh: Do the leukaemic inhibitory factor (LIF) knockout mice have astrocytes that either do not have glial fibrillary acidic protein (GFAP) or do not have astrocytes at all?

Bartlett: That's an interesting point, which depends on your definition of an astrocyte. Unfortunately, LIF receptor knockout mice die at birth, as do the ciliary neurotrophic factor (CNTF) receptor and the gp130 knockout mice, so it is impossible to answer the question unequivocally. As far as we can tell, however, newborn mice do not contain GFAP-positive cells in the brain. But, the first GFAP-positive cell does not appear until embryonic day (E) 17, so development may be delayed rather than abrogated. We haven't looked at other markers. The number of neurons looks normal, although some degeneration of the neuronal population is apparent.

Ghosh: Is there a LIF knockout mouse that does not die at birth?

Bartlett: The LIF knockout mice do not die at birth, nor do the CNTF knockout mice. We have some evidence (S. S. Cheema & P. Bartlett, unpublished observations) that LIF knockout mice also have a reduction in the number of GFAP-positive cells (about 30–40%). If one looked at histological sections stained with anti-GFAP antibody, one wouldn't pick it up unless one did a detailed cell count. What we're looking at with the LIF-β receptor mutant is the cumulative effect of the loss of LIF, CNTF, oncostatin M and possibly another CNTF-like ligand (G. Yancopoulis, personal communication).

J. Price: Why don't they generate oligodendrocytes?

Bartlett: It is possible that stimulation with fibroblast growth factor (FGF) causes clones to remain in a more primitive state. Alternatively, it may do the opposite, as Jack Price suggested in the discussion following his presentation (Price et al 1995, this volume), i.e. it may promote the proliferation of an astrocyte neuronal precursor which looks like, but is different to, an epithelial cell. We should use other stimulants, like platelet-derived growth factor (PDGF), to attempt to stimulate oligodendrocyte production.

J. Price: Have you tried stimulating with just serum?

Bartlett: It doesn't work with single cells.

Parnavelas: What is the age of the embryos that these cells are taken from?

Bartlett: E10.

Parnavelas: Is it possible that oligodendrocyte precursors are not present at that stage?

Bartlett: We showed many years ago (Abney et al 1981) that oligodendrocytes can be generated *in vitro* from single cell suspensions of rat brains taken from as early as E10.

Parnavelas: We have looked at over 300 neuronal and glial clones in adult rats following retroviral injections in the cerebral ventricles of embryos between E14 and E20 (Parnavelas et al 1995, this volume). We found few oligodendrocyte clones in those samples. In contrast, when we injected retroviruses in the subventricular zone in postnatal rats, we found many oligodendrocyte clones throughout the cortical white and grey matter.

Bartlett: Another step, possibly another environmental signal, seems to be required for the generation of oligodendrocytes. If we looked at clones generated from E12 or E14, we may see oligodendrocytes but they may be derived from the subventricular zone.

LaMantia: Factors are present at this age that determine whether these cells become neurons or not. Cortical neurons are not generated until four or five days after this, so are factors present in the cortical rudiment's neuroepithelium that keep the cells in a precursor state?

Bartlett: Yes, and it is possible that FGF-2 is that factor. We have to remove the stimulus of FGF-2 in order for differentiation to occur in the proliferating neuroepithelial cells. One way of initiating differentiation of precursor cells is for them to move away from the ventricular zone—the area that has the highest levels of heparan sulfate proteoglycan (HSPG) and FGF-2. We understand how the presentation of both those molecules is important in the right context, but we don't understand how the cell actually gets fed one or the other.

LaMantia: FGF-3 is expressed dorsally in the telencephalon at E10. Where are FGF-1 and FGF-2 expressed?

Bartlett: FGF-2 is expressed throughout the neuroepithelium of the telencephalon at E10. FGF-1 is not expressed until E11 or E12, but again it is fairly widespread throughout the neuroepithelium.

Rubenstein: There is evidence that in the forebrain, oligodendocytes may not come from a cortical source but may migrate to the subventricular zone later on (Timsit et al 1995). Is it possible that you don't observe oligodendrocytes when you make your dissections at E10 or E12 because they simply aren't there at that stage?

Bartlett: This is an interesting proposition, although recent results using cortical tissue from E12 rats (Davis & Temple 1994) suggest that clones can generate all three principal cell types (neurons, astrocytes and oligodendrocytes), so it is unlikely that this is the explanation. These experiments were done under serum-free conditions with conditioned medium from neuronal cells. It is possible that stimulation with FGF is the primary reason why we're not generating oligodendrocytes.

Innocenti: What is left of the old hypothesis that microglia and neuroglia share the same origin (reviewed in Fujita et al 1981)?

Bartlett: There is no evidence to suggest that neuroepithelial cells have the potential to become microglial cells. All the studies to date, using cell-surface markers to follow microglial precursor cells, show that they migrate in.

J. Price: I agree only with the proviso that microglial precursor cells migrate early. For example, if you culture the E16 ventricular zone, microglial cells appear in the cultures because they have already moved into the brain by that stage.

Bartlett: Progenitors of microglia are circulating in the blood at E11–E12, at a time when the thymus is also being invaded by dendritic precursors. Presumably, this is the time when they invade the brain as well. Their presence

Factors affecting differentiation 97

at the commencement of neurogenesis has made interpretations problematic. There is no evidence of microglia in cultures obtained from E10 neuroepithelium.

Innocenti: However, the possibility that at least the ramified form of microglia is of neuroectodermal origin has only recently been advocated and discussed (reviewed by Ling & Wong 1993). I was curious why the people who described the clones of cortical cells did not mention whether or not they saw microglia. Perhaps this would imply that microglia were not of neuroectodermal origin.

Levitt: This is only a minor controversy. There is clearly a population that does migrate, but that doesn't mean to say that there isn't a population that's intrinsic to the neuroepithelium. Almost all the evidence suggests that microglia are formed from monocytes migrating from the periphery (Ling & Wong 1993). Even after injury in embryogenesis or early in development, the cell that responds is the peripheral monocyte and not the intrinsic microglia (Milligan et al 1991).

Is your neuroepithelium generating a glial cell type and a neuronal cell type at the same time?

Bartlett: Yes, a clone can generate both cell types.

Levitt: But this does not mean that *in vivo* the epithelium generates both cell types at the same time as the wall expands. We showed that there are large numbers of radial glial cells at the beginning of neurogenesis in the monkey, although I don't know if this is true for the rat (Levitt & Rakic 1980, Levitt et al 1983).

LaMantia: There are not many radial glial cells present at the stage that Perry Bartlett is looking at (E10 in the mouse). Instead, the columnar cells may just be epithelial cells.

Levitt: We tend to use conventional markers to decide whether a cell is present. If stained cells are observed in a species in which the conventional marker works early, then one can be confident that they are present. However, if stained cells are not observed, they may still be present. The important question is perhaps, if one assumes an early coexistence of neurons and glia, what is behind the simultaneous expansion of the glial cell population and the neuronal population?

Bartlett: The generation of neurons and astrocytes are temporally different both *in vitro* and *in vivo*. Neurons are generated virtually within a day of plating E10 neuroepithelium, whereas the GFAP-positive population is not apparent until about one week after plating (equivalent to E17, the time at which GFAP first appears in the mouse). The clones show a similar temporal difference.

Levitt: Are the glial cells making intermediate filaments?

Bartlett: Yes. GFAP is present, but not until the clone is at least a week old. One question I would like to raise concerns the origin of radial glial cells. How do we identify radial glial cells *in vitro* and what is their relationship to a stem cell? Leber et al (1990) suggest that radial glial cells and stem cells may be the same cell. The studies on radial glial cells in the chick also support that interpretation (Alvarez-Buylla et al 1990), but I would like to hear Pasko Rakic comment on what is known about the generation of radial glial cells.

Rakic: It is probably a question of the timing of cellular events, which is different in different species. For example, there is an early commitment to glial and neuronal lineages in the monkey and human cerebrum. In both species glial cells express vimentin and GFAP early. In rats the same type of cells may be present but they don't express GFAP. People using GFAP came to the conclusion that radial glial cells appear in the rat only after birth. However, these cells are actually there before birth. Therefore, these markers are not good enough to detect cell differences.

Bartlett: Do we have any good radial glial markers? It seems to me that most of the radial glial markers also mark neuroepithelial cells.

Rakic: Then they are not good markers!

References

Abney ER, Bartlett PF, Raff MC 1981 Astrocytes, ependymal cells, and oligodendrocytes develop on schedule in dissociated cell cultures of embryonic rat brain. Dev Biol 83:301–310

Alvarez-Buylla A, Theelen M, Nottebohm F 1990 Proliferation hot spots in adult avian ventricular zone reveal radial cell division. Neuron 5:101–109

Davis AA, Temple S 1994 A self-renewing multipotential stem cell in embryonic rat cerebral cortex. Nature 372:263–266

Fujita S, Tsuchihashi Y, Kitamura T 1981 Origin, morphology and function of the microglia. In: Vidorio EA, Fedoroff S (eds) Glial and neuronal cell biology. Alan R. Liss, Inc, New York, p141–169

Leber SM, Breedlove SM, Sanes JR 1990 Lineage, arrangement, and death of clonally related motoneurons in chick spinal cord. J Neurosci 10:2451–2462

Levitt P, Rakic P 1980 Immunoperoxidase localization of glial fibrillary acidic protein in radial glial cells and astrocytes of the developing Rhesus monkey brain. J Comp Neurol 193:815–840

Levitt P, Cooper ML, Rakic P 1983 Early divergence and changing proportions of neuronal and glial precursor cells in the primate cerebral ventricular zone. Dev Biol 96:472–484

Ling E-A, Wong W-C 1993 The origins and nature of ramified and amoeboid microglia: a historical review and current concepts. Glia 7:9–18

Milligan CE, Levitt P, Cunningham TJ 1991 Brain macrophages and microglia respond differently to lesions of the developing and adult visual system. J Comp Neurol 314:136–146

Parvavelas JG, Mione MC, Lavdas A 1995 The cell lineage of neuronal subtypes in the mammalian cerebral cortex. In: Development of the cerebral cortex. Wiley, Chichester (Ciba Found Symp 193) p 41–58

Price J, Williams BP, Götz M 1995 The generation of cellular diversity in the cerebral cortex. In: Development of the cerebral cortex. Wiley, Chichester (Ciba Found Symp 193) p 71–84

Timsit S, Martinez S, Allinquant B, Peyron F, Puelles L, Zalc B 1995 Oligodendrocytes originate in a restricted zone of the embryonic ventral neural tube defined by DM-20 messenger RNA expression. J Neurosci 15:1012–1024

Emx and *Otx* gene expression in the developing mouse brain

Edoardo Boncinelli*, Massimo Gulisano†, Fabio Spada and Vania Broccoli‡

DIBIT, Istituto Scientifico H.S. Raffaele, Via Olgettina 60, 20132 Milano and *Centro Infrastrutture Cellulari, Via Vanvitelli 32, 20129 Milano, †Istituto di Biologia Generale, Via Androne 81, 95124 Catania, ‡Istituto di Genetica, Dipartimento di Biologia Evoluzionistica Sperimentale, Via Belmeloro 8, 40126 Bologna, Italy

>*Abstract.* The homeobox genes *Emx1*, *Emx2*, *Otx1* and *Otx2* are all expressed in the rostral brain of embryos at E10. Their expression domains are continuous regions of the developing brain contained within each other, such that the expression domain of *Otx2* contains that of the other three genes, the expression domain of *Otx1* contains that of *Emx1* and *Emx2*, and the expression domain of *Emx2* contains that of *Emx1*. The *Emx1* expression domain includes the dorsal telencephalon and it has a posterior boundary slightly anterior to that between the presumptive diencephalon and telencephalon, whereas the *Otx2* expression domain covers almost the entire forebrain and midbrain. Starting from E10.75, *Otx2* expression disappears progressively from the presumptive cerebral cortex, whereas *Emx1*, *Emx2* and *Otx1* are expressed in this structure until late gestation. In particular, *Emx2* appears to be expressed exclusively in the germinal ventricular zone of the developing cerebral cortex.
>
>*1995 Development of the cerebral cortex. Wiley, Chichester (Ciba Foundation Symposium 193) p 100–116*

The study of vertebrate genes partially homologous to regulatory genes controlling the development of the *Drosophila* embryo (Lawrence & Morata 1994) has provided invaluable information about the genetic control of positional values in development. Many of these genes contain a DNA motif termed the homeobox and are thus called homeobox genes. A number of homeobox genes control cell identity in specific regions or segments both in invertebrates and vertebrates (McGinnis & Krumlauf 1992). The homeobox is a DNA sequence coding for a 61 amino acid residue protein domain called the homeodomain. The homeodomain is able to recognize and bind specific DNA sequences. Proteins containing homeodomains (homeoproteins) are transcription factors that act as activators or repressors of groups of downstream target genes.

There are several families of homeobox genes. These include the *Hox* genes (Krumlauf 1994) which are the vertebrate homologues of *Drosophila* homeotic genes. They control the specification of body regions along the vertebrate axis and provide positional cues for the developing neural tube, in particular the rhombencephalon and spinal cord from the branchial area down to the tail. In contrast, the development of the anteriormost body domain corresponding to the anterior head has remained relatively obscure in both invertebrates and vertebrates (Finkelstein & Boncinelli 1994). In the insect embryo the nature of anterior head segmentation has been controversial and the genes that govern it are mostly unknown. In vertebrates the existence of compartments or segments in the forebrain and midbrain is controversial and the underlying molecular mechanisms of pattern formation have not been determined.

A recent breakthrough has been the identification of genes in *Drosophila* and their homologues in vertebrates that appear critical to anterior head and brain specification. We focus here primarily on four vertebrate homologues of the two *Drosophila* genes *empty spiracles* (*ems*) and *orthodenticle* (*otd*) (see Finkelstein & Perrimon 1991, Cohen & Jürgens 1991 for reviews). These four genes are: *Emx1* and *Emx2* (Simeone et al 1992a,b), which are related to *ems*; and *Otx1* and *Otx2* (Simeone et al 1992a, 1993, Finkelstein & Boncinelli 1994), which are related to *otd*. The two *Otx* genes code for homeoproteins containing a homeodomain of the *bicoid* class. This class has a characteristic lysine residue at position 50 which corresponds to position 9 of the recognition helix and has been reported to confer DNA binding specificity. In addition, the substitution of this lysine by glutamine in the *bicoid* homeodomain leads to the replacement of its DNA binding specificity with that of the *antennapedia*-like homeoproteins (Simeone et al 1993).

The four *Emx* and *Otx* vertebrate genes are expressed in extended regions of the developing rostral brain of mouse mid-gestation embryos, including the presumptive cerebral cortex and olfactory bulbs. Here we summarize the mRNA expression results, and we discuss a possible role of the four genes in establishing the boundaries of the various embryonic brain regions. We will focus on three main aspects of these patterns: expression of the four genes in mouse embryos at embryonic day (E) 10; expression of *Otx* genes in the diencephalon at mid-gestation; and expression of the four genes in the developing cerebral cortex.

Expression of the four genes in mouse embryos at E10

The developing neural tube in mouse embryos at E10 shows recognizable presumptive regions corresponding to future anatomical subdivisions. The entire neural tube consists of neuroepithelial cells in active proliferation and most of the specific differentiating events have not yet occurred. All four genes are expressed at this stage in development. Their expression domains (Simeone

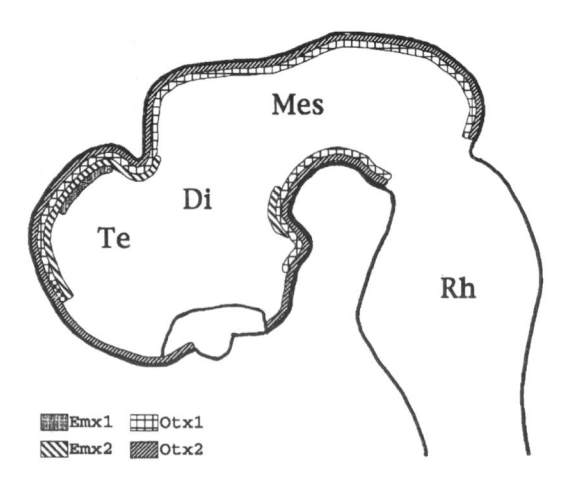

FIG. 1. The expression domains of *Emx* and *Otx* genes in the developing central nervous system of mouse embryos at embryonic day 10. A small checked area at the anterior boundary of the *Otx1* expression domain indicates an anterior region of dorsal telencephalon where the limits of expression of *Emx2* and *Otx1* are less well defined. Di, diencephalon; Mes, mesencephalon; Rh, rhomboencephalon; Te, telencephalon.

et al 1992a) are continuous regions of the developing brain contained within each other, such that the expression domain of *Otx2* contains that of the other three genes, the expression domain of *Otx1* contains that of *Emx2* and *Emx1* and the expression domain of *Emx2* contains that of *Emx1* (Fig. 1). The *Emx1* expression domain includes the dorsal telencephalon with a posterior boundary slightly anterior to that between the presumptive diencephalon and telencephalon. *Emx2* is expressed in dorsal and ventral neurectoderm with an anterior boundary slightly anterior to that of *Emx1* and a posterior boundary within the roof of the presumptive diencephalon. This boundary probably coincides with the boundary between the first and second thalamic segment (Kuhlenbeck 1973), which subsequently gives rise to ventral thalamus and dorsal thalamus, respectively. The *Otx1* expression domain contains the *Emx2* domain. It covers a continuous region including part of the telencephalon, the diencephalon and the mesencephalon, with an anterior boundary approximately coincident with that of *Emx2*. Laterally, the posterior boundary of the *Otx1* domain coincides with that of the mesencephalon. In median sections a strong hybridization signal extends only half way along the mesencephalon, dividing the mesencephalic dorsal midline into two domains. Finally, the *Otx2* expression domain contains the *Otx1* domain, both dorsally and ventrally, and it covers almost the entire forebrain and midbrain, to the exclusion of the early optic area.

The expression of *Emx* and *Otx* genes identifies several regions in the forebrain (Fig. 1). Some of these regions correspond to presumptive anatomical subdivisions, whereas the significance of others remains to be assessed. For example, it is clear that the dorsal expression of the two *Emx* genes identifies a presumptive cortical region, part of which will form the neocortex and archicortex. The expression of *Emx2* also appears to define the boundary between the future dorsal and ventral thalamus. On the other hand, the expression of these genes does not offer an unambiguous cue for the boundary between the presumptive ventral thalamus and the posterior dorsal telencephalon.

In summary, analysis of the brain at E10 shows a nested pattern of expression of the four genes in brain regions defining an embryonic rostral, or preisthmic, brain as opposed to hindbrain and spinal cord. The appearance of the four genes is also sequential: *Otx2* is expressed as early as E5.5 (Simeone et al 1993), followed by *Otx1* and *Emx2* at E8–8.5 then *Emx1* at E9.5 (Simeone et al 1992a). It seems reasonable to postulate a role for the four homeobox genes in establishing the identity of the various embryonic brain regions. The specification of the regions of the early rostral brain seems to be a discrete process with its centre in the dorsal telencephalon.

Expression of *Otx* genes in the brain of mouse embryos at mid-gestation

The two *Otx* genes are expressed in specific restricted regions of the developing brain in mouse embryos at mid-gestation (Simeone et al 1993). They are expressed in the dorsal and basal telencephalon, diencephalon and mesencephalon, but not in the spinal cord. Their expression domains in the mesencephalon show a sharp posterior boundary, both dorsally and ventrally, approximately at the level of the rhombic isthmus. From E9.25 onward, the expression of both genes marks the posterior boundary of mesencephalon to the exclusion of presumptive anterior cerebellar domains.

Otx1 and *Otx2* are also expressed in restricted regions of the diencephalon—the epithalamus, dorsal thalamus and mammillary region of the posterior hypothalamus. In these regions, the hybridization signal is almost exclusively confined to cells of the ventricular zone. Their expression domains do not include the ventral thalamus. A two-layered narrow stripe of expression is detectable at the boundary between the dorsal and ventral thalamus, i.e. the zona limitans intrathalamica, which is the precursor of the lamina medullaris externa and mammillothalamic tract. Other localizations are the fasciculus retroflexus, the precursor of the habenulointerpenduncular tract, the stria medullaris, including the region surrounding the posterior commissure, the primordium of mammillotegmental tract, the epiphysis, the fornix and the sulcus lateralis hypothalami posterioris. Posterior to the diencephalon, *Otx1* and *Otx2* are expressed in mesencephalic regions of the tectum and tegmentum, possibly at the level of presumptive periventricular bundles.

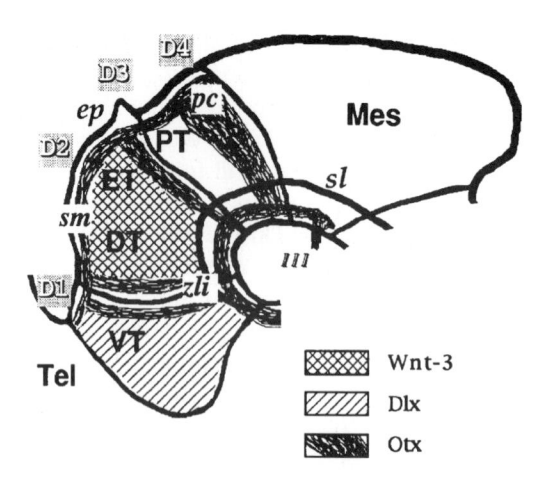

FIG. 2. Schematic representation of the expression domains of *Otx1* and *Otx2*, *Wnt3*, and *Dlx* genes in the diencephalon of embryos at embryonic day 12.5. Within the diencephalic regions, bold letters designate the columnar nomenclature: DT, dorsal thalamus; ET, epithalamus; PT, pretectum; and VT, ventral thalamus. Outside the profile, the proposed new subdivision into four neuromeres, D1 to D4 (Figdor & Stern 1993), is indicated. *ep*, epiphysis; Mes, mesencephalon; *pc*, posterior commissure; *sl*, sulcus limitans; *sm*, stria medullaris; Tel, telencephalon; *zli*, zona limitans intrathalamica; *III*, third cranial nerve.

Both *Otx* genes are expressed in the olfactory epithelium and in the developing inner ear—from early expression in the otic vesicle to expression in the epithelia of auricular ducts of the sacculus and cochlea. They are also expressed in the developing eye, including the external sheaths of the optic nerve.

Areas and boundaries in the diencephalon

The expression of *Otx* genes in the diencephalon and mesencephalon of embryos at E12.5–14.5 co-localizes with boundary regions and presumptive axon tracts, including the posterior commissure (Fig. 2). This expression is confined to precursor cells surrounding these structures as if these cells could be used as borders of pathways for the pioneer axon tracts. This is particularly evident in the posterior commissure and along the zona limitans intrathalamica (see also Fig. 3D). *Otx* gene expression in the posterior commissure is limited to cells of the ventricular epithelium. The primary fibres running along the surface of the posterior commissure are not labelled. Expression of *Otx* genes along the zona limitans intrathalamica might constitute a framework for the axon patterning of the lamina medullaris and other structures that physically separate the dorsal thalamus from the ventral thalamus. The existence of this barrier might account for the sharp dorsal boundary of the expression domain

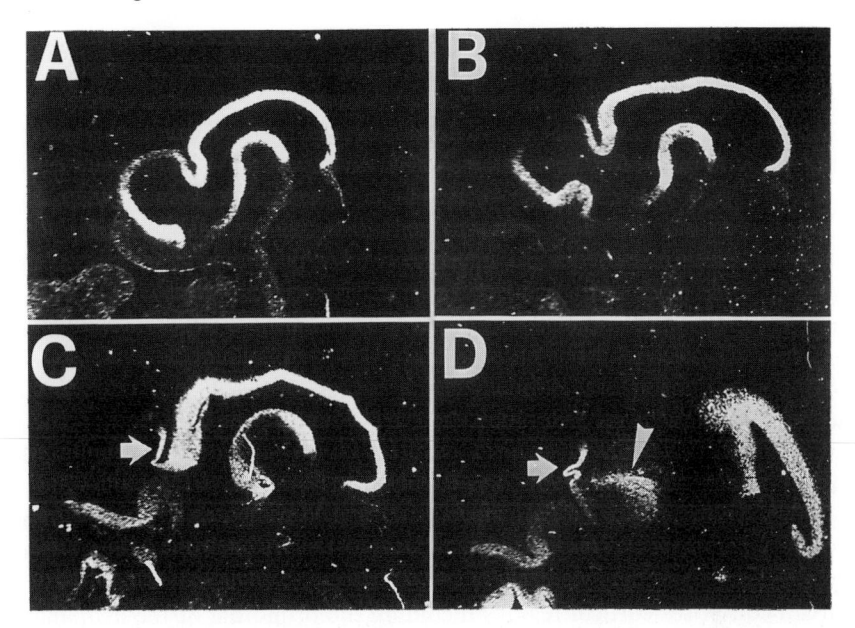

FIG. 3. *Otx2* expression in the developing brain of mouse embryos at embryonic day (E) 10.75 (A), E11 (B), E12 (C) and E12.75 (D). Arrows in C and D point to the developing choroid plexus. An arrowhead in D indicates the zona limitans intrathalamica in the diencephalon.

in the ventral thalamus of the *Dlx* homeobox genes (Simeone et al 1994) and for the sharp ventral boundary of the expression domain in the dorsal thalamus of the *Wnt3* gene, which encodes a putative differentiation factor (see Boncinelli 1994 for review). Both *Otx* genes are also expressed around the developing optic nerve. This localization provides clues to axon pathfinding and patterning in a similar manner to that along the zona limitans intrathalamica. Expression of *Otx* genes may, therefore, provide a global framework for the primary scaffold of specific axon pathways in the early neuroepithelium of the forebrain. We are currently testing this hypothesis with *in vitro* analysis of axon growth and propagation in *Otx*-transfected cells.

Otx genes may play different roles in the development of the head in at least two different stages. They first specify territories or areas in the rostral brain of mouse embryos at E8–10, and then they provide a set of positional cues required for growing axons to follow specific pathways within the embryonic central nervous system. It is not clear whether these two functions are independent. It is also of interest to consider that in the *Drosophila* mutant for *orthodenticle*, pioneer axons of the posterior commissures fail to develop normally, as if appropriate positional cues were missing (Tessier-Lavigne 1992).

Expression of the four genes in the developing cerebral cortex

All four genes are expressed between E9.5 and E10.5 in the presumptive cerebral cortex (Simeone et al 1992a), but *Otx2* expression disappears progressively from this region at E10.75 (Fig. 3). This process is relatively quick and it is initiated in the central areas. *Otx2* expression persists in the forming choroid plexus (Figs 3C and 3D), a characteristic *Otx2* localization (Boncinelli et al 1993), and in the septum and ganglionic eminence. Expression along the zona limitans intrathalamica is also observable (Fig. 3D).

Emx1 and *Emx2* are expressed in the presumptive cerebral cortex in a developmental period (E9.5–16) corresponding to major events in cortical neurogenesis (Simeone et al 1992b). From E12.5 to E13.5, the *Emx1* expression domain comprises cortical regions including the primordia of the neopallium, hippocampal and parahippocampal archipallium. *Emx1* expression seems mainly, but not exclusively, characteristic of cortical regions that are hexalaminar in nature (Kuhlenbeck 1973). During the same period, the *Emx2* expression domain comprises presumptive cortical regions including the neopallium, hippocampal and parahippocampal archipallium and selected paleopallial localizations, but not the basal internal grisea (Simeone et al 1992b). The *Otx1* expression domain in the forebrain (Simeone et al 1993, Frantz et al 1994) includes the dorsal telencephalon but also extends to basal regions.

It is interesting to consider the temporal pattern of expression of the *Emx* genes in various zones of the forming cerebral cortex. At least three zones can be considered: the germinal neuroepithelium or ventricular zone, where cortical neurons proliferate; the transitional field; and the forming cortical plate, from which the cortical grey matter will subsequently develop (Bayer & Altman 1991). The germinal neuroepithelium is practically the sole component of the prosencephalic wall up to E12.5. A transitional field then appears which includes the subventricular zone, the nature of which is not known, and the intermediate zone, where differentiating cortical cells are translocated before migrating to the cortical plate. As development proceeds, the thickness of the cortical plate increases progressively. Both the transitional field and the cortical plate develop at the expense of the neuroepithelium, according to specific spatial and temporal gradients within the forming cortex. Two major neurogenetic and morphogenetic gradients can be observed: one progressing anterior to posterior, and a second progressing ventrolateral to dorsomedial. The various cortical layers are formed in an inside out pattern (Bayer &

Emx and *Otx* genes 107

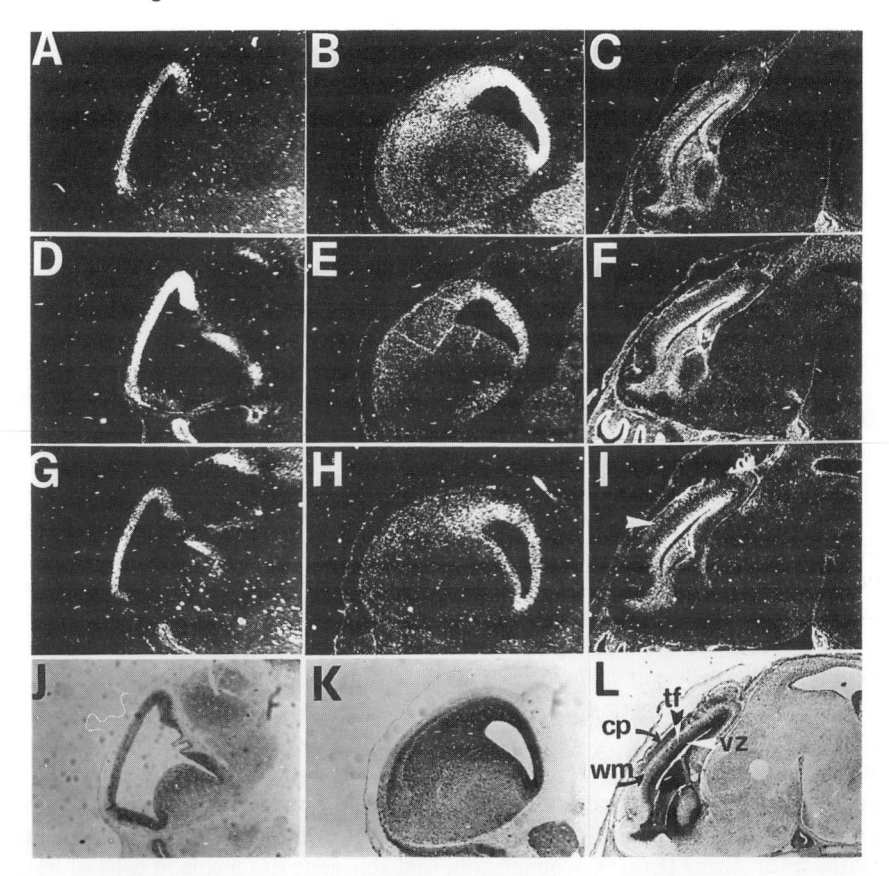

FIG. 4. Expression of *Emx1* (A–C), *Emx2* (D–F) and *Otx1* (G–I) in sagittal sections of the developing cerebral cortex of mouse embryos at embryonic day (E) 12.5 (A, D and G), E14.5 (B, E and H) and E17 (C, F and I). J–L show the corresponding bright fields. An arrowhead in I indicates the early cortical plate where *Otx1* expression is detectable. cp, cortical plate; tf, transitional field; wm, white matter; vz, ventricular zone.

Altman 1991). Autoradiographic experiments using [^{3}H]thymidine to label neurons at their birth have shown that neurons destined to occupy the depth of the cortex are generated first, and that the waves of neurons generated subsequently bypass the earlier ones by active migration and settle above them.

Figure 4 shows the expression of *Emx1*, *Emx2* and *Otx1* in the cortex of mouse embryos at E12.5, E14.5 and E17. *Emx2* is expressed within the germinal neuroepithelium at all stages. At E12.5, the *Emx2* hybridization signal is distributed uniformly across the cortex without major differences, but from E13.5 onwards it is confined to the germinal neuroepithelium of the ventricular zone, excluding both the transitional field and the cortical plate. This is most apparent at E14.5 and later stages (Figs 4E and 4F). The coincidence between *Emx2*-expressing cells and proliferating cells is confirmed by bromodeoxyuridine labelling (not shown). From E14.5 onwards, *Emx2* cortical expression declines progressively in the anterior and ventrolateral regions and at E19 it is confined solely to specific cell layers in the hippocampus.

Emx1 expression is not as specific and extends to other zones. It is expressed in nearly every cortical neuron. A weaker *Emx1* signal is present only in the late intermediate zone just prior to its evolution into white matter. *Otx1* is expressed in the deeper layers of the telencephalic cortex following migration. This was reported previously in a study on the rat cortex (Frantz et al 1994). When the cortex is organized into layers, i.e. from postnatal day (P) 16 to adult, *Otx1* is expressed in a subpopulation of neurons in layer V and uniformly in layer VI, with the exception of the frontal area (Frantz et al 1994 and our unpublished results). The cortical plate is not evident in mouse embryos at E12.5, and the expression of *Emx1* and *Otx1* is practically coincident in the telencephalon (Figs 4A and 4G). At E14.5 the first cortical cells migrate to the external layers (Bayer & Altman 1991) and are clearly positive for both genes (Figs 4B and 4H). These cortical cells are fated to become cells of deeper layers of the cortex. Figure 4I shows a clear signal for *Otx1* in the cortical plate (arrowhead) at a stage (E17 in the mouse) when cells fated to become layers V and VI are born and have migrated to the cortical plate. No detectable signal is found at this stage in the intermediate zone, suggesting that the majority of *Otx1*-positive cells have already migrated to the cortical plate. In contrast, both the cortical plate and the intermediate zone are positive for *Emx1* (Fig. 4C), suggesting that this gene has a role in cells of both the upper and deeper layers during cortical development.

Concluding remarks

Emx and *Otx* genes represent good markers for the study of the developing brain, and in particular for the developing cerebral cortex. Their expression patterns during mouse embryogenesis allow us to follow major events in cortical neurogenesis and differentiation, and they are also useful for following the phylogenetic evolution of cortical structures. With this in mind, we have cloned and characterized the cognates in other species, including man, frog (Boncinelli et al 1993), chick (Bally-Cuif et al 1995), zebrafish and lamprey (our unpublished results). The analysis of expression patterns in these species promises interesting conclusions, on both the phylogenetic and the ontogenetic aspects of neocortex formation.

Acknowledgements

We thank everyone in our lab for comments and helpful suggestions. This work was supported by grants from the EC BIOTECH Programme, the Telethon-Italia Programme and the Italian Association for Cancer Research.

References

Bally-Cuif L, Gulisano M, Broccoli V, Boncinelli E 1995 c-*otx2* is expressed in two different phases of gastrulation and is sensitive to retinoic acid treatment in chick embryo. Mech Dev 49:49–63
Bayer SA, Altman J 1991 Neocortical development. Raven Press, New York
Boncinelli E 1994 Early CNS development: *distal-less* related genes and forebrain development. Curr Opin Neurobiol 4:29–36
Boncinelli E, Gulisano M, Pannese M 1993 Conserved homeobox genes in the developing brain. C R Acad Sci Ser III Sci Vie 316:972–984
Cohen S, Jürgens G 1991 *Drosophila* headlines. Trends Genet 7:267–272
Figdor M, Stern C 1993 Segmental organization of embryonic diencephalon. Nature 363:630–634
Finkelstein R, Boncinelli E 1994 From fly head to mammalian forebrain: the story of *otd* and *Otx*. Trends Genet 10:310–315
Finkelstein R, Perrimon N 1991 The molecular genetics of head development in *Drosophila melanogaster*. Development 112:899–912
Frantz GD, Weimann JM, Levin ME, McConnell SK 1994 Otx1 and Otx2 define layers and regions in developing cerebral cortex and cerebellum. J Neurosci 14:5725–5740
Krumlauf R 1994 *Hox* genes in vertebrate development. Cell 70:191–201
Kuhlenbeck H 1973 Central nervous system of vertebrates, vol 3, part 2: Overall morphology pattern. S. Karger, Basle
Lawrence PA, Morata G 1994 Homeobox genes: their function in Drosophila segmentation and pattern formation. Cell 70:181–189
McGinnis W, Krumlauf R 1992 Homeobox genes and axial patterning. Cell 68:283–302
Simeone A, Acampora D, Gulisano M, Stornaiuolo A, Boncinelli E 1992a Nested expression domains of four homeobox genes in the developing rostral brain. Nature 358:687–690
Simeone A, Gulisano M, Acampora D, Stornaiuolo A, Rambaldi M, Boncinelli E 1992b Two vertebrate homeobox genes related to the *Drosophila empty spiracles* gene are expressed in the embryonic cerebral cortex. EMBO J 11:2541–2550
Simeone A, Acampora D, Mallamaci A, Stornaiuolo A, Nigro V, Boncinelli E 1993 A vertebrate gene related to *orthodenticle* contains a homeodomain of the *bicoid* class and demarcates anterior neuroectoderm in the gastrulating mouse embryo. EMBO J 12:2735–2747
Simeone A, Acampora D, Pannese M et al 1994 Cloning and characterization of two members of the vertebrate *Dlx* gene family. Proc Natl Acad Sci USA 91:2250–2254
Tessier-Lavigne M 1992 Axon guidance by molecular gradients. Curr Opin Neurobiol 2:60–65

DISCUSSION

Walsh: Are *Emx* homologues present in *Xenopus*? And if so, do alterations in their expression have any effect?

Boncinelli: *Emx* homologues are present in *Xenopus laevis*. They are expressed early in development in a similar region as in the mouse embryo, and in *Xenopus Emx1* is probably specifying some sort of positional information. Teleosts don't have an *Emx1* homologue, they only have an *Emx2* homologue. This could be interesting in evolutionary terms because lung fish have both *Emx1* and *Emx2* homologues. It is possible that *Emx1* arose as a gene duplication of *Emx2*.

LaMantia: There is a shift in *Otx1* expression from a broad to a restricted pattern in the diencephalon. What is known about the regulation of *Otx1* expression?

Boncinelli: We are now transfecting cells to pinpoint the genes that are both downstream and upstream. The *Otx2* regulatory regions are a mess. We have been struggling with this region for two years and we have been unable to reproduce its expression pattern in transgenic mice. We haven't looked at the regulation of *Otx1* yet.

LaMantia: You have shown that both *Otx1* and *Otx2* are expressed at different times in distinct areas. Therefore, it's possible that the downstream targets of both genes are the same, and that these targets are performing the same functions in different tissues. Could you elaborate on what you think those same things are and what kind of downstream targets they may have.

Boncinelli: I could elaborate, but it would be purely speculation. Two years ago I thought they were upstream of about 10 different genes, but now I think there are probably hundreds. The results of transfection experiments (E. Boncinelli & S. Guazzi, unpublished results) suggest that *Otx2* may regulate cell shape. Obviously, there are numerous factors involved in determining cell shape, intermediate filaments and adhesion molecules for instance, but cells transfected with *Otx2* certainly change their cell shape dramatically—they become flat and resemble tiles.

Bolz: Do transfected neurons have this shape?

Boncinelli: I don't know.

D. Price: Do you have any evidence for interactions between these different homeobox genes?

Boncinelli: No. When we overexpress *Otx2* in *Xenopus* embryos, the expression of *Otx1*, *Emx1* and *Emx2* is neither up-regulated nor down-regulated.

Rubenstein: How did you reduce the expression of the genes in *Xenopus*?

Boncinelli: We reduced the expression by using antibodies, antisense oligonucleotides or ribozymes.

Rubenstein: Do helicases prevent the action of antisense oligonucleotides in *Xenopus*?

Boncinelli: No, in our hands the oligonucleotides work beautifully. Obviously, one has to be very patient and success depends on the oligonucleotides that are chosen.

Rubenstein: Do you think that prechordal mesoderm or anterior mesoderm are necessary for generating the head?

Boncinelli: Yes.

Rubenstein: How can you reconcile this with the planar induction studies that suggest that underlying mesoderm is not required to induce anterior genes (Doniach 1992).

Boncinelli: It is not clear whether positional information occurs by radial or planar induction. The process of differentiation requires other factors in addition to positional information. For example, radial induction is necessary for the development of eyes. A few markers have been analysed, but we must be careful when drawing conclusions (Altaba 1994). When we produce exogastrulae, we obtain ectoderm and mesoderm, which begs the question as to whether radial induction is necessary for the expression of certain genes in the neuroectoderm. All our evidence suggests that the expression of *Otx2* requires radial induction.

Rubenstein: It has been suggested that the transforming growth factor (TGF) family and the fibroblast growth factor (FGF) family may be candidates for early induction molecules. Is the expression of *Otx1*, *Otx2*, *Emx1* and *Emx2* changed by TGF or FGF?

Boncinelli: In *Xenopus* embryos all factors that promote dorsalization enhance the expression of *Otx2*, and factors that promote ventralization down-regulate the expression of *Otx2*. For example, retinoic acid causes the complete ventralization of the embryo (the embryos develop without brains), and *Otx2* expression is totally switched off in these embryos (Pannese et al 1995).

Rubenstein: One of the reasons that Edoardo Boncinelli and I are here is to discuss the genes that are likely to be involved in the patterning of different regional areas of the cerebral cortex. Pasko Rakic's proto-map (Rakic 1988) hypothesizes that certain genes are expressed in the ventricular zone which divide the neuroepithelium of the cerebral cortex into domains, giving rise to different functional and structural areas. This led me to search for these genes, but as yet there is no strong evidence that they exist.

Boncinelli: There is a sharp boundary between the expression of *Emx1* and *Emx2* in the cortex and the basal ganglia throughout development, so there must be a predetermined difference in the cell identity of cortical neurons compared to ganglionic neurons.

A relevant question to ask is what *Emx1* is actually doing, because it is expressed almost exclusively in the cortex, suggesting that there must be something in the nature of cortical neurons that is specified by *Emx1*.

Walsh: There's an interesting parallel in the lineage results, where we also observe a sharp border between the cortex and the striatum (Reid et al 1995). Glial cells in the striatum are rarely related to cortical cells, but neurons don't seem to cross the border between the cortex and the striatum.

On the other hand, we have observed clones that are dispersed between the cortex and the hippocampus (C. Walsh & C. Reid, unpublished observations). This is in parallel with the pattern of expression of *Emx1*, which also crosses the boundary between the cortex and the hippocampus, or the boundary between the cortex and olfactory bulb. Is it possible that the two regions represent one developmental unit?

LaMantia: No. Several lines of evidence suggest that the olfactory targets and the ventricular telencephalic olfactory targets are within separate inductive fields which are induced by a neural crest-derived population in the lateral cranial mesenchyme (Stout & Graziadei 1980, LaMantia et al 1993, Drake et al 1994).

Walsh: But that's not to say that they do not have different external influences or inductive effects outside the brain which cause regionally distinct effects on an initially uniform developmental structure.

Rubenstein: We have identified some genes that have distinct patterns of expression within the cerebral cortex (Porteus et al 1994). Typically, they're expressed in the postmitotic layers (the mantle). Discontinuities in their expression suggest that they subdivide the cortex into paleo, limb and cortical domains. As yet, we have no evidence for transverse subdivisions within the neocortex.

Rakic: Can you comment on whether some markers display gradients of expression.

Boncinelli: Gradients are definitely involved, and there may be anterior-to-posterior gradients and lateral-to-medial gradients, but what looks like a spatial gradient may actually be a temporal gradient.

Rakic: That doesn't matter because if two gradients intersect at one point, they can provide useful coordinates.

Kennedy: We have seen differences in proliferation rates in the germinal zone, which gives rise to individual areas of the neocortex (Dehay et al 1993). In the monkey the proliferation rate in the germinal zone that gives rise to area 17 is higher than that which gives rise to area 18. This difference in proliferation constitutes a gradient. In the mouse there is also a mesolateral gradient of proliferation in the region that gives rise to the motor and the somatosensory cortex (F. Polleux, C. Dehay & H. Kennedy, unpublished observations). The ventricular zone that gives rise to motor cortex has a higher rate of proliferation than the adjacent somatosensory region. It's also on a gradient, but not a moving gradient and it is, therefore, not to be confused with the rostrocaudal gradient of neurogenesis. The mediolateral gradient that was described in one of your papers (Simeone et al 1992) is pronounced and could underline a regional difference.

LaMantia: How should we define regional differences? During neurogenesis, precursors from the cortical rudiment make several types of neurons and glia. However, although there are some regional variations in the neurons, it's impossible to determine which group of cells in the embryonic cortex will have which particular characteristic. Is it possible that regional variation isn't a neurogenic feature? Regional identity may be assigned after the time when cortical neurons are finally generated, and it may involve an intermediate differentiation step which produces, for example, adhesive differences that affect the attraction of thalamic axons. It is difficult to imagine how transcriptional regulation is involved in regional differentiation in the absence of a target. No one has been able to find anything that marks one region as being different from the next at the time when the neurons are generated.

Levitt: Not exactly. There are molecular differences in the archicortex, allocortex and mesocortex early in development. For example, we have identified an adhesive protein in the rodent that is expressed immediately after neurogenesis when the cells are in the process of migrating to the final position (Levitt et al 1995, this volume, Horton & Levitt 1988). It is expressed before the neurons have settled in the cortical plate. We don't know what regulates its transcription, but there is a gradient of expression, i.e. there are cortical areas that have high levels and cortical areas that have low levels.

LaMantia: But this is more the issue of the differences between areas 17 and 18, the differences between areas 3 and 4, and the differences between areas 1 and 2, rather than the differences between the hippocampus and the cortex. Even though there is some sort of historical pre-patterning that gives this result, is there a time when the ventricular zone of the cortex is similar to the situation in *Drosophila* where there are bands of gene expression that specify regional identity of distinct neocortical fields? If this is a situation where position and identity are established by a series of steps, is the identification of transcription factors that mark area 17 as being different to other areas, for example, going to involve a fruitless search?

Ghosh: In the H-2Z1 transgenic mouse line (Cohen-Tannoudji et al 1994), cells express β-galactosidase only in layer IV of the somatosensory cortex. If you move those cells to a different location before thalamic innervation, they still express β-galactosidase at the right time. This experiment suggests that determination occurs early on.

Kennedy: However, in these experiments expression of the marker gene is not completely restricted to the somatosensory cortex but is merely more frequent in that area. Hence, the experiments of Cohen-Tannoudji et al (1994) demonstrate that early determination could involve some sort of gradient in the ventricular zone.

Krubitzer: If differential gene expression marked the functional boundaries of developing fields in the cerebral cortex, how would this change in different species? Different species may retain some fields from a common ancestor and also develop new ones. Would you expect to see common patterns of gene expression for the fields that they have in common, and different patterns of expression for the new ones?

Boncinelli: I expect that differences between species exist which are correlated with modulation of gene expression, but it would be premature to say anything more than that because we have not yet found localized expression of *Otx1*, *Otx2*, *Emx1* or *Emx2*.

Bartlett: Looking at localized expression of mRNA may give a good indication as to whether a gradient exists, but this may not always reflect the amount of protein that is present. Would we see different patterns or gradients if we looked at protein levels?

Rubenstein: For the homeobox genes that we have studied, protein distribution is the same as mRNA distribution (Porteus et al 1994).

Boncinelli: As far as we can tell, this is also true for *Otx2*. There are at least three different *Otx2* transcripts, so when I say expression, I mean the 3'-region that is common to all three.

The more we study in terms of molecular genetics, the earlier we seem to find that everything is decided. Therefore, I would not be surprised to find that the subdivision of the cortex is a very early event, but this is simply a prediction on the basis of what is known about other tissues.

Rakic: Smooth molecular gradients expressed in the cortical plate may eventually be converted into stepwise, sharp cellular changes by the interaction of cortical neurons and different afferent systems. For example, there may be an initial smooth anteroposterior gradient of molecules that make the occipital pole of the cerebral vesicle more attractive to the axons from the lateral geniculate nucleus (LGN) and the frontal pole more attractive to axons from the medial dorsal nucleus of the thalamus. These gradients may be sufficient to attract a separate set of axons preferentially to the opposite poles of the cerebrum, but it may not be sufficient to accomplish full differentiation of the cortex into distinct cytoarchitectonic fields. For example, if the prospective area 24 of the embryonic monkey frontal lobe was exposed to the LGN axons, I doubt that it would differentiate into area 17, which has an elaborate sublaminae in layer IV and characteristic cytochrome oxidase blobs in layers II and III that are specialized for colour vision. There would not be enough cortical neurons to accomplish this because the cortex of area 24 is very thin and does not have a layer IV. Also, in the ventricular zone of the posterior part of the occipital lobe, neurogenesis lasts one month longer and, as a result, the cortex of area 17 becomes almost twice as thick as that of cortex in area 24 (Rakic 1982). These differences in cell allocation show up early. For example, Henry Kennedy mentioned that in the macaque monkey the kinetics of cell proliferation in the ventricular zone subjacent to area 17 is different from that subjacent to area 18 (Kennedy & DeHay 1993). Thus, there is some areal specialization before thalamic afferents arrive at the cortex. It is likely that a population of cortical neurons of the occipital pole in the monkey has a capacity to respond in a particular way to the input from the magnocellular and parvocellular moieties of the LGN in order to create species-specific features of this area that are known to subserve separate colour and non-colour vision. However, cortical neurons can accomplish this only in response to the information from the appropriate input.

D. Price: I have a question for John Rubenstein and Edoardo Boncinelli. Joliot et al (1991) suggest that the *antennapedia* gene product is taken up by cells, transported to the nucleus and regulates neural morphogenesis. Can you comment on that work, in view of the possibility that homeodomain proteins could act as intercellular signals?

Boncinelli: This is conceivable, but probably not the only mechanism, or even the major mechanism, by which cells exchange this type of information. Obviously, what is true for the Antennapedia complex may not be for *Otx2* and vice versa. It's an interesting mechanism that must be taken into consideration, but it has to be demonstrated *in vivo*.

References

Altaba ARI 1994 Pattern formation in the vertebrate neural plate. Trends Neurosci 17:233–243

Cohen-Tannoudji M, Babinet C, Wassef M 1994 Early determination of a mouse somatosensory cortex marker. Nature 368:460–463

Dehay C, Giroud P, Berland M, Smart I, Kennedy H 1993 Modulation of the cell cycle contributes to the parcelation of the primate visual cortex. Nature 366:464–466

Doniach T 1992 Induction of anteroposterior neural pattern in *Xenopus* by planar signals. Development (suppl) 114:183–193

Drake DP, Gerwe EA, Linney E, LaMantia A-S 1994 Role of retinoid signalling in abnormal forebrain differential in the developing mutant mouse *Small Eye*. Soc Neurosci Abstr 20:254

Horton HL, Levitt P 1988 A unique membrane protein is expressed on early developing limbic system axons and cortical targets. J Neurosci 8:4653–4661

Joliot A, Pernelle C, Deagostini-Bazin H, Prochiantz A 1991 Antennapedia homeobox peptide regulates neural morphogenesis. Proc Natl Acad Sci USA 88:1864–1868

Kennedy H, DeHay C 1993 Cortical specification of mice and men. Cereb Cortex 3:171–186

LaMantia A-S, Colbert MC, Linney E 1993 Retinoic acid induction and regional differentiation prefigure olfactory pathway formation in the mammalian forebrain. Neuron 10:1035–1048

Levitt P, Ferri R, Eagleson K 1995 Molecular contributions to cerebral cortical specification. In: Development of the cerebral cortex. Wiley, Chichester (Ciba Found Symp 193) p 200–213

Pannese M, Polo C, Andreazzoli M et al 1995 The *Xenopus* homologue of *Otx2* is a maternal homeobox gene that demarcates and specifies anterior body regions. Development 121:707–720

Porteus MH, Bulfone A, Liu K, Puelles L, Lo LC, Rubenstein JLR 1994 Dlx-2, Mash-1, and Map-5 expression and bromodioxyuridine incorporation define molecularly distinct cell populations in the embryonic mouse forebrain. J Neurosci 14:6370–6383

Rakic P 1982 Early developmental events: cell lineages, acquisitions of neuronal positions, and areal and laminar development. Neurosci Res Prog Bull 20:439–451

Rakic P 1988 Specification of cerebral cortical areas. Science 241:170–176

Reid CB, Liang I, Walsh C 1995 Systematic widespread clonal organization in cerebral cortex. Neuron 15:1–20

Simeone A, Gulisano M, Acampora D, Stornaiuolo A, Rambaldi M, Boncinelli E 1992 Two vertebrate homeobox genes related to the *Drosophila empty spiracles* gene are expressed in the embryonic cerebral cortex. EMBO J 11:2541–2550

Stout RP, Graziadei PPC 1980 Influence of the olfactory placode on the development of the brain in *Xenopus laevis*. Neurosci 5:2175–2186

General discussion II

Teleology for tangential migration

Daw: I would like to bring up the subject of the teleology for tangential migration, in terms of the final organization of the cortex. Pasko Rakic (1972) proposed that the teleology for radial migration is related to the columnar system. One could propose, for example, that tangential migration in the proliferative zone might be related to areas that are topographically connected, and that the migration which takes place later on within the cortical plate might be related to columns that have similar properties. Alternatively, one could propose that tangential migration is simply Brownian motion, and that it doesn't matter where the neurons end up tangentially, in terms of the final organization of the cortex.

Walsh: There are two different ways of looking at this. One is an argument similar to the proposal by Dennis O'Leary (1989) that it may be evolutionarily advantageous to have a cortex which is specified primarily by its afferents, as opposed to being specified by intrinsic patterns of cell division. The advantages for that may be that a simple mutation in a homeobox gene may, for example, change the limb of an animal from a paw to a fin. The brain will, therefore, have to deal with different peripheral inputs over a relatively short number of generations. Consequently, the greater the extent to which the functional organization of the cortex is regulated by its afferents, the more flexibility the cortex has to adapt to a changed periphery.

This is simply an argument for having a relatively uninformed cortex that is specified by its afferents and that will allow the cortex and the periphery to evolve independently. The cortex, for example, could enlarge by simply adding more cells, but it could still maintain its functional organization with the periphery because that functional organization can be dictated and inscribed by the peripheral afferents to the cortex.

There's a second evolutionary argument for why tangential dispersion may be selected. Somatic mutation must be a common event during the development of an organism. DNA polymerases have an error rate of at least one in every 10^9 bases and the size of the eukaryotic genome is about 10^9 bp. This means that every mitotic division may involve mutation. Bernards & Gusella (1994) suggested that there are cells in everyone's body that carry mutations for any given gene. If a somatic mutation occurred in a neural gene in a dividing cell, the mutation would be inherited in its progeny. In the absence of tangential migration, the resulting clone would then produce a focus of cells that had a defective neural function. On the other hand, if you just sprinkled the abnormal cells over the cortex they might have less functional impact.

Bonhoeffer: But there is no advantage for cells to follow a pattern rather than being dispersed throughout the cortex by Brownian motion.

Walsh: The idea of cells following a pattern may not mean a plan to disperse cells, but may mean that cells have to wait a certain amount of time before they can divide again. The cells may actively disperse from each other or it may just be difficult to keep them together in an epithelial sheet.

Bonhoeffer: But then you're back to teleology.

O'Leary: If one assumes that the progenitors start off at different points in the epithelium, then a scattered distribution of cells will be generated, although within a given clone there will be periodicity.

J. Price: It is a mistake to assume that the scattered dispersion of cells contradicts the idea of a proto-map, even though this is probably not what Pasko Rakic had in mind when he proposed that model (Rakic 1988). A certain degree of dispersion can be accommodated within the idea of areal specification in the ventricular zone. For example, if just 10% of the cells, which would have been missed in everyone's analyses, stayed put in an absolute radial fashion, they could provide the scaffold for a proto-map.

Walsh: Microenvironmental differences may occur throughout the cortex and would also not be inconsistent with tangential clonal dispersion.

LaMantia: Also, cells that are specified and that migrate can still carry their positional information with them. An example of this is the migration of neural crest cells.

Blakemore: According to your evidence, progenitor cells are multipotential while they're migrating, so it is difficult to imagine how they could constitute a stable, committed proto-map at that stage. Only the static progenitors that give rise to local clusters appear to be more or less committed in the type of cell that they generate. Therefore, if there is a proto-map, it must surely be represented by the immobile progenitors that give rise to local clusters, which may lie in the subventricular zone. It is possible that the microenvironment in that layer determines their commitment.

J. Price: Do you mean commitment to cell type?

Blakemore: Yes.

J. Price: Why is this relevant?

Blakemore: Because some of the areal differences in the neocortex are defined as differences in the proportions and laminar distributions of different cell types.

J. Price: The process whereby ventricular zone cells acquire positional identity and that by which cells become allocated to a particular type could be totally orthogonal to one another.

Molnár: We're talking about the cortex as though it was a completely liquid tissue during development, but there are planar sheets in the developing cortex with a constant topography; for example, the bottom of layer VI and the subplate. These are stable sheets with topographies that are defined from embryonic day (E) 15–16 in the rodent. The deployment of thalamocortical projections is complete by this stage, and there is an initial matching between the different thalamic nuclei within the sheet that establishes a relatively constant map at the peak of the cell generation. Consequently, even if there is a widespread dispersion of the cortical plate cells, there is a constant stable sheet onto which the afferents can be deployed, lined up and sorted out.

Bolz: This is a chicken and egg situation because the stable sheet needs to recognize something to organize itself.

Blakemore: Zoltán Molnár, are you suggesting that the stable sheets formed by the lowest layers of the cortex can influence the migration of subsequent generations of neurons?

Molnár: I am suggesting that it is possible. Different growth factors or surface molecules may be involved. The regulation of cell divisions in the ventricular and subventricular zones by the ingrowing thalamic axons is an uncharted territory. One can only speculate about the possible ways of regulation. Voltage-sensitive dyes (Crair et al 1993, Higashi et al 1993) or current source–density analysis techniques (K. Toyama, T. Kurotani, N. Yamamoto & Z. Molnár, unpublished results 1993) can be used to show that the activation patterns elicited by thalamic stimulation do not reach the subventricular zone. It is possible that the signals are too small to be detected, even if one had more sensitive techniques. We still have to find the way that thalamic fibres exert their influence, if any, on the dividing cells in the ventricular zone. Radial glia would be an ideal substrate to transmit information from thalamic afferents down to the mitotic factory in the ventricular zone.

Krubitzer: It is possible that the events that we observe during development are not relevant to the final function of the organism, and that some of these events represent ways of moving around obstacles, which are laid down early in evolution and may constrain or channel future development, rather than being necessary steps that ultimately generate a particular function.

Blakemore: If the pattern of lateral migration were similar in absolute dimensions in the monkey and in the mouse, then the consequences would be different. In the mouse, clones can be distributed over virtually the whole of the neocortex, contributing to somatosensory, motor, olfactory and visual systems. Similar lateral dispersion, in absolute distance, in the monkey would only spread the clone across a small fraction of a single cortical field.

Kennedy: Yes, but it takes 10 times longer in the monkey to do the same thing. If you had the same migration rates, you would end up with the same relative degree of scatter.

Rakic: Menezes & Luskin (1994) have shown recently that some cells migrate from the proliferative subventricular zone of the telencephalon to the olfactory bulb. Apparently, these cells continue to divide as they move perpendicularly to the radial glial fibres. However, just because they migrate laterally does not mean that they are not committed. In fact, they are immunoreactive to cell class-specific antibodies and their movement is restricted. Therefore, lateral migration in the telencephalic wall may actually signify cell commitment rather than lack of it. If a similar population of cells exist in primates, they would have to migrate at least 10 times further than those in the mouse. Likewise, if some cells in the monkey, as suggested by Chris Walsh for the mouse, migrate from the temporal or parietal lobes to the frontal lobe, the total length of their journey in the middle of gestation must be measured in centimetres rather than in millimetres. The cell size in both species is approximately the same, so this would pose a considerable transportation problem in a large and partially convoluted cerebrum. The problem of distance and pathfinding in humans would be even greater. If, however, in primates this long distance allocation of postmitotic cells does not exist, this would signify to me that the dispersed cells are irrelevant for the specification of neocortex. They may represent a class of neurons that subserve the same function in all areas or brain structures, unlike neuronal classes which migrate strictly radially and may carry gradient molecules or other signals important for regional specificity.

Kennedy: The distance in the monkey is 10 times greater than the mouse. That's fine because the time period is 10 times longer—60 days for the monkey and six days for the rat. If they are migrating at the same rate in both species, they would migrate the same equivalent distance in terms of cortical space.

Rakic: But cells don't migrate for 60 days in the monkey.

Bonhoeffer: What do we actually mean by the term committed? Do we mean sensory as opposed to motor neurons, visual as opposed to auditory neurons or neurons in area 17 as opposed to area 18?

Blakemore: That's an important point. Some of the most important features of the differentiation of an area probably involve the connections and relationships between neurons, established after they're in position. The neurons themselves may not differ very much in their genetic expression. For example, a layer V pyramidal cell in area 18 may not be very different initially from one in area 3.

Bonhoeffer: There is also evidence that some sensory areas can be turned into other ones (Sur et al 1988). Consequently, how stringent should we be with a definition of commitment?

Blakemore: It seems clear that the kind of commitment we've been talking about—commitment of a progenitor to produce a certain type of glial cell or pyramidal neuron, rather than non-pyramidal, is only the first step in establishing the ultimate fate of cells. An entirely different form of commitment might determine many aspects of the internal connectivity of distinct areas, and hence their cytoarchitecture and their computational function.

General discussion II

J. Price: It's possible that the periodicity of dispersion is not that important. We studied pyramidal neurons in the pyramidal areas of the hippocampus (Grove et al 1992). The clones here were regularly spaced, 150 μm on average, so we were able to trace them with regard to the hippocampal fields. We found that they don't respect the boundaries between fields, but march straight through the presubiculum, the subiculum, CA1, CA2 and CA3.

Bonhoeffer: What about the dentate gyrus?

J. Price: Our evidence suggested that clones included CA4 neurons and dentate gyrus neurons, but we obtained few examples of these, so I can't be completely certain. I can't explain why we see equal spacing in the hippocampus or the neocortex, but I doubt that this has any functional consequence.

Daw: Could the periodicity have anything to do with the possibility that there are integral velocities of migration?

Walsh: That did not occur to me. I have no results that it has or that it hasn't.

Rubenstein: Do you think that the timing of the cell cycle is involved?

Walsh: The timing of the cell cycle has to be important because the cells have to wait a certain amount of time before they can divide again. But Nigel Daw's question was whether some cells migrate twice as fast as others per cell cycle and I don't know the answer to that.

Parnavelas: In what percentage of your clones do you see periodicity?

Walsh: In all the clones that we analysed, fewer that 1% of sibling cells are spaced 1–1.5 mm apart (Reid et al 1995). That is most remarkable because there are large numbers of siblings that are either spaced 0–0.5 mm or 2–3 mm apart, but there is a striking lack of sibling cells spaced 1–1.5 mm apart.

Rakic: The developing cerebellum may be an instructive model system for understanding the relationship between cell lineages and connectivity in other areas. In the cerebellar rhombic lip, which is a small mitotically active centre situated on each side of the rhombencephalon, cells proliferate and then migrate in two directions: one population leads to the pons, and the other forms the external granular layer of the cerebellar hemispheres. The pontine cohort eventually generates neurons of the pontine grey nucleus, whereas the external granular layer generates granule cells of the cerebellar cortex. The interesting point is that the cells of the pons later form long axons that enter the cerebellar hemisphere and find their way to the cortex where they contact the granule cells. Thus, the clonally related cells that originate from the same progenitor pool, and later have a different history, eventually become connected to each other.

Rubenstein: Gray & Sanes (1991) used retroviral labelling on the tectum of the chick. They showed that radial migration is not random—the cells migrate to the subventricular layer, then some migrate longitudinally and others migrate radially.

LaMantia: That's a very specific cell class which takes a specific migratory route.

Rubenstein: It is possible that the cortex is so complicated that some of the tangential movement we're observing may not be random but may be part of a directed migration.

Boncinelli: Cells may move freely within compartments, but may get trapped at their boundaries. Therefore, it is possible that migration does fit within the proto-map hypothesis.

Walsh: The existence of compartments would have been tested for by analysing the distances between non-sibling cells. If compartments were present, there may be places in the cortex where cells divide preferentially. Such 'hot-spots' of cell division would be an alternative explanation for the preferential spacing between sibling cells because a cell divides in one hot-spot, then goes to another at a set distance, and so on. However, if there were compartments or hot-spots for cell division, then there would be a similar periodic spacing between sibling cells and between non-sibling cells. We did not see this periodic spacing between non-sibling cells (Reid et al 1995) which argues against the existence of compartments, although we can't rule it out entirely.

Blakemore: Also, if there are specific zones where migratory progenitors tend to divide hence where non-migrating progenitors accumulate, one would expect the cortex to be thicker above those zones, but in fact it is fairly uniform in thickness.

LaMantia: Chris Walsh, have you looked at the intermediate stages of cortical neuroblast migration to see whether there are cells that migrate radially then disperse tangentially? It is possible to use [^{3}H]thymidine labelling to show that cells migrate radially from the ventricular zone at the place where they undergo their final division. If it is the ventricular cells that are migrating tangentially, they could be displaced randomly if they were constrained in the lateral dimension. However, the known migrations that are tangential are also directed in some way (Domesick & Morest 1977, Gray & Sanes 1991).

Walsh: It's difficult to do that experiment because the clones are not coherent, and so it is difficult to infer mechanisms of migration. The retroviral method is not the optimum method for studying migration because you can only infer patterns of migration based on the patterns of dispersion. However, if we inject at E15 and analyse at E21, we observe that about a third of the dispersed clones have dispersed sibling cells within the ventricular zone (Walsh & Cepko 1993). The presence of dispersed cells within the proliferative zones before radial migration occurs is strong evidence that some of the progenitor cells themselves disperse.

Rubenstein: Are you sure they're not in the subventricular zone?

Walsh: The material was not always counter stained so it may be difficult to determine whether they're in the ventricular zone or subventricular zone. However, they were in the proliferative zones.

General discussion II

About another third of the clones have one cell in the proliferative zone in one place, and one cell in the intermediate zone of the cortex in another place (Walsh & Cepko 1993). How did they separate like this? It may be the same mechanism, but it is possible that one cell left the proliferative zone and migrated non-radially through the intermediate zone.

The last third of the clones contained cells within the cortical plate that are dispersed widely. The retroviral method cannot determine how these cells migrated apart. We can, therefore, be confident that there's dispersion in the proliferative zone, but we can't say what other mechanisms may also be operating.

Bartlett: This argument can be turned around the other way. There are many areas where progenitors are moving in relation to the rostrocaudal axis. We know that boundary formation, even in the hindbrain, is probably not due to clonal restriction, but is due to the regulated expression of certain genes either side of the boundary. Do we really have to explain why dispersion occurs? We're really interested in the other mechanisms, which involve gene regulation, that form these functional boundaries.

LaMantia: There are two studies (Fraser et al 1990, Wingate & Lumsden 1994) that show that there is a restriction of cell movement in the hindbrain which obeys rhombomere boundaries.

Bartlett: That's not entirely true. Fraser et al (1990) are saying that even after the boundaries are formed in the hindbrain, there is still a lot of dispersion across these boundaries.

LaMantia: They claimed that the numbers of cells transgressing rhombomere boundaries was between 5 and 10%.

Rubenstein: That may be similar to what we're seeing here.

Bartlett: The boundaries of expression of the *Hox* genes aren't initially well-defined, they're fuzzy and then they sharpen up probably because of differential gene regulation.

J. Price: David Wilkinson (unpublished observations) also sees this with *Kroz-20* expression. Expression is initially fuzzy, but by the time the rhombomere boundaries are morphologically apparent, the pattern of gene expression has sharpened up.

Boncinelli: Gene expression produces boundaries, and boundaries constrain the movement, so it depends on exactly when you ask the question.

J. Price: Exactly. Dispersion occurs within the rhombomeres, but the boundaries stop dispersion between rhombomeres.

Molnár: I agree that boundaries exist in the developing and adult cerebral cortex. These boundaries may be rigid or modifiable. The boundaries of different areas in the adult cerebral cortex can shift, and there is a spectacular plasticity in the system (Merzenich et al 1983, Kaas 1991). There are signs of movement of these cytoarchitectonic areas well after the completion of cell generation, migration and differentiation.

Differentiation factors in culture

Ghosh: We have taken cells from the rat cortex at E13 or E14, dissociated them and looked at differentiation over time in culture (Ghosh & Greenberg 1995). At the time of plating in serum-free medium, all of the cells are nestin positive. It is difficult to get these single dissociated cells to divide without serum. However, clusters of cells in serum-free medium will continue to divide. When we added fibroblast growth factor 2 (FGF-2) to the serum-free medium, we found that it is a potent mitogen for these cells. However, we also found that differentiated neurons appear in these cultures over time in the presence of FGF-2. This suggests that in the presence of a proliferative signal, signals that regulate neuronal differentiation can continue to function. We were interested in the possibility that endogenous factors made by these cells play a role in mediating their differentiation. In our hands the best candidate for the differentiating factor is neurotrophic factor 3 (NT-3) because when we added anti-NT-3 antiserum to these cells, there was a significant reduction in the number of microtubule-associated protein 2-positive cells in each cluster. This is not a survival effect because the anti-NT-3 antiserum does not have this effect on cultures of postmitotic neurons. Also, there is no reduction in the total cell number in these clusters in the presence of anti-NT-3. NT-3 probably causes the cells to leave the cell cycle because if you add excess NT-3, there is a reduction in the number of bromodeoxyuridine-positive cells. This is consistent with the finding that during development, there's a loss of FGF responsiveness in some cells. It is not known whether NT-3 regulates this step, but NT-3 does seem to be a good candidate for a neuronal differentiation factor.

We have also studied the generation of oligodendrocytes in culture. We have found that neurons are generated only in the first few days in culture. They are not generated after about five or six days. However, there is a pool of cells that continues to divide and undergoes a morphological transformation after about 10 to 12 days in culture. These cells stain with anti-galactocerebroside antibodies suggesting that they have adopted an oligodendrocyte-like fate.

I would like to point out that, although these cell culture experiments are useful for trying to identify differentiation factors, there are probably a lot of factors that influence differentiation. Consequently, the danger is that cell culture experiments will give us an over-simplified view.

Bartlett: I have two points. FGF-2 is expressed at the right time in development, but I'm not sure that NT-3 is expressed at the right time, although something else could be acting through the TrkC receptor. Secondly, there is a vast difference between culturing a single cell and culturing a clump of cells. We had the same result three or four years ago (Murphy et al 1990). Many of the cells become neuronal in the presence of FGF-2. We first thought that the difference between proliferation and differentiation was just a matter of FGF dose, but it's not. A secondary message is involved. Transforming growth factor β-like molecules seem to be able to induce neuronal differentiation, but in our hands FGF-1 is far more effective. FGF-1 can induce a robust single-celled clone to make 200–500 neurons. This is not to say that other factors are not involved. These factors may induce the differentiation of the neuronal subtypes and may be responsible for the apparent homogeneity of the progeny of individual precursors. NT-3 may well play a part in the selection of these subtypes.

Ghosh: These experiments have been done in culture, so it's difficult to know whether this is exactly what's happening *in vivo*.

Bartlett: NT-3 might also be involved in neurite outgrowth. What you might be measuring is a secondary differentiation event.

Ghosh: NT-3 has an influence on the cell proliferation, which suggests that it can act on precursors while they are still dividing.

Jones: We looked at the expression of nerve growth factor, brain-derived neurotrophic factor (BDNF), NT-3 and NT4/5 in the frontal cortex of fetal monkeys and rats. We found that their expression was turned on after lamination is fully formed in the cortex and at a time when process formation is probably at its maximum, including the outgrowth of axons and the finding of their targets (Huntley et al 1992). For example, in the monkey, BDNF expression is turned on at about E100. I can't rule out that these, or other members of the neurotrophin family, are not operating earlier, but in terms of what we see *in vivo*, NT-3 is not expressed at the correct time.

Ghosh: But there is published evidence that both NT-3 and the TrkC receptor are expressed in the embryonic cortex (Maissonpierre et al 1990, Tessarollo et al 1993) and, therefore, could play a role in neuronal differentiation. Nevertheless, it is worth considering that an unidentified molecule related to NT-3 may be involved in regulating neurogenesis *in vivo*.

References

Bernards A, Gusella JF 1994 The importance of genetic mosaicism in human disease. N Engl J Med 331:1447–1449
Crair MC, Molnár Z, Higashi S, Kurotani T, Toyama K 1993 The development of thalamocortical and intracortical connectivity in rat somatosensory 'barrel' cortex imaged by optical recording. Soc Neurosci Abstr 19:702.5
Domesick VB, Morest DK 1977 Migration and differentiation of ganglion cells in the optic tectum of the chick embryo. Neurosci 2:459–475
Fraser S, Keynes R, Lumsden A 1990 Segmentation in the chick embryo hindbrain is defined by cell lineage restrictions. Nature 344:431–435
Ghosh A, Greenberg ME 1995 Distinct roles for bFGF and NT-3 in the regulation of cortical neurogenesis. Neuron 15:89–104
Gray GE, Sanes JR 1991 Migratory paths and phenotypic choices of clonally related cells in the avian optic tectum. Neuron 6:211–225

Grove EA, Kirkwood TBL, Price J 1992 Neuronal precursors in the rat hippocampal formation contribute to more than one cytoarchitectonic plate. Neuron 8:217–229

Higashi S, Crair MC, Kurotani T, Molnár Z, Toyama K 1993 Imaging neural excitation propagating from thalamus to somatosensory 'barrel' cortex of the rat. Soc Neurosci Abstr 19:49.15

Huntley GW, Benson DL, Jones EG, Isackson PJ 1992 Developmental expression of brain derived neurotrophic factor mRNA by neurons of fetal and adult monkey prefrontal cortex. Dev Brain Res 70:53–64

Kaas JH 1991 Plasticity of sensory and motor maps in adult mammals. Annu Rev Neurosci 14:137–167

Maisonpierre PC, Belluscio L, Friedman B et al 1990 NT3, BDNF and NGF in the developing rat neuron system: parallel as well as reciprocal patterns of expression. Neuron 5:5101–5109

Menezes JRL, Luskin MB 1994 Expression of neuron-specific tubulin defines a novel population in the proliferative layers of the developing telencephalon. J Neurosci 14:5399–5416

Merzenich MM, Kaas JH, Wall JT, Sur M, Nelson RJ, Felleman DJ 1983 Progression of change following median nerve section in the cortical representation of the hand in areas-3B and area-1 in adult owl and squirrel monkeys. Neuroscience 10:639–666

Murphy M, Drago J, Bartlett PF 1990 Fibroblast growth factor stimulates the proliferation and differentiation of neural precursor cells *in vitro*. J Neurosci Res 25:463–475

Rakic P 1972 Mode of cell migration to the superficial layers of fetal monkey neocortex. J Comp Neurol 145:61–83

Rakic P 1988 Specification of cerebral cortical areas. Science 241:170–176

Reid CB, Liang I, Walsh C 1995 Systematic widespread clonal organization in cerebral cortex. Neuron 15:1–20

Sur M, Garraghty PE, Roe AW 1988 Experimentally induced visual projections into auditory thalamus and cortex. Science 242:1437–1441

Tessarollo L, Tsoulfas P, Martin-Zanca D et al 1993 TrkC, a receptor for NT3, is widely expressed in the developing neuron system and in non-neuronal tissues. Development 118:463–475

Walsh C, Cepko CL 1993 Widespread clonal dispersion in proliferative layers of cerebral cortex. Nature 362:632–635

Wingate RJT, Lumsden AGS 1994 Where do rhombomeres go? Fate maps in the chick embryo. Soc Neurosci Abstr 20:254

Guidance of thalamocortical innervation

Zoltán Molnár and Colin Blakemore

Department of Physiology, University of Oxford, South Parks Road, Oxford OX1 3PT, UK

Abstract. We propose that a sequence of individually simple mechanisms influences the pattern of thalamocortical innervation, which itself contributes to the determination of regional differentiation of the neocortex. In co-culture, the cortex appears to exert a remote growth-promoting influence on thalamic axons from E15, becomes growth-permissive to axon invasion at about E20 and expresses a stop signal, causing termination in layer IV, from P2–3. This cascade of cortical signals may determine the timing of events *in vivo*. However, any part of the thalamus will innervate any region of the developing cortex in culture, without obvious preference, suggesting that the topographic distribution of thalamic fibres *in vivo* does not depend on regional chemospecificity. The initial extension of axons from the cortical preplate and the thalamus starts at about E14, and the topography of both may be influenced by their temporal sequences of outgrowth (chronotopy). The axon arrays meet in the basal telencephalon, after which the preplate scaffold may guide thalamic axons and ensure both their 'capture' within the subplate layer and the establishment of the waiting period. The unusual pattern of innervation in the Reeler mutant mouse supports the hypothesis that thalamic axons grow over preplate fibres to find the waiting compartment.

1995 The development of the cerebral cortex. Wiley, Chichester (Ciba Foundation Symposium 193) p 127–149

From the perspective of developmental neuroscience, the most perplexing feature of the cortex is its parcellation into distinct regions, each recognizable on the basis of its particular input and output connections, its functional role and (usually) its local cytoarchitecture. There is compelling evidence that the input from the thalamus to the cortex can act as a local extrinsic signal, setting the boundaries of certain cortical fields and influencing their regional differentiation (O'Leary et al 1995, this volume). At the very least, the thalamic input itself becomes part of the cytoarchitectural signature by contributing to the neuropil of the cortex, and it influences the functional status of each area by delivering afferent inputs on which the cortex acts.

All areas of the rodent neocortex receive a thalamic projection (Caviness & Frost 1980). Occasionally, in the adult, non-adjacent thalamic nuclei project to contiguous cortical regions; but in general, adjacent neocortical fields receive their projections from neighbouring thalamic nuclei, and proximity relationships correspond in the cortex and the thalamus (Caviness 1988). The capacity of the thalamic input to serve as a local signal depends on it being guided to the appropriate territory in the cortex, so it is important to ask how axons from the specific nuclei of the thalamus find their way to their cortical targets.

In the brains of eutherian mammals, the crucial steps in thalamocortical development take place *in utero*, at stages when manipulative interventions are difficult. In the rat the first cells of the cortical plate itself are born on embryonic day (E) 14 (defining the date of conception as E0), and the last at E21 (parturition is at E22). Thalamic axons are distributed in a remarkable topographic order, directly below the cortical plate, shortly after the arrival of the first true neurons of the cortical plate, and before any obvious regional differentiation has occurred. Lund & Mustari (1977), using degeneration techniques, described the early outgrowth of fibres from the fetal rat thalamus, and observed that they accumulate in the subplate layer, under the cortical plate, for a waiting period of a few days before invading the cortex itself.

A cascade of signals from the cortex may control the timing of thalamic innervation

Organotypic co-culture offers the opportunity of studying, in a simplified system, the expression of signals that might play a part in the initiation, guidance and termination of the thalamic innervation of the developing cortex (Yamamoto et al 1989, 1992a, Blakemore & Molnár 1990, Molnár & Blakemore 1991, this volume: Price et al 1995, Bolz et al 1995).

We cultured small explants from the diencephalon of fetal Sprague–Dawley rats and slices of neocortex, 350–400 μm thick, in a serum-free (N2) medium (Romijn 1988), on collagen-coated microporous membranes in stationary Costar Transwell-COL culture chambers (Molnár & Blakemore 1991). Cortical slices survive well and remain about 150–200 μm thick even after many weeks in culture. This is in contrast to the roller-tube method, where they collapse to virtual monolayers (Gähwiler 1988). Thalamic explants cultured alone also maintain their integrity quite well, although they tend to flatten somewhat more than cortical slices.

We placed thalamic explants from rat embryos (between E14.5 and E21, but most often at E16) close to the ventricular aspect of slices of neocortex (from other rats between E14.5 and postnatal day (P) 11, which spans the natural period of thalamic fibre outgrowth, arrival, accumulation and ingrowth). For most of our experiments, we obtained thalamic explants from the region of the putative lateral geniculate nucleus (LGN), the visual nucleus in the posterodorsal thalamus. We dissected cortical slices from the occipital region assumed to be the precursor of the primary visual cortex (area 17). We preincubated the thalamic explants with the carbocyanine dye DiI (1,1'dioctadecyl-3,3,3',3'-tetramethylindocarbocyanine perchlorate) to label outgrowing axons (Molnár & Blakemore 1991).

The cortex exerts a remote growth-promoting influence on the thalamus from very early stages

Under our culture conditions, there was little neurite extension from a thalamic explant cultured alone or with a cortical slice placed more than a few millimetres away. However, if the gap between the two explants was less than about 2 mm when cultured on a collagen-coated membrane (or 4 mm on laminin), axons streamed out of the thalamic explant on all sides and the gap between the explants was filled with either a mass of fibres, or one or two large fibre bundles.

This target-dependent axon outgrowth was observed for cortical slices at all ages tested, indicating that a diffusible growth-promoting substance was produced from the beginning of formation of the cortical plate until after the age at which innervation of the cortex is complete. We saw no evidence of a direct effect on the direction of growth of individual axons indicative of a true trophic influence (Lumsden 1992), but if the culture chambers were left completely undisturbed, neurites usually extended further from the explant and more densely on the side facing the cortical slice. This implies that, even in a stationary liquid medium, a sufficient concentration gradient is established to influence the overall pattern of growth. Therefore, it is possible that *in vivo*, the diffusible substance might not only initiate axon outgrowth from the thalamus, but also determine its direction, perhaps enticing axons towards the primitive internal capsule.

Although these results have been confirmed and extended by Price et al (1995, this volume), Bolz et al (1992) suggested that the cortex does not have a remote influence because they did not see a difference in thalamic axon growth with and without a neighbouring piece of cortex. However, they used a serum-enriched medium that stimulates considerable (and variable) axon outgrowth even from isolated thalamic explants. In such conditions it is difficult to interpret the effect of an added cortical slice.

The cortex becomes growth permissive at about E20

When an E15–16 fetal thalamic explant is cultured with a cortical slice taken before about E20, thalamic axons grow out across the floor of the culture chamber, initially at a speed of about 1 mm/day. When they reach the slice most of them encircle it and some run over its surface. Few axons invade the

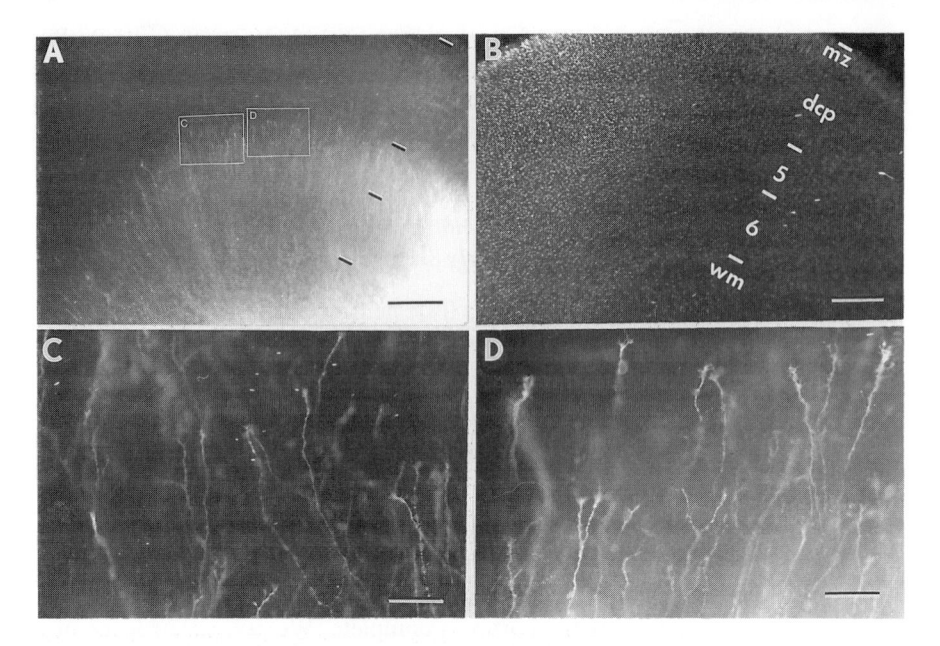

FIG. 1. Darkfield micrographs under epifluorescent illumination. DiI (1,1'dioctadecyl-3,3,3',3'-tetramethylindocarbocyanine perchlorate)-labelled axons from an embryonic day 16 lateral geniculate nucleus explant invade an occipital cortical slice, taken at postnatal day 3, after 1.5 days in culture. The plane of focus was adjusted to about the middle of the thickness of the section to show the axons within the tissue. (A) The dense array of thalamic fibres has grown in a radial direction through the cortical slice from its ventricular surface and has just reached layer IV. Cortical lamination is indicated (see B). The majority of the fibres in such co-cultures reach and begin to terminate in layer IV after two days *in vitro*. Bar = 250 μm. (B) The same as A but viewed under ultraviolet illumination to reveal the cortical layering and the pial surface, after staining with bisbenzimide. dcp, dense cortical plate; mz, marginal zone; wm, white matter; 5, layer V; 6, layer VI. Bar = 250 μm. (C, D) High-power views of the regions indicated by the inset squares in A. Growth cones are seen at the tips of many axons, and some of them have branched as they approach layer IV. Bars = 50 μm.

slice suggesting that thalamic fibres do not find early embryonic cortex an attractive environment for growth.

At about E20 (for the occipital cortex) the tissue of the cortical slice becomes growth permissive and thalamic axons penetrate it readily (Fig. 1). This is in agreement with Bolz et al (1992, 1995, this volume). When confronted with cortex taken at E20–P2, fibres grow out profusely from an E16 thalamic explant, pass through the ventricular surface of the slice and onward, in a radial direction, through the intermediate zone and into the cortical plate itself.

Growing at approximately 1 mm/day, they pass through the layer of immature neurons that constitutes the dense cortical plate and reach the marginal zone after about 3–4 days in culture without terminating or hesitating. Some axons then turn and run tangentially through the marginal zone. However, most axons penetrate the surface, turn 90° and run over it, directly below the pia itself, forming a fasciculated band of fibres similar to the circling axons seen at younger stages.

A stop signal is expressed in layer IV after about P3

In cortical explants taken at P3 or older, thalamic axons grow somewhat less quickly than in younger cortex. After only four days in culture the majority of them have branched locally, lost their growth cones and terminated around 300 μm below the pial surface, at a depth corresponding to layer IV. Just as *in vivo*, some axons (less than 10%) extend up to the marginal zone and appear to terminate there. Yamamoto et al (1989) were the first to observe such lamina-specific termination in co-culture. Yamamoto et al (1992b) used time-lapse confocal microscopy of co-cultures of this sort to observe thalamic axons slowing down, branching and terminating in layer IV.

We have suggested that, at around P3–4 (the age at which thalamic axons are completing their invasion of the cortex and terminating in layer IV *in vivo*), the cortex starts to express a stop signal, which makes thalamic fibres branch and terminate in layer IV (Molnár & Blakemore 1991). However, the failure of most thalamic axons to grow beyond layer IV could merely indicate that the upper layers of the cortex, which form later, lack growth-permissive properties or become positively inhibitory to axon growth. We excluded these possibilities by culturing E16 thalamus adjacent to the pial surface of cortex older than P3. We observed that the axons grew down readily through the upper layers and still terminated in the presumptive layer IV (Fig. 2).

Thalamic innervation *in vitro* is not obviously regionally selective

The cascade of molecular influences revealed by these experiments (growth-promoting signal, growth-permissive stage and stop signal) is synchronous with the natural sequence of outgrowth, waiting, innervation and termination. In principle, regional differences in the nature of either the early growth-promoting signal or the growth-permissive property could exert a chemospecific influence (Sperry 1963) on the innervation of target regions in the cortex by thalamic axons. However, this seems unlikely. It is difficult to imagine how regional differences in the diffusible growth-promoting substance could establish the topographically precise patterns of thalamic outgrowth that are actually seen *in vivo* (see below). In addition, the growth-permissive property is not expressed until about E20, which is three or four days after

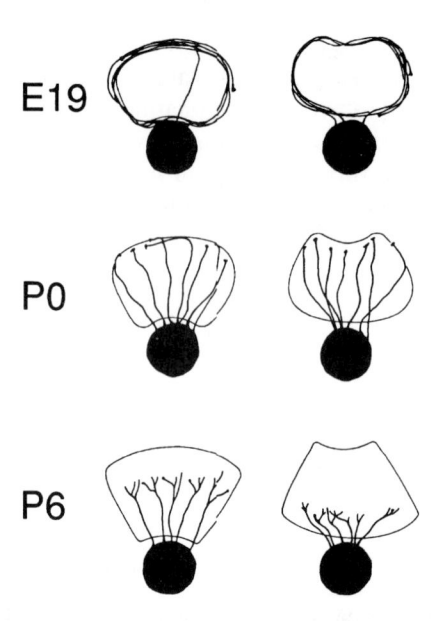

FIG. 2. Summary diagram of the patterns of thalamic fibre growth from embryonic day (E) 16 lateral geniculate nucleus explants (filled circles) after five days of co-culture with rat occipital cortical slices, taken at E19, postnatal day (P) 0 and P6. Drawings in the left and right columns illustrate typical patterns of ingrowth when the thalamic explant is placed against the ventricular and pial surface, respectively. Only a small fraction of the usual number of outgrowing fibres are represented. Cortical axons are not drawn. Different patterns are observed depending on the age of the cortical slice. (E19) Cortical slices at E19 or younger, promote the outgrowth of axons from the thalamic block independent of their orientation, but very few of them penetrate the slice. Most grow around its margin and a few run over its surface. (P0) For cortical slices from rats older than E20, thalamic fibres grow into the ventricular surface and run radially through the cortical plate to the marginal zone, where some turn laterally to run over or beneath the pial surface. When confronted with the pial surface, thalamic fibres also enter and grow all the way through the slice to the opposite side. (P6) When cultured next to the ventricular surface of cortical slices from rats older than about P2, most thalamic fibres branch and terminate some 250–300 μm below the pial surface, in what is presumed to be layer IV, and very few invade the upper layers. This behaviour is presumably due to a specific stop signal expressed in layer IV rather than a lack of growth permissiveness in the supragranular layers because a similar pattern of ingrowth and termination occurs if the thalamic explant is placed against the pial surface.

thalamic axons have distributed themselves in a precise pattern under the cortical plate.

Despite these prima facie doubts, we tested the possibility of regional chemospecificity either by culturing thalamic blocks with single slices from regions of cortex that would not normally be innervated *in vivo*, or in a choice paradigm, in which a central thalamic explant was flanked with two slices, one from the appropriate target area and the other from an inappropriate region (Molnár & Blakemore 1991). These procedures were capable of revealing target preferences. When confronted with a slice of E18–P8 cerebellum, some thalamic axons encircled the slice but very few penetrated it. However, there were no obvious differences in the patterns of innervation when the thalamic explant was flanked with two different explants of neocortex, one from the correct target and the other from a region that the chosen region of thalamus would never approach *in vivo*. This experiment does not exclude the possibility of a subtle gradient of signals across the cortex, nor does it rule out the expression of regionally specific factors at the tips of early corticofugal axons (see below). However, it does suggest that different regions of the cortical plate itself are not identified by thalamic axons on the basis of chemoselective signals.

Do early corticofugal projections from the preplate guide the orderly deployment of thalamocortical fibres?

Although the cascade of cortical signals defined *in vitro* may help to determine the timing of events *in vivo*, we must seek other explanations for the topographic pattern of thalamic axon guidance.

McConnell et al (1989), using axon tracing with carbocyanine dyes in fetal fixed brains, reported that neurons of the subplate (the early-generated preplate cells that remain below the cortical plate as true cortical neurons migrate past them; see Blakemore 1995, this volume) send pioneering axons to the diencephalon in the cat. They, and Shatz et al (1990), suggested that these axons might play a part in the establishment of ascending and other descending projections.

What evidence would support the hypothesis that corticofugal axons from subplate cells guide thalamic axons to their correct cortical target regions? First, it would be necessary to show that the array of subplate axons is topographically organized, each cortical region pioneering a route to the corresponding thalamic nucleus. Second, subplate axons ought to form their guidance projections before axons start to leave the thalamus. Third, we would need to establish that thalamic axons grow over the subplate scaffold, and follow it to their targets. We have traced axons with carbocyanine dyes in the fetal rat to test these predictions.

Early projections from the subplate are topographically organized

We placed small crystals of carbocyanine dye into the cortex of perfused fetal rats of known gestational age and, after sufficient incubation to allow diffusion of the dye along axons, we examined sections by fluorescence and confocal microscopy. The first axons leave the occipital cortex at E14, immediately after the first postmitotic neurons have migrated to the outer edge of the cerebral wall to form the preplate. The axons turn laterally and start to grow down as a discrete bundle through the intermediate zone. They reach the basal telencephalon by late E15 and curve medially beneath and through the developing corpus striatum towards the primitive internal capsule. At this stage, they have particularly large and elaborate growth cones at their tips.

The fact that these early projections arise from preplate (mainly subplate) cells can be confirmed by back-labelling the cell bodies. Placement of a dye crystal in the internal capsule at E15 labels multiform and bipolar cells, which lie mainly below the first cells of the true cortical plate. A similar crystal placement a day or so later also begins to label pyramidal neurons of the cortical plate, presumably cells of the future layers V and VI, whose axons are bound for the thalamus and the brainstem, respectively.

Labelling with single crystals in other regions of the hemisphere, and with parasagittal or coronal rows of dye placement of different colour, shows that each region sends a tightly organized bundle of axons towards the internal capsule, with little or no intermixing of axons from neighbouring cortical areas, even when the array narrows as it approaches the internal capsule (Blakemore & Molnár 1990). At this early stage, each labelled bundle has an uncomplicated geometry, simply curving in an arc without twists or sharp bends, taking the shortest route through the intermediate zone. Thus, pioneering axons from the preplate do seem to establish a topographic scaffold of strictly ordered fibres.

Thalamic axon growth is synchronized with preplate outgrowth

If dye crystals are placed in the cortex after E15, the pattern of axon labelling gives the impression that the early corticofugal projections grow through the internal capsule and penetrate the thalamus itself. All cortical areas appear to connect to their corresponding thalamic nuclei by about E16 (Fig. 3), the age at which Lund & Mustari (1977) suggested that axon outgrowth from the LGN begins. However, even as early as E16, cortical crystal implantations lead to labelling of cell bodies in the matched thalamic nucleus, presumably due to retrograde diffusion along thalamic axons that have already reached the cortex at that age. Indeed, DiI crystal placements in the LGN itself show that axons begin to grow out at E14. This is two days earlier than Lund and Mustari's original estimates, which are based on less sensitive degeneration methods. These thalamic axons reach the occipital pole at about E16 (Molnár & Blakemore 1990, Blakemore & Molnár 1990, De Carlos & O'Leary 1990, 1992) and they gather within the subplate layer before invading at about E19–20. Catalano et al (1991) have also reported that fibres from the ventral thalamus arrive under the somatosensory cortex early and enter the cortical plate before birth. They demonstrated that some axons penetrate the lower layers of the

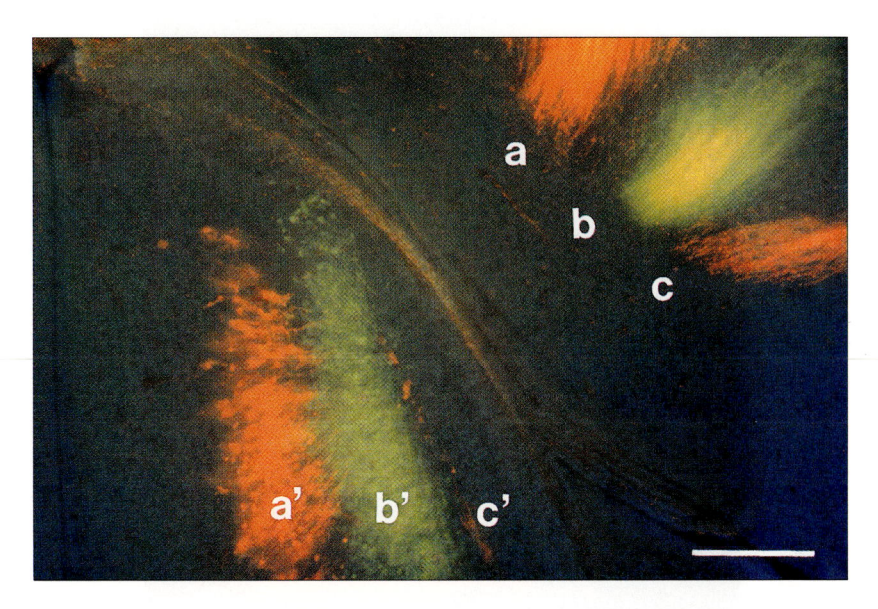

FIG. 3. Fluorescent photomicrograph taken from a 100 μm thick horizontal section of an embryonic day (E) 16 brain (rostral, up). Three different carbocyanine dyes were used: one DiA (4-[4-dihexadecylaminostyryl]-N-methylpyridinium iodide) dye and two DiI (1,1'dioctadecyl-3,3,3',3'-tetramethylindocarbocyanine perchlorate) dyes. A crystal of each dye was placed in a parasagittal row along the right hemisphere. Thalamic fibres had already arrived at the cortex at E16, so each placement labelled a mixed bundle of thalamocortical and corticofugal axons linking cortex and diencephalon. Five weeks incubation at room temperature allowed full anterograde and retrograde diffusion. The film was exposed three times using rhodamine, fluorescein and ultraviolet filters, to reveal the three dyes and the bisbenzimide counter-staining. Three distinct bundles are visible within the basal telencephalon (a, b and c to the right) approaching the primitive internal capsule. The corresponding labelled thalamic cells line up along anteroposterior lines forming slabs (a', b' and c' on the left). The most anterior cortical crystal placement (a) revealed the most medial thalamic slab (a') and more posterior cortical points (b and c) were connected to more lateral slabs (b' and c'). Bar = 300 μm.

to each other, as in the results of Miller et al (1993). However, we attribute this to misalignment of the two labelled points because in many experiments, we saw a definite intermingling of thalamic and subplate axons. Figure 4 illustrates thalamic and subplate axons running in opposite directions through the intermediate zone but contributing to the same fascicles. The theory of co-fasciculation is also supported by the fact that discrete crystal implantation at any point in the cortex at E16 (after the arrival of thalamic axons, therefore labelling both descending and ascending projections) leads to the staining of a single, tight bundle of fibres rather than two separate ones.

Evidence from the Reeler mouse

Independent and compelling evidence for the growth of thalamic axons over the preplate scaffold comes from studies of the Reeler mouse. In this mutation, the entire cortex forms in a roughly outside-in sequence, leaving the preplate stranded as a 'superplate' above the inverted layers of the cortical plate itself (Caviness et al 1988). Despite this gross aberration of cortical layering and the absence of a normal subplate, thalamic fibres innervate the appropriate cortical regions, although they have a bizarre laminar distribution in the adult, running obliquely up to the top and looping down to terminate deep in the cortex (Caviness 1976).

We have examined the entire process of corticogenesis and thalamic innervation in the Reeler mouse (Molnár & Blakemore 1992). The earliest stages of preplate formation, the outgrowth of a topographically ordered array of descending pioneer axons from the preplate cells, the synchronous outgrowth of thalamic axons through the primitive internal capsule and their tangential distribution in association with the scaffold of corticofugal fibres towards the appropriate cortical regions all occur normally in the Reeler mouse. Only after cells of the cortical plate itself start to arrive (at about E15 for the occipital cortex), does the Reeler phenotype diverge in appearance from the normal phenotype. In the Reeler mouse, cortical plate cells accumulate entirely below the 'superplate' cells, whose pioneer axons subsequently come to lie in oblique fascicles, surrounded by the thickening cortical plate.

The evidence from co-culture experiments suggests that the cortex does not become growth permissive (at least in the rat) until a few days after thalamic axons arrive at the cortex *in vivo* (see Fig. 2 and Götz et al 1992). When thalamic axons in the Reeler mouse reach the cortex, however, they do not accumulate below, but run up diagonally in fascicles through the thickening cortical plate. They then gather over the 'superplate' above, where they wait for a few days before growing downwards and arborizing towards the bottom of the plate among cells that presumably correspond to the normal layer IV (Fig. 5). It seems likely that thalamic fibres grow through the hostile environment of the cortical plate by following the oblique fascicles of preplate axons laid down earlier. The unusual pattern of axon growth in the

Thalamocortical innervation 137

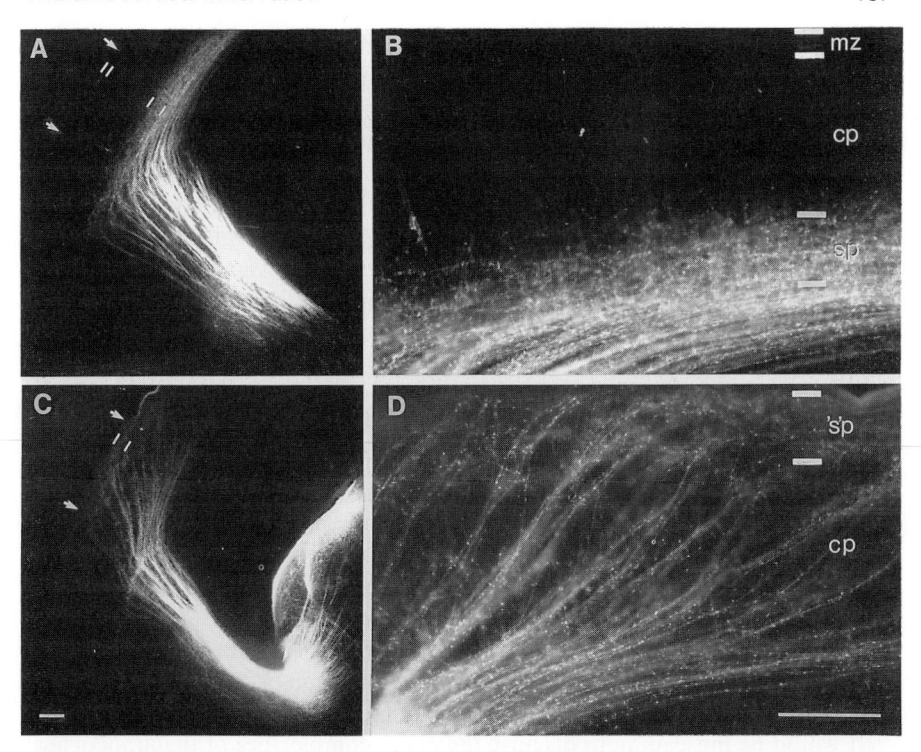

FIG. 5. Coronal sections showing the thalamocortical projection in embryonic day 18 fetal mice. (A, B) Normal mice. (C, D) Reeler mice from the same litter of a heterozygous female (rl/+) that had been mated with a homozygous (rl/rl) male. Axons were revealed by a single DiI (1,1'dioctadecyl-3,3,3',3'-tetramethylindocarbocyanine perchlorate) crystal placed into the dorsal thalamus, which did not back-label cortical neurons. The low-power micrographs (A, C) of the dorsolateral segments of the left hemispheres show the thalamic fibres streaming out of the internal capsule, through the lower part of the corpus striatum and up through the intermediate zone to reach the occipital cortex. The region of cortex between the arrows in A and C is magnified in B and D. The laminar boundaries marked in A and B correspond to the marginal zone (mz) and the bottom of the cortical plate (cp), as indicated in B and D. In the normal mouse, thalamic axons accumulate just below the cortical plate, in the subplate zone (sp). In the Reeler mouse, the fibres run in oblique fascicles through the cortical plate, all the way up to the 'superplate' ('s'p) where they accumulate. Bars = 100 μm.

Reeler mouse can be explained by precisely the same algorithm of interaction and guidance that we propose for the normal mouse.

Conclusions

We suggest that the timing of outgrowth, accumulation, invasion and termination of thalamic axons is regulated by a temporal cascade of simple signals expressed by the cortex. However, it is unlikely that the precise tangential topography of the thalamic projection results from regional chemospecificity of such signals.

The intimate association of preplate axons and thalamic fibres in normal and Reeler mice implies that thalamic axons follow the preplate scaffold to the cell bodies of preplate cells, where they subsequently wait for the cortical plate to become growth permissive. However, the scaffold of preplate axons is not available to act as a guidance structure when fibres are first leaving the thalamus. The initial ordered outgrowths from both preplate and thalamus may depend on interaction with extracellular guidance cues. However, they may be influenced simply by the order in which fibres leave the cortex and the thalamus, each wave of axons being laid down on the surface of the pre-existing axon array. In rodents, where the whole process of cortical neurogenesis is completed within a week, the anteroventral areas of the hemisphere antecede the caudodorsal in maturity by at least a day (Berry & Rogers 1965, Lund & Mustari 1977). It is likely that a similar temporal gradient exists in the developing thalamus. These gradients may impose chronotopic patterns on the outgrowth of axons, and may determine the sequence in which the two fibre arrays arrive and interact in the basal telencephalon. Time is unidimensional and cannot determine the entire two-dimensional pattern of thalamocortical innervation. The other axis of the map may depend on a gradient of molecular signals on the preplate and thalamic axon arrays, or merely on the spatial separation of parts of the growing scaffolds.

It would be misleading to disguise the many complexities of thalamocortical innervation that must be explained, especially in animals with a larger and more complex neocortex than that of rodents. Single cortical areas can be innervated by more than one thalamic nucleus and individual thalamic nuclei can innervate several cortical regions. For regions other than primary sensory areas, corticocortical connectivity is often more important than thalamic input. The reversals of mapping that often occur at the borders of cortical areas also need to be explained. However, we are beginning to see how the basic aspects of thalamocortical innervation might be generated by the sequential combination of a number of individually simple mechanisms.

Acknowledgements

We thank André M. Goffinet for providing Reeler mouse specimens and Richard Adams for help with the confocal microscopy. This study was supported by grants from the Wellcome Trust, the Medical Research Council (MRC), the Soros-Hungarian Academy of Sciences Foundation and the Human Frontier Science Program. It forms part of the work of the McDonnell–Pew Centre for Cognitive Neuroscience and the MRC Research Centre in Brain and Behaviour. The excellent technical assistance of Pat Cordery and William Hinkes is gratefully acknowledged. Z. M. is a Junior Research Fellow at Merton College and currently holds an MRC Fellowship.

References

Berry M, Rogers AW 1965 The migration of neuroblasts in the developing cerebral cortex. J Anat 99:691–709

Bicknese AR, Sheppard AM, O'Leary DDM, Pearlman AL 1994 Thalamocortical axons extend along a chondroitin sulfate proteoglycan-enriched pathway coincident with the neocortical subplate and distinct from the efferent path. J Neurosci 14:3500–3510

Blakemore C 1995 Introduction: mysteries in the making of the cerebral cortex. In: Development of the cerebral cortex. Wiley, Chichester (Ciba Found Symp 193) p 1–20

Blakemore C, Molnár Z 1990 Factors involved in the establishment of specific interconnections between thalamus and cerebral cortex. Cold Spring Harbor Symp Quant Biol 55:491–504

Bolz J, Novak N, Staiger V 1992 Formation of specific afferent connections in organotypic slice cultures from rat visual cortex cocultured with lateral geniculate nucleus. J Neurosci 12:3054–3070

Bolz J, Kossel A, Bagnard D 1995 The specificity of interactions between the cortex and the thalamus. In: Development of the cerebral cortex. Wiley, Chichester (Ciba Found Symp 193) p 173–191

Catalano S, Robertson RT, Killackey HP 1991 Early ingrowth of thalamocortical afferents to the neocortex of the prenatal rat. Proc Natl Acad Sci USA 88:2999–3003

Caviness VS Jr 1976 Patterns of cell and fibre distribution in the neocortex of the reeler mutant mouse. J Comp Neurol 170:435–488

Caviness VS Jr 1988 Architecture and development of the thalamocortical projection in the mouse. In: Bentivoglio M, Spreafico R (eds) Cellular thalamic mechanisms. Elsevier Science, Amsterdam

Caviness VS Jr, Frost DO 1980 Tangential organization of thalamic projections of the neocortex in the mouse. J Comp Neurol 194:355–367

Caviness VS Jr, Crandall JE, Edwards MA 1988 The Reeler malformation: implications for neocortical histogenesis. In: Peters A, Jones EG (eds) Cerebral cortex, vol 7: Development and maturation of cerebral cortex. Plenum Press, New York, p 489–499

De Carlos JA, O'Leary DDM 1990 Subplate neurons 'pioneer' the output pathway of rat cortex but not pathways to brainstem or spinal targets. Soc Neurosci Abstr 16:139

De Carlos JA, O'Leary DDM 1992 Growth and targeting of subplate axons and establishment of major cortical pathways. J Neurosci 12:1194–1211

Gähwiler BH 1988 Organotypic cultures of neural tissue. Trends Neurosci 11:484–489

Götz M, Novak N, Bastmayer M, Bolz J 1992 Membrane-bound molecules in rat cerebral cortex regulate thalamic innervation. Development 116:507–519

Lumsden AGS 1992 Chemotaxis in the developing nervous system. In: Kater S, Letorneau P, Macagno E (eds) The nerve growth cone. Raven Press, New York, p 167–180

Lund RD, Mustari MJ 1977 Development of the geniculocortical pathway in rats. J Comp Neurol 173:289–305

McConnell SK, Ghosh A, Shatz CJ 1989 Subplate neurons pioneer the first axon pathway from the cerebral cortex. Science 245:978–982

Miller B, Chou L, Finlay BL 1993 The early development of thalamocortical and corticothalamic projections. J Comp Neurol 335:16–41

Molnár Z, Blakemore C 1990 Relationship of corticofugal and corticopetal projections in the prenatal establishment of projections from thalamic nuclei to specific cortical areas of the rat. J Physiol 430:104P

Molnár Z, Blakemore C 1991 Lack of regional specificity for connections formed between thalamus and cortex in coculture. Nature 351:475–477

Molnár Z, Blakemore C 1992 How are thalamocortical axons guided in the Reeler mouse? Soc Neurosci Abstr 18:330
Molnár Z, Adams R, Blakemore C 1993 3-D confocal microscopic study on the early development of thalamocortical projections in the rat. Soc Neurosci Abstr 19:450
O'Leary DDM, Borngasser DJ, Fox K, Schlaggar BL 1995 Plasticity in the development of neocortical areas. In: Development of the cerebral cortex. Wiley, Chichester (Ciba Found Symp 193) p 214–230
Price DJ, Lotto RB, Warren N, Magowan G, Clausen J 1995 The roles of growth factors and neural activity in the development of the neocortex. In: Development of the cerebral cortex. Wiley, Chichester (Ciba Found Symp 193) p 231–250
Rakic P 1977 Prenatal development of the visual system in rhesus monkey. Philos Trans R Soc Lond B Biol Sci 278:245–260
Romijn HJ 1988 Development and advantages of serum-free, chemically defined nutrient media for culturing of nerve tissue. Biol Cell 63:263–268
Shatz CJ, Luskin MB 1986 The relationship between the geniculocortical afferents and their cortical target cells during development of the cat's primary visual cortex. J Neurosci 6:3655–3668
Shatz CJ, Ghosh A, McConnell SK, Allendoerfer KL, Friauf E, Antonini A 1990 Pioneer neurons and target selection in cerebral cortical development. Cold Spring Harbor Symp Quant Biol 55:469–480
Sperry RW 1963 Chemoaffinity in the orderly growth of nerve fibre patterns and connections. Proc Natl Acac Sci USA 50:703–709
Yamamoto N, Kurotani T, Toyama K 1989 Neural connections between the lateral geniculate nucleus and visual cortex in vitro. Science 245:192–194
Yamamoto N, Yamada K, Kurotani T, Toyama K 1992a Laminar specificity of extrinsic cortical connections studied in culture preparations. Neuron 9:217–228
Yamamoto N, Higashi S, Sugihara H, Toyama K 1992b Axonal growth and branch formation of geniculate fibres in visual cortex studied in coculture preparations. Soc Neurosci Abstr 18:105

DISCUSSION

D. Price: Did you place the crystals quite laterally within the cortex?

Molnár: I have not examined this question systematically, but I certainly did not restrict my carbocyanine crystal placements to the lateral cortex. I also did extensive dye injections in dorsal and ventral parts; indeed, the majority were quite dorsal.

D. Price: The reason I mention this is because Clasca et al (1994) suggested that axons from the subplate cells in the medial part of the ferret cortex do not make it to the internal capsule. Consequently, I wondered if you had difficulty obtaining corticothalamic labelling with the medial injections.

Molnár: When I placed crystals in the internal capsule I found that the back-labelled cells did extend dorsally. I could not get back-labelled cells in the cingulum, but the entire lateral cortex and most of the dorsal regions contained back-labelled subplate cells. Interestingly, when I placed carbocyanine crystals into the dorsal thalamus itself, I saw very few back-labelled subplate cells at any age suggesting that in the rodent their projections either do not penetrate deep into the thalamus or that they reach the thalamus but do not arborize extensively and therefore they did not take up the dye. Slightly larger crystals, intruding into the ventrolateral diencephalon, or small crystals placed directly in the region of the primitive internal capsule produce considerable back labelling of subplate cell bodies as well as of layer VI and V cells, depending on age and crystal position (Molnár 1994). That is why I tried labelling from the internal capsule. It's an interesting hypothesis that these subplate projections might themselves 'wait' outside the thalamus. Recently, John Mitrofanis described transient cell populations outside the thalamus in the thalamic reticular and perireticular nuclei (Mitrofanis 1992). I suspect that this may be the compartment where the subplate and perhaps even layer VI projections accumulate temporarily before they enter deep into the thalamus itself. Shatz & Rakic (1981) looked at these early corticofugal projections in the fetal rhesus monkey using a purely anterograde tracer, [^{3}H]proline, and they found that the orthogradely transported label appears in the surrounding of the lateral geniculate nucleus (LGN) weeks before it is detected within the LGN itself. This suggests that the corticofugal projections wait outside the thalamus. This waiting compartment might be identical with the transient cell populations that John Mitrofanis observed in the thalamic reticular and perireticular nuclei. At present, what these early corticofugal projections are doing is a mystery.

Bolz: You showed in your labelling experiment that thalamic axons only invade the cortex late in development. We have similar results from co-culture experiments in the rat (Götz et al 1992). You mentioned that in the Reeler mouse, they have to grow through the cortical plate very early in development. Have you tried a co-culture experiment with the Reeler mouse at the age at which the cortex in normal mice is not permissive for thalamic fibres?

Molnár: No, but it is an important experiment. I suspect that a membrane preparation from embryonic Reeler cortex would be hostile.

Bolz: You could do a regular co-culture experiment instead of a membrane preparation.

Molnár: The thalamic fibres would probably not enter *in vitro*.

Bolz: Why do you think that thalamic fibres *in vivo* would be able to grow through the cortical plate?

Molnár: Because *in vivo*, the earliest corticofugal projections of the preplate and the thalamic projections find each other and fasciculate with each other.

Bolz: It is possible that the axons of subplate cells provide the substrate for thalamic fibres to enter the cortical plate.

Molnár: The initial association of the two fibre systems both in normal and Reeler mice occurs below the cortical plate at an early stage. The extending thalamic fibres are then indifferent to the position of these preplate cells. This mechanism could explain why the thalamic fibres are deployed into the superplate in the Reeler mouse. There are at least three distinct patterns of thalamic fibre organization in different parts of the pathway from the thalamus to the cortex. Adjacent axons remain parallel and are closely associated throughout the primitive internal capsule. On entering the corpus striatum, the tight, homogeneous bundles break up and the fibres reorganize into small fascicles. In normal mice the fibres defasciculate at the lateral edge of the striatum, and individual axons and small fascicles course towards the subplate along parallel paths in the white matter. In Reeler mice they remain in large fascicles throughout the cortical plate. The different fibre organizations might reflect the differences in the adhesive properties of the environment. It may be possible to tell what was happening during their establishment from these patterns, in the same way as you can tell whether someone was skiing fast or slow, or whether they were professionals or inexperienced, from traces left behind in the snow. I suspect that the striatum is a hostile environment for thalamic fibre ingrowth. In the Reeler mouse these fibres do not break up from their fascicles at the lateral edge of the striatum, but run through the cortical plate, and I suggest that they keep their association with their corticofugal counterparts from the superplate.

Levitt: Do they break up because they're part of a large bundle that comes from different thalamic nuclei and then has to make different decisions about where to turn? This may have nothing to do with the hostile environment, but may instead represent a decision point for these fibres.

Molnár: The decision is probably not made at the striatum–intermediate zone boundary because when the fibres leave the striatum, they seem to know what they're doing. It's an important point. In the Reeler mouse the fibres don't break up into smaller fibres or smaller fascicles. There is a border between the striatum and the cortical plate, but the fibres change their direction slightly and run up to the superplate in oblique fascicles. The steepness of the intracortical part of these fascicles depends on the age and thickness of the cortex. A sharp change in the direction of the fascicles is not observed initially, but it is more apparent when the cortex becomes thicker.

Levitt: You may be labelling a more diverse population of thalamic fibres later in gestation.

Blakemore: The apparent change in trajectory of thalamic axons at the boundary between the intermediate zone and the cortical plate in the Reeler mouse can be accounted for by the thickening of the cortical plate itself, displacing all the preplate cells upwards and enveloping their descending axons to form oblique fascicles, which are then followed by the ingrowing thalamic fibres. It may be confusing, in both Reeler and normal mice, to look at the picture in the older mouse where fibre pathways often have sharp bends or inflections, which give the axons the appearance of having made deliberate choices and changed their direction of growth. For instance, efferent fibres that leave the cortex appear to have grown down and turned 90° in the white matter, but the neurons may actually have spun their axons off in the intermediate zone during migration, and then continued to migrate upwards, leaving a 90° bend in the axon.

Levitt: How can you tell the difference between that situation and the addition of groups of fibres from different thalamic nuclei which start to grow later on and are making choices? Are you saying that they all grow out straight from these different thalamic areas and then make decisions about where to go as they interact with the different populations of subplate axons that are coming down?

Molnár: There is a sequential outgrowth of different thalamic fibres, but the difference is only a maximum of two days.

Blakemore: But we don't know enough about the timing of outgrowth from the different segments of the thalamus and we don't know if this growth is temporally matched with the wave of maturation that sweeps across the cortex.

Levitt: It's difficult to believe that it is synchronous because the time of origin of those different neuronal populations spans (in the rodent) at least four days.

You are suggesting that there are some specific interactions among different populations in the internal capsule between these different fibres. Also, you dismiss the possibility that specification occurs within the cortex itself because the posterior thalamic region that you grew with different areas of the cortex grew into the frontal cortex. How can you exclude the possibility that there are a number of general thalamic regions that grow into all areas of the cortex, and that what you're seeing is a combination of both specific and general thalamic nuclei that are growing into your explants? Your dissections are early, i.e. at embryonic day (E) 16, and don't include, for example, populations of the dorsal lateral geniculate neurons that were generated the day before and haven't migrated out to the nuclear region of the LGN. In addition, the non-specific posterior thalamic nucleus (PO) is located in the same dissected tissue and will grow into any cortical area.

Molnár: We dissected at E16 in order to study cells that were generated at E12–14. E16 is also the first day at which lateral geniculate nuclei can be distinguished from the rest of the thalamus by nissl staining.

Levitt: But in this situation, when you're talking about specificity and dismissing one major mechanism, it's important to determine that there is not a complement of neurons in that thalamic explant that would project to every area of the cortex. You have to show that what you're culturing are dorsal lateral geniculate neurons and nothing else because if a component of PO is present, then it becomes more difficult to dismiss the possibility that the fibres innervating frontal cortex explants arise from thalamic nuclei in the explant other than the LGN.

Molnár: We are confident that lateral geniculate nuclei are in the explants. When we didn't find any specificity, we studied different thalamic nuclei, excluding lateral geniculate nuclei, but we still didn't see preferential ingrowth.

Blakemore: Pat Levitt is saying that it would be difficult to cut an explant from the thalamus that did not include some non-specific thalamic nucleus.

There are two arguments that we could present. One concerns the density and uniformity of outgrowth—the whole explant is surrounded by large numbers of growing neurites. Secondly, the pattern of termination of these fibres in the cortical explant, at least in explants taken after postnatal day (P) 2, is so obviously typical of input from the specific relay nuclei: there is dense arborization in layer IV, which is quite different from the normal termination pattern for the non-specific thalamic afferents.

O'Leary: If you did the same experiment with E18 rat tissue, you could probably dissect out the LGN free of other thalamic nuclei.

Molnár: But the problem then is that at E18, they are not virgin fibres because they have seen the cortex.

O'Leary: That's true, but this would probably not have a major impact on the interpretations of that experiment.

Jones: If the outgrowth reflects the situation in the adult, the number of axons coming from the general non-specific thalamic nuclei would be less than those coming from the lateral geniculate nuclei or from some of the other major relay nuclei. This would suggest that the conditions are tilted in favour of there being a major complement of specific afferents.

Levitt: The large outgrowth from the explant could originate from a relatively small number of neurons in culture, and we don't know which neural population in the thalamus is dominant in the culture system. For example, dorsal lateral geniculate neurons are sensitive to target deprivation and might die preferentially in the culture system, whereas general thalamic neurons are not nearly as sensitive to target deprivation.

Jones: Yes, but it is possible that the PO neurons survive because their projections are widespread, so it's difficult to compromise their terminations in the same way that you can compromise the terminations of a lateral geniculate neuron. I agree that you cannot predict what's going to happen in culture, but if this reflects the situation in the adult, then you would expect there to be quantitatively less outgrowth.

O'Leary: I expect that you would get the same result if you dissected at E18, but I would hesitate to use that finding to rule out the possibility of molecular or genetic specification of different regions of neocortex. For instance, the *in vitro* experiments of Bonhoeffer and his colleagues on the development of retinotopically ordered connections in the retinotectal system of chicks (Walter et al 1987a,b) showed that temporal axons will grow on caudal tectal membranes, but when temporal axons are given a choice between the correct (rostral) and the incorrect (caudal) part of the tectum, they will grow preferentially on rostral membranes. Therefore, there is clearly a molecular distinction between rostral and caudal tectum.

Molnár: We were initially interested in looking at the nature of the ingrowth pattern to the cortical plate. There wasn't an obvious specific affinity for the subplate *in vitro*, so it is possible that we were not asking the right question in the right system. If the cortical plate itself is not responsible for the deployment of the initial thalamocortical projections, then it has to be the subplate. We should now look at the different regions within the subplate. It is possible that the organotypic slice culture preparation, where the surface is only a few cell layers thick compared to a flatmount, is not the best system for addressing these questions. Nevertheless, these *in vitro* experiments suggest that the cortical plate itself does not have a strict, inherent chemical matching tag to guide the deployment of fibres from appropriate nuclei.

LaMantia: Classical experiments have been performed that address the same question, albeit at the back part of the brain or in the peripheral nervous system (Langley 1897, Landmesser 1980, Purves et al 1981). The results suggested that specificity is apparent over a short distance; but when a cell is removed from its normal location and confronted with any neuron or muscle target, if its normal target is not present, it will innervate any available neuronal or muscle target. Consequently, the specificity that you're looking at for thalamocortical pathway formation works only over short distances in the intact developing cortex. Apparent lack of specificity is only observed when the neuron is confronted with a different target other than its normal target. The level of neuronal differentiation at P6 in the cerebellum is different from the cortex because cerebellar development is not complete. Consequently, if you had looked at the cerebellum, you would have probably observed little in the way of explant outgrowth or *in vitro* innervation. You seem to have the same results in all of your cultures, i.e. the thalamic explant will grow into any neuronal target tissue.

Molnár: I did co-culture experiments with younger cerebellum, but the results were similar to the P6 experiments. I do not see why you find the *in vitro* results, where we manipulated the system, contradictory rather than supportive. I would like to use your argument to support mine.

LaMantia: The argument is that in both the normal mouse and in the Reeler mouse, there is remarkable specificity regardless of how the axons get there. This seems to arise because the route of the axons reflects the availability of the substrates on which to grow. Rather than dismissing biochemical specificity, because it doesn't reside in gross differences that one can pick up in a culture experiment, what do you think is causing this specificity in the mouse at an early stage?

Molnár: Relationships between neighbouring neurons may be important in the initial delivery of their axons. If you want to show that this is really important, you have to challenge the system by moving the cells and giving them the chance to interact with other neighbours. If you don't see specificity, then you have two options; you either disregard your results and consider them to be *in vitro* artefacts, or you accept them.

LaMantia: In which region did you not see specificity?

Molnár: We did not see specificity between different cortical and thalamic regions.

LaMantia: Yes, but this is not *in vivo*.

Levitt: There is also some appropriate matching of thalamic projections and cortical targets in transplant experiments (Barbe & Levitt 1992, 1995, Castro et al 1991). Transplant experiments can be viewed as regenerative explant experiments where thalamic fibres are being challenged with a new piece of embryonic tissue to determine whether they can tell the difference between one embryonic tissue and another. For some thalamic projections the answer is yes, but for other projections it seems to be no. However, there are differences in thalamic tissue that the fibres can perceive. Perhaps the ultimate test, in terms of connectivity, is the behavioural study of Barth & Stansfield (1994) which shows that the ingrown thalamic fibres can tell the difference between pieces of embryonic cortical tissue that are transplanted. If there was no specificity, then behavioural recovery could be mediated by an area of cortex.

Blakemore: But you're talking about events that are happening *in vivo* at the time of thalamic axon invasion of the cortical plate. Even in the rat, this is several days after the thalamic fibres have been deployed in fairly precise topographical order in the subplate layer under the cortex. There are several facts that are clearly relevant. One is that thalamic axons appear to grow in a topographically ordered pattern through the whole of the diencephalon and the ventral telencephalon before they contact preplate axons. It's difficult to believe that a remote influence from the cortex, which at this stage has few, if any, cortical plate cells, could be responsible for organizing such tight fibre–fibre ordering.

Levitt: There is probably also some chemospecificity, either quantitative or qualitative, between those different axons.

Blakemore: Both thalamic and preplate axons grow in close topographic relationship to neighbouring fibres of the same class and with fibres of the other group. They might be interacting chemospecifically. However, we have put forward the idea that the timing of outgrowth is also critical. The first fibres that are laid down, originating from the preplate of the rostroventral cortex and from the corresponding region of the thalamus, appear to take a virtually straight path towards the internal capsule (there might be something attractive about the internal capsule that determines the overall direction of growth). The others that follow in sequential order are simply laid down on the top of these first pioneers. It seems inconceivable that the immediate topographic ordering of thalamic axons, even within the diencephalon, could depend on some sort of chemospecific interaction with cells of the cortical plate, because cells of the cortical plate are not even in place at the time of thalamic outgrowth; and the growth permissive property (which is, in any case, membrane bound and, therefore, incapable of remote influence) is not expressed until some six days later in the rat. Chemospecific interaction with the surface properties of preplate axons, during the 'handshake' in the ventral telencephalon, might play a part in the guidance of thalamic axons over the distal phase of their growth.

However, the results of co-culture experiments suggest that even cells of the subplate do not become growth permissive for thalamic axons until E19–20 in the rat. If a gradient of chemospecific properties does play a part in the final guidance of thalamic axons, it must be expressed selectively on the growing tips of the descending preplate axon array at a very early stage.

Levitt: Is chemospecificity apparent in the subplate?

Blakemore: The 'choice' experiments in co-culture yield no evidence of regional specificity, but there remains the possibility that a chemospecific gradient is expressed *in vivo* on the axons of the preplate scaffold.

Molnár: Anthony LaMantia supports chemospecificity in the early pathways.

LaMantia: Yes, the Reeler mouse experiments show this.

Molnár: But there is an alternative explanation to chemospecificity. The ordering of fibres can be explained by the timing of their establishment with chronotopic fibre ordering.

LaMantia: That's a trap that a lot of people fall into. A negative biological clock does not exist. Differences in timing suggest that there are molecular differences that set the clock which makes the axons recognize different target regions at distinct times giving regional order.

Blakemore: That's different from a specific tag that enables one axon to interact preferentially with another.

LaMantia: But the outcome is the same. This is an important point because when people refer to timing it seems to lift it out of the realm of molecular specificity, but a timing phenomenon has to be specified in a molecular way. If those axons grow out at earlier, intermediate or later times, then something molecular is tagging them temporally rather than spatially or absolutely.

Molnár: If neurons and their projections line up in a queue, according to a first come, first served basis, strict, specific chemical tags may not be necessary.

Rakic: There may be some problems with the interpretation of these experiments. When thalamic nuclei are injected in Reeler mice, it is difficult to be sure that they are actually injected.

Molnár: That's why we did the cortical carbocyanine crystal placement experiments in these early stages (E15 and E18) in both normal and Reeler mice. In the back-labelled thalamic nuclei and fibre bundles we didn't see an obvious difference in the two genotypes.

Innocenti: If you look at subcortical white matter, where corticocortical axons cross thalamocortical axons, there is a zigzag pattern of trajectories suggesting that some axons may fasciculate temporarily with one axonal system, then another and another until they finally decide what to do and where to go (Innocenti 1987). Do you have any evidence which suggests that, at the site of interaction between thalamic fibres and cortical fibres, the trajectory of one population of fibres is modified by the presence of the other? If this is the case, it is possible that they are being guided rather than bumping into each other.

Molnár: Some people believe that studies of growth cone morphology can demonstrate various forms of interactions (see Mason & Godement 1992). The growth cone morphology of an exploring axon is different to the growth cone morphology of an axon growing quickly. There are numerous time-lapse video studies in different systems, including callosal axons (Halloran & Kalil 1994) and retinal axon pathfinding in the optic chiasm (see Stretavan & Reichardt 1993) or the optic tectum (see Stuermer 1992), which show that growth cone morphology is informative about growth cone behaviour. To the best of my knowledge, similar studies have not been done on the thalamocortical fibres. Therefore, I don't know if one axon searches specifically for other axons.

Ghosh: We published observations of the growth cone morphology of growing thalamic axons (Ghosh & Shatz 1992). The first thalamic axons that grow into the cortex have a more complicated growth cone morphology than those which grow in later. I'm not sure what the significance is of these differences in morphology.

References

Barbe MF, Levitt P 1992 Attraction of specific thalamic input by cerebral grafts depends on the molecular identity of the implant. Proc Natl Acad Sci USA 89: 3706–3710

Barbe MF, Levitt P 1995 Age-dependent specification of the cortico-cortical connections of cerebral grafts. J Neurosci 15:1819–1834

Barth TM, Stanfield BB 1994 Homotopic, but not heterotopic, fetal cortical transplants can result in functional sparing following neonatal damage to the frontal cortex. Cereb Cortex 4:271–278

Castro AJ, Hogan TP, Sorensen JC et al 1991 Heterotopic neocortical transplants. An anatomical and electrophysiological analysis of host projections to occipital cortical grafts placed into sensorimotor cortical lesions made in newborn rats. Dev Brain Res 52:231–236

Clasca F, Angelucci A, Sur M 1994 Layer 5 neurons establish the first cortical projection to the dorsal thalamus in ferrets. Soc Neurosci Abstr 20:98

Ghosh A, Shatz CJ 1992 Pathfinding and target selection by developing geniculocortical axons. J Neurosci 12:39–55

Götz M, Novak N, Bastmeyer M, Bolz J 1992 Membrane bound molecules in rat cerebral cortex regulate thalamic innervation. Development 116:507–519

Halloran MC, Kalil K 1994 Dynamic behaviours of growth cones extending in the corpus callosum of living brain slices observed with video microscopy. J Neurosci 14:2161–2177

Innocenti GM 1987 Structure, specificity and discontinuities in the development of corticocortical connections. Pontif Acad Sci Scr Varia 59:91–108

Landmesser L 1980 The generation of neuromuscular specificity. Ann Rev Neurosci 3:279–302

Langley JN 1897 On the regeneration of pre-ganglionic and post-ganglionic visceral nerve fibres. J Physiol 22:215–230

Mason CA, Godement P 1992 Growth cone form reflects interactions in visual pathways and cerebral targets. In: Letourneau PC, Kater SB, Macagno ER (eds) The nerve growth cone, Raven press, New York, p 405–423

Mitrofanis J 1992 Patterns of antigenic expression in the thalamic reticular nucleus of developing rats. J Comp Neurol 320:161–181

Molnár Z 1994 Multiple mechanisms in the establishment of thalamocortical innervation. DPhil thesis, Oxford University, Oxford, UK

Purves D, Thompson WJ, Yip JW 1981 Reinnervation of ganglia transplanted to the neck from different levels of the guinea pig sympathetic chain. J Physiol 313:49–63

Shatz CJ, Rakic P 1981 The genesis of efferent connections from the visual cortex of the fetal rhesus monkey. J Comp Neurol 196:287–307

Stretvan DB, Reichardt LF 1993 Time-lapse video analysis of retinal ganglion cell axon pathfinding at the mammalian optic chiasm: growth cone guidance using intrinsic chiasm cues. Neuron 10:761–777

Stuermer CAO 1992 The formation of topographically ordered connections during development and regeneration of the vertebrate visual system. In: Cronly-Dillon JR (ed) Vision and visual dysfunction, vol 11: Development and plasticity of the visual system. CRC Press, Boca Raton, CA, p 88–111

Walter J, Kern-Veits B, Huf J, Stolze B, Bonhoeffer F 1987a Recognition of position-specific properties of tectal cell membranes by retinal axons *in vitro*. Development 101:685–696

Walter J, Henke-Fahle S, Bonhoeffer F 1987b Avoidance of posterior tectal membranes by temporal retinal axons. Development 101:909–913

Subplate neurons and the patterning of thalamocortial connections

Anirvan Ghosh

Division of Neuroscience, Enders 250, Children's Hospital and Department of Microbiology and Molecular Genetics, Harvard Medical School, 300 Longwood Avenue, Boston, MA 02115, USA

Abstract. The patterning of the cerebral cortex into functionally distinct domains relies on the formation of appropriate connections between the thalamus and the cortex during development. To identify the mechanisms that underlie cortical target selection by thalamic axons, we have examined the role of cellular interactions in the formation of connections between the lateral geniculate nucleus (LGN) and the visual cortex during development of the cat visual system. The morphology of LGN axons as they grow towards the visual cortex suggests that interactions within the subplate zone may be important in the development of geniculocortical connections. The requirement for subplate neurons in this process was examined by ablating subplate neurons underlying the visual cortex at various developmental stages. When subplate cells are deleted between E38 and E42, prior to target innervation by LGN axons, these axons fail to select the visual cortex as their correct target and instead grow past it, staying restricted to the white matter. Deletion of subplate cells at later stages, between P2 and P7, does not affect target selection, but instead it prevents the segregation of LGN axons into ocular dominance columns within layer IV of the cortex. The effects of subplate neuron ablation suggest that interactions between thalamic axons and subplate cells are of critical importance in the specification of thalamocortical connections during development.

1995 Development of the cerebral cortex. Wiley, Chichester (Ciba Foundation Symposium 193) p 150–172

A central organizational feature of the cerebral cortex is that different regions of the cortex mediate distinct functions. For example, the occipital cortex is devoted to the processing of visual information and regions of the temporal cortex are involved in auditory information processing. This spatial ordering in the cortex comes about, in part, from the patterning of connections between the thalamus and cortex: different thalamic nuclei receive input from distinct peripheral organs, and in turn send axonal projections to restricted input regions of the cortex. Thus, the lateral geniculate nucleus (LGN) of the thalamus projects to the primary visual cortex (occipital cortex) whereas the medial geniculate nucleus (MGN) projects to the auditory cortex (temporal cortex). This review focuses on cellular interactions that are important for establishing appropriate connections between the thalamus and the cortex, and it discusses the evidence that interactions involving a transient class of neurons, the subplate neurons, are critical for the patterning of thalamocortical projections.

To determine the role of cellular interactions in thalamocortical development, we must first consider the architecture of the cerebral cortex early in development. (Developmental times refer to the cat because most experiments discussed here were performed in the cat; however, these developmental events are central features of cortical organization among all mammals.) The developing cerebral wall consists of three separate zones: the ventricular zone; the intermediate zone; and the cortical plate. The ventricular zone is the region within which the neuronal and glial precursor cells proliferate. As cells become postmitotic, they migrate through the intermediate zone to take up their final positions within the cortical plate. The intermediate zone contains the axons that project to and from the cortex, and it develops into the cortical white matter. As summarized in Fig. 1, cells that will form the adult cerebral cortex (layers II–VI) are generated in the ventricular zone between embryonic day (E) 30 and E55 (Luskin & Shatz 1985a). Gestation is 65 days. Cells of layer IV, which receive the major ascending input from the thalamus, are generated between E38 and E42 and they do not finish migrating to the cortical plate until after E50 (Luskin & Shatz 1985a, Shatz & Luskin 1986). In addition to the cells of layers II–VI, there is another population of neurons in the cortex that is present only transiently during development. These are the subplate neurons, which are generated between E26 and E30 and are the first postmitotic neurons of the cortex. They occupy the subplate region, which is immediately below the cortical plate (approximately the upper half of the intermediate zone), during much of development (Luskin & Shatz 1985b).

The possibility that subplate neurons may be important in cortical development was initially suggested on the basis of two sets of observations. Firstly, transneuronal labelling experiments indicated that LGN axons grow into the telencephalon relatively early and they reach their cortical target region before their target layer IV cells have migrated into position (Lund & Mustari 1977, Rakic 1977, Shatz & Luskin 1986). During this period, the LGN axons are restricted to the subplate, suggesting that interactions within this zone may be important for the development of the geniculocortical projection. Secondly, a series of anatomical and biochemical studies suggested that subplate neurons achieve a high degree of phenotypic maturity during fetal development and that subplate neurons participate in synaptic circuits before thalamic axons grow into the cortical plate (Chun & Shatz 1988, 1989a). I will discuss a series of more recent studies which support the proposal that interactions involving subplate neurons are critical for the patterning of thalamocortical connections.

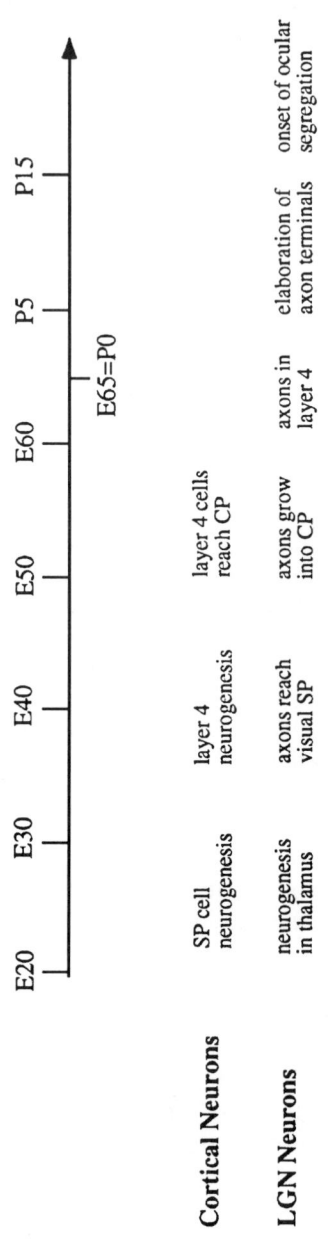

	E20	E30	E40	E50	E60	P5	P15
					E65=P0		
Cortical Neurons	SP cell neurogenesis		layer 4 neurogenesis	layer 4 cells reach CP			
LGN Neurons	neurogenesis in thalamus		axons reach visual SP	axons grow into CP		axons in layer 4	elaboration of axon terminals	onset of ocular segregation

FIG. 1. Timing of events associated with development of the thalamocortical pathway in the cat visual system. Gestation in the cat is 65 days. CP, cortical plate; E, embryonic day; LGN, lateral geniculate nucleus; P, postnatal day; SP, subplate.

Development of geniculocortical axons and the 'waiting period'

[^{3}H]Thymidine labelling indicates that neurons in the LGN are generated between E26 and E30. Shortly after they become postmitotic, thalamic neurons extend axons into the developing telencephalon. Labelling of thalamic axons with the lipophilic tracer DiI (1,1′dioctadecyl-3,3,3′,3′-tetramethylindocarbocyanine perchlorate) indicates that the first thalamic axons have grown into the telencephalon by as early as E30 (Ghosh & Shatz 1992a). These axons, which pioneer the ascending projections into the cortex, are tipped with growth cones that are larger and more elaborate than those of later growing axons. It is noteworthy that at E30, when thalamic axons first grow into the telencephalon, the only postmitotic neurons in the cortex are the preplate neurons (Luskin & Shatz 1985b). The preplate consists of neurons that occupy the marginal zone and subplate later in development, and the trajectory of thalamic axons suggests that they may interact with preplate cells during this period of development.

LGN axons continue to grow dorsally and posteriorly towards the presumptive visual cortex between E30 and E36. A tight fascicle of axons, identifiable as the optic radiations, can be seen coursing towards the visual cortex by E36. Inspection of individual axon morphologies indicates that many axons travelling within the optic radiation extend axon collaterals as they traverse past non-visual cortical areas (Ghosh & Shatz 1992a). Strikingly, these collaterals are invariably restricted to the subplate and they do not invade the cortical plate. Furthermore, these collaterals are no longer present once the LGN axons invade the visual cortex.

Although the first LGN axons arrive within the presumptive visual cortex by E36, they do not invade the cortical plate at this time, but instead they are restricted to the subplate zone immediately below the cortical plate (Shatz & Luskin 1986, Ghosh & Shatz 1992a). This early arrival of LGN axons within the visual subplate has also been reported in rodents and primates (Lund & Mustari 1977, Rakic 1977, Bicknese et al 1994). The period during which LGN axons stay restricted to the subplate before growing into the cortical plate has been referred to as the 'waiting period' (Rakic 1977, Shatz & Luskin 1986). The morphology of LGN axons during this period, however, suggests that they are not simply waiting: these axons have extensive branches within the subplates suggesting that they may participate in cell–cell interactions. In support of this, Herrmann et al (1994) demonstrated that LGN axons make synaptic contacts within the subplate, and that at least some of these contacts are made with subplate neurons. Therefore, LGN axons appear to participate in synaptic interactions with subplate neurons before they invade the cortical plate.

Between E50 and E55, the LGN axons grow into the cortical plate where they form connections with layer IV neurons. Most of these axons grow past layers V and VI, and they turn within layer IV, which is their ultimate target

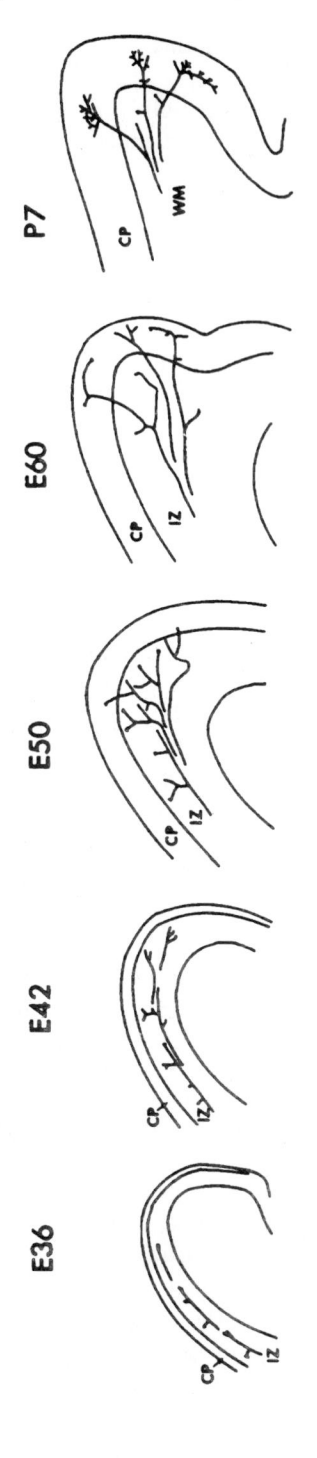

FIG. 2. Diagrammatic representation of the morphological changes within lateral geniculate nucleus axons as they grow toward visual cortex. CP, cortical plate; E, embryonic day; IZ, intermediate zone; P, postnatal day; WM, white matter. (Reproduced with permission from Ghosh & Shatz 1992a.)

layer (Ghosh & Shatz 1992a). LGN axons branch extensively within layer IV during the next few weeks and they segregate to give rise to the pattern of ocular dominance columns by the sixth postnatal week (Ghosh & Shatz 1992a, LeVay et al 1978, Antonini & Stryker 1993).

These observations (summarized in Fig. 2) indicate that thalamic axons invade the telencephalon relatively early in development and that axons from different thalamic nuclei navigate to their appropriate target regions before most cells of the cortical plate are generated. The morphology of the growing thalamic axons indicates that interactions important for their guidance must take place in a restricted zone of developing cerebral wall called the subplate. This zone is remarkable because it contains a population of neurons that displays a high degree of phenotypic complexity during early cortical development but it is largely eliminated by programmed cell death once the permanent connections of the cortex are formed. These neurons play a critical role in the development of thalamocortical connections because the patterning of thalamic projections is severely disrupted when they are absent.

The developmental history of subplate neurons

The subplate contains the first postmitotic neurons of the developing cerebral cortex. In the cat these cells are generated between E26 and E30 (Luskin & Shatz 1985b). Morphological, biochemical and electrophysiological studies indicate that these cells acquire many features of mature neurons early in development and they seem to participate in synaptic circuits before their counterparts in the cortical plate (reviewed in Shatz et al 1990). Studies on the development of cortical projections indicate that axons of subplate neurons pioneer the formation of efferent projections from the cortex (McConnell et al 1989, 1994). Subplate axons from various cortical regions can be seen growing towards the future internal capsule as early as E30. Thalamic axons and subplate axons together form the axon tract, which later develops into the internal capsule. Retrograde labelling experiments indicate that subplate neurons project to the superior colliculus and the thalamus, which are the major subcortical targets of the cells of layers V and VI (McConnell et al 1989). At a later stage, some subplate cells also project across the corpus callosum, which is a major target of the cells of layers II and III (Chun & Shatz 1989a). Subplate cells, therefore, extend axons to several structures that are later innervated by neurons of the cortical plate.

Biochemical studies indicate that subplate neurons are a heterogeneous class of cells. One of the first indications of this was the observation that many of these cells express neuropeptides and neurotransmitters long before cells in the cortical plate (Chun et al 1987, Chun & Shatz 1989a). Labelling subplate neurons when they become postmitotic with a pulse of [^{3}H]thymidine and immunostaining sections with specific antibodies reveals that subsets of subplate neurons express neuropeptide Y, cholecystokinin, somatostatin and glutamic acid decarboxylase (GAD). Subplate neurons that have long distance projections can transport [^{3}H]aspartate (retrograde transport), suggesting that some of these cells may also use excitatory amino acids as neurotransmitters (Antonini & Shatz 1990). In contrast, GAD immunoreactivity (which serves as an indicator of inhibitory neurons) in subplate neurons is restricted to those cells that have local projections (Antonini & Shatz 1990).

Several lines of evidence indicate that subplate neurons can participate in synaptic circuits during embryonic development. The first synapses of the cortex were found in the subplate region by Cragg (1975). Then, by identifying subplate neurons with [^{3}H]thymidine, it was shown that subplate neurons receive synaptic inputs specifically during embryonic development. At least some of these inputs are from thalamic axons (Chun et al 1987, Chun & Shatz 1988, Herrmann et al 1994). These findings are supported by electrophysiological studies of subplate neurons in acute slices, which indicate that subplate neurons receive functional excitatory inputs from fibres coursing through the optic radiations (Friauf et al 1990). The location of these fibres corresponds to the location of growing LGN axons, so it is likely that at least some of the synapses on subplate neurons are from LGN axon terminals.

Perhaps the most striking feature of subplate neurons is that they are a transient population of neurons (Valverde & Facal-Valverde 1988, Chun & Shatz 1989b). Although subplate neurons are present in large numbers during fetal and early postnatal development, most are lost by programmed cell death by the end of the fourth postnatal week. The observations mentioned above, however, suggest that subplate neurons are a highly developed class of neurons which can participate in cellular interactions in the developing cortex. Their location within the developing cerebral wall, moreover, suggests that interactions between subplate neurons and thalamic axons may have important consequences for cortical development.

Requirement for subplate neurons in the formation of thalamocortical projections

The role of subplate neurons in various aspects of thalamocortical development has been explored by examining the consequences of ablating subplate neurons at different developmental ages. These experiments involve intracortical injections of kainic acid; therefore, the specificity of these ablations deserves comment. Kainic acid is a glutamate receptor agonist and causes excitotoxic cell death of neurons expressing kainate receptors (Coyle et al 1981). Intracortical white matter injections of kainic acid between E38 and postnatal day (P) 7 leads to cell death of subplate neurons as determined by the loss of microtubule-associated protein 2-immunoreactive cells in the subplate. However, these injections do not affect the viability of neurons in the overlying

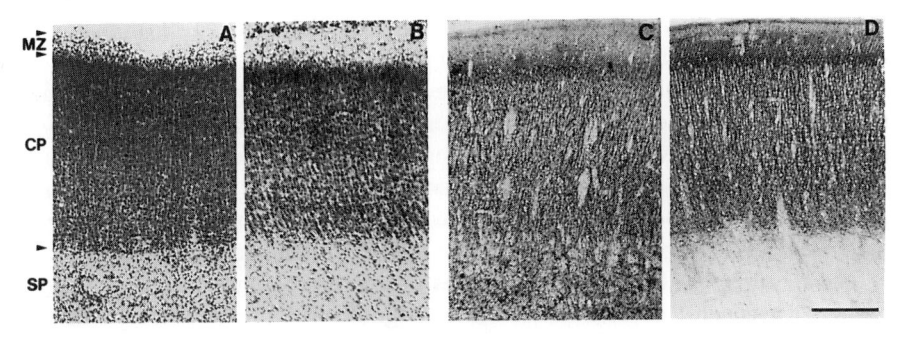

FIG. 3. Effects of white matter kainic acid injections at embryonic day (E) 42 on the cellular constituents of the developing cortex examined at E60. (A, B) Cresyl violet stained sections indicate that the cellular organization of the cortical plate in kainic acid treated cats (B) is not notably different from controls (A). (C, D) Sections immunostained with an antibody against the neuron-specific microtubule-associated protein 2 show that in contrast to control injections (C), kainic acid injections lead to the elimination of subplate neurons but not cortical plate cells (D). CP, cortical plate; MZ, marginal zone; SP, subplate. Bar = 300 μm (applicable to A, B, C, D).

cortical plate (Fig. 3) (Chun & Shatz 1988, Ghosh et al 1990, Ghosh & Shatz 1992b, Ghosh & Shatz 1993). Other cellular elements in the region appear unaffected by kainic acid injections; for example, [^{3}H]thymidine labelling experiments indicate that neurogenesis and migration of layer IV cells are not affected in kainic acid-injected regions. Similarly, radial glial cells also appear unaffected, based on DiI labelling and vimentin immunostaining of radial glial fibres. These observations suggest that the primary effect of such kainic acid injections is the selective elimination of subplate neurons. We believe that this selectivity may be due to the early maturation of subplate neurons, which leads to an increased expression of glutamate receptors in these cells.

Subplate neurons underlying the visual cortex were ablated at E42 or E43 and the effects on geniculocortical development were examined by labelling of LGN axons with DiI days to weeks after the lesion. LGN axons are normally restricted to the subplate between E43 and E50, and they extend branches within the region. At comparable times in subplate-ablated cats, branching of axons within the subplate was lost, but there was no precocious invasion of the cortical plate (Ghosh & Shatz 1993). This suggests that during the 'waiting period', subplate neurons do not simply act as a barrier to cortical plate innervation and that ingrowth of thalamic axons into the cortical plate may be restricted by other signals. In addition, there is recent evidence that, at least in the rat, inhibitory signals present in the cortical plate may promote the ingrowth of thalamic axons at later developmental times (Götz et al 1992).

Thalamic axons normally invade the cortical plate after E50, and it was during this time that the subplate ablations had their most dramatic effects. In normal (and control injected) cats there is a robust innervation of the visual cortex by LGN axons between E55 and P7. In contrast, LGN axons fail completely to innervate the primary visual cortex when subplate neurons are ablated and instead they grow past it, remaining restricted to the white matter (Fig. 4). These experiments suggest that the geniculocortical axons have a severe defect in target recognition in the absence of subplate neurons.

The requirement for subplate neurons in the patterning of thalamocortical axons is a general effect because an analogous situation is observed in the auditory cortex following subplate ablations, i.e. MGN axons fail to invade the auditory cortex in subplate-ablated cats and instead grow past it. These observations in the visual cortex and the auditory cortex suggest that thalamic axons interact with subplate neurons, and that subplate neurons play a critical role in the guidance of thalamic axons to their appropriate cortical targets.

Subplate neurons and the formation of ocular dominance columns

Although LGN axons invade the visual cortex between E50 and E55, subplate neurons continue to persist in the white matter for the next several weeks. During this period, many of the subplate neurons project their axons into the cortical plate and receive synaptic inputs from white matter fibres (Friauf et al 1990). Therefore, it appears that although subplate neurons play a critical role in target selection by thalamic axons, they may continue to participate in later developmental events.

One of the major postnatal developmental events in the visual cortex is the segregation of geniculocortical axons into ocular dominance columns. Axons of LGN neurons receiving inputs from the two eyes are initially intermixed within layer IV of the visual cortex. At about the third postnatal week, axons from the two eyes begin to segregate; and by the sixth postnatal week, LGN axons have segregated into the typical eye-specific patches of the adult (LeVay et al 1978). The anatomical segregation of LGN axons within layer IV is the basis for the physiologically defined ocular dominance columns in visual cortex.

The possibility that subplate neurons may influence interactions between LGN axons and layer IV cells was tested in a series of experiments in which subplate neurons were ablated during the first or third postnatal weeks by kainic acid injections into the white matter. The pattern of geniculocortical projections was examined by transneuronal transport of [^{3}H]proline following a unilateral eye injection. Segregation of LGN axons is complete by the sixth postnatal week in normal cats, and transneuronal labelling of geniculocortical axons reveals the distinct pattern of ocular dominance patches in layer IV of

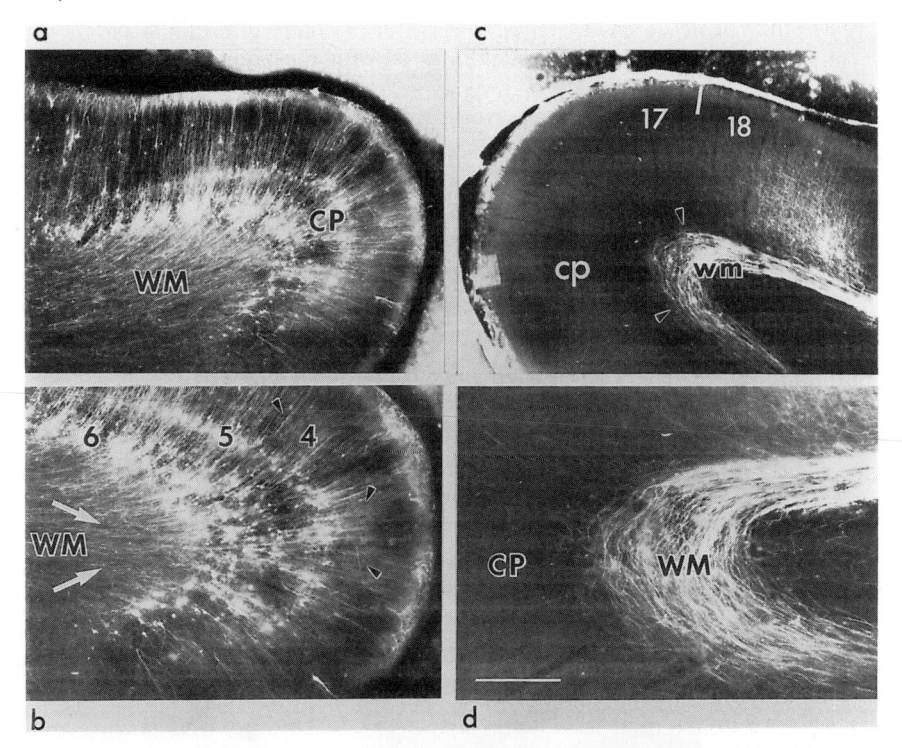

FIG. 4. Aberrant pathfinding by thalamic axons in cats with embryonic ablations of subplate neurons. In this experiment, geniculocortical axons were labelled with an injection of DiI (1,1'dioctadecyl-3,3,3',3'-tetramethylindocarbocyanine perchlorate) into the lateral geniculate nucleus (LGN) at postnatal day 5 following a control (a, b) or kainic acid (c, d) injection into the white matter underlying visual cortex at embryonic day 42. In the control experiment, as in normal cats, LGN axons can be seen invading (white arrows in b) and branching in primary visual cortex (arrowheads in b). In contrast, following subplate ablation, LGN axons fail to grow into primary visual cortex and instead grow past it, staying restricted to the white matter (arrowheads in c). Layers IV–VI shown as 4–6. 17, area 17; 18, area 18; CP, cortical plate; WM, white matter. Bar (shown in d) = 750 μm for a and c, and 250 μm for b and d.

the visual cortex (Fig. 5A). In striking contrast, if the subplate is ablated during the first postnatal week, the LGN axon terminals within layer IV are uniformly labelled, indicating that the ocular dominance columns have failed to form (Fig. 5B). Retinal axons in these cats are normally segregated within the LGN, so the defect in ocular dominance column formation reflects a failure of eye-specific segregation of LGN axons within layer IV (Ghosh & Shatz 1992b).

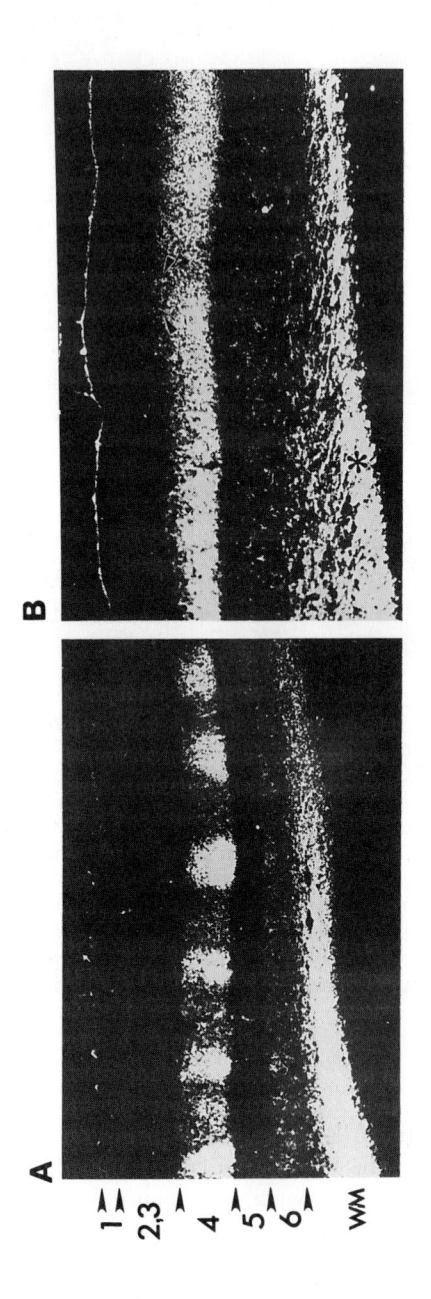

FIG. 5. Failure of ocular dominance column formation in the visual cortex of cats with postnatal ablations of subplate neurons. Comparison of the geniculocortical projection to the visual cortex of normal (A) and subplate-ablated (B) cats visualized by transneuronal transport of [³H]proline. In normal cats the periodic patchy distribution of transported label indicates that distinct columns are present within layer IV by postnatal day (P) 44. In contrast, following ablation of subplate neurons at P6 there is a complete failure of lateral geniculate nucleus axon segregation within layer IV as indicated by the uniform distribution of label (shown here at P55). Asterisk in B indicates the location of the kainic acid injection site. Layers I–VI shown as 1–6. WM, white matter. (Reproduced with permission from Ghosh & Shatz 1994.)

Ablation of subplate neurons during the third postnatal week leads to a similar defect in ocular dominance column formation. There is also an additional alteration in the patterning of thalamocortical axons: LGN axons are normally restricted to layer IV of the visual cortex with a smaller projection to layer VI; but subplate ablation at P18 or P20 causes LGN axons to lose their laminar restrictions to layer IV, and instead also project to layers II and III (Ghosh & Shatz 1994). This last set of experiments warrants cautious evaluation because at these later stages, kainic acid injections lead to some loss of layer VI cells in addition to subplate neurons. Nevertheless, these experiments suggest that the laminar restriction of thalamic axons to layer IV can be modulated by inputs from the subplate (and perhaps layer VI).

These postnatal subplate ablation experiments indicate that subplate neurons continue to participate in thalamocortical development even after LGN axons have grown into layer IV and can influence the eye-specific segregation of geniculocortical axons. A number of previous observations suggest that activity-dependent interactions are required for ocular dominance column formation (Stryker & Harris 1986, Antonini & Stryker 1993). It is possible that subplate neurons influence the process of ocular segregation by modulating synaptic interactions between LGN axons and layer IV neurons.

Subplate neurons as pioneers of cortical efferent projections

Recent evidence suggests that subplate neurons are not only involved in the patterning of thalamic inputs to the cortex, but may also be involved in the development of cortical efferent projections. Subplate neurons extend axons towards the internal capsule shortly after they become postmitotic. These axons, which are the first axons of the cerebral cortex, separate the preplate (which includes the future subplate neurons) from the ventricular zone and they form the structure that will develop into the white matter in the adult. The subplate pioneer axons project to multiple distant targets, including the thalamus, the superior colliculus and the contralateral cortex (McConnell et al 1989, Chun & Shatz 1988). Thus, structures that receive input from cortical plate neurons later in development receive their first cortical innervation from subplate axons. These observations suggest that subsets of subplate neuron axons are involved in guiding the cortical axons that will later grow to their appropriate targets.

The pioneering role of subplate axons and its functional consequences have been examined by two lines of investigation. In one set of experiments, the relationship between the early growing subplate axons and late growing cortical axons has been examined by labelling the respective axonal populations at various times during development. In addition, the requirement for subplate neurons in guiding cortical efferents to their targets has been examined by ablating subplate neurons prior to target selection by cortical axons.

Axon labelling experiments indicate that the subplate neurons extend their axons before their cortical plate counterparts (McConnell et al 1994). Subplate axons grow across the internal capsule to subcortical structures by E35, whereas cortical plate axons do not reach the same targets until about E50. This delay is unexpected because layer VI cells, which form the major projection to the thalamus, are generated between E30 and E35 (immediately after subplate neurons) and they may have been expected to extend axons towards the thalamus shortly thereafter. The location of the descending axon tracts indicates that axons from cortical plate neurons grow adjacent to subplate neuron axons. This suggests that these later growing axons may indeed use the subplate pioneer axons for guidance. This possibility is also supported by morphological differences between growth cones of subplate axons and those from cortical axons that grow later, i.e. the growth cones of subplate axons are relatively large and elaborate, and consist of multiple filopodia; whereas the cortical axons tend to be smaller and of much simpler morphology (Kim et al 1991).

The ideal test of the pioneering role of subplate axons would be to delete subplate neurons very early in development, immediately after they become postmitotic. Unfortunately, the only technique that ablates subplate neurons effectively, i.e. kainic acid injections into the intermediate zone, is not effective until about E37. I suspect that this delay in susceptibility reflects a developmental increase in the expression of kainate receptors in the subplate, although there is a lack of direct experimental evidence. Nevertheless, deletion of subplate neurons between E37 and E43 affects the formation of cortical efferent projections (McConnell et al 1994). The effect, however, is more variable than the effect of subplate ablations on the targeting of thalamocortical axons. Deep layer projections to the LGN and superior colliculus are absent in 50% of subplate-ablated cats. The subcortical projections look surprisingly normal in the remaining 50%. The reason for these subplate ablation differences is not clear. It is not related to the age of subplate ablation or the extent of subplate neuron loss because there are examples where one finds completely normal subcortical projections even in cats in which the subplate is deleted along the entire pathway for descending axons. Thus, subplate neurons are not absolutely required for the formation of cortical efferent projections. However, cortical efferents fail to invade their targets in a large percentage of cases, suggesting that subplate neurons can influence the patterning of descending cortical projections. In this sense, it appears that subplate neurons have a significant role in the formation of cortical efferent projections.

Conclusions

Although the selection of appropriate targets by growing axons is of critical importance in the development of the nervous system, the mechanisms by which precise patterns of connectivity are generated are not understood. The problem is even more perplexing in the developing cortex because the cortical target cells of thalamic axons (the layer IV neurons) are not present when the thalamic axons enter the telencephalon. How can thalamic axons from different nuclei recognize their cortical target areas when the target neurons aren't present? My observations indicate that at least part of the solution seems to be that the initial patterning of thalamocortical projections is determined largely by interactions between thalamic axons and subplate neurons, a population of cortical neurons that are generated early and present only transiently.

Anatomical evidence from the cat and several other species indicates that growing thalamic axons are only in a position to interact with subplate neurons (Bayer & Altman 1990, Bicknese et al 1994, Kostovic & Rakic 1990). The most compelling evidence in support of a role for subplate neurons in cortical target selection comes from studies which show that LGN and MGN axons fail to recognize and innervate their appropriate cortical targets when subplate neurons from these cortical regions are ablated. Subplate neurons are also involved in other aspects of thalamocortical organization. Ablation of subplate neurons affects the formation of cortical efferent projections, and the deletion of subplate neurons after LGN axons have grown into layer IV prevents their segregation into ocular dominance columns. Thus, this population of neurons seems to be involved in regulating multiple aspects of thalamocortical development.

A major challenge ahead will be to understand the mechanisms by which subplate neurons mediate their different functions. For instance, it will be important to determine whether subplate neurons in different cortical areas are unique (perhaps by the expression of distinct cell surface molecules) in a way that allows specific recognition by thalamic axons. It is also possible that timing or activity-dependent mechanisms may play an important role in the patterning of thalamic axons, given that thalamic axons and subplate neurons participate in synaptic interactions.

A few years ago, subplate neurons were considered as a minor neuronal population of little apparent significance. It now appears that this transient population of neurons may be central to the patterning of thalamocortical connections in various mammalian species. It will be of interest to determine the roles of various subpopulations of subplate neurons, and to identify the mechanisms by which subplate neurons regulate key aspects of cortical development.

Acknowledgements

All the experiments described here were carried out in the laboratory of Dr Carla Shatz at Stanford University. I would like to thank her for her guidance and involvement in this work. I would also like to thank Drs Marla Luskin, Jerold Chun, Susan McConnell,

Antonella Antonini and Eckhardt Friauf, whose important contributions I have discussed here.

References

Antonini A, Shatz CJ 1990 Relation between putative transmitter phenotypes and connectivity of subplate neurons during cerebral cortical development. Eur J Neurosci 2:744–761

Antonini A, Stryker MP 1993 Development of individual geniculocortical arbors in cat striate cortex and effects of binocular impulse blockade. J Neurosci 13:3549–3573

Bayer SA, Altman J 1990 Development of layer 1 and the subplate in the rat neocortex. Exp Neurol 107:48–62

Bicknese AR, Sheppard AM, O'Leary DDM, Pearlman AL 1994 Thalamocortical axons extend along a chondroitin sulfate proteoglycan-enriched pathway coincident with the neocortical subplate and distinct from the efferent path. J Neurosci 14:3500–3510

Chun JJM, Shatz CJ 1988 Distribution of synaptic vesicle antigens is correlated with the disappearance of a transient synaptic zone in the developing cerebral cortex. Neuron 1:297–310

Chun JJM, Shatz CJ 1989a The earliest-generated neurons of the cat cerebral cortex: characterization by MAP2 and neurotransmitter immunohistochemistry during fetal life. J Neurosci 9:1648–1667

Chun JJM, Shatz CJ 1989b Interstitial cells of the adult neocortical white matter are the remnant of the early generated subplate neuron population. J Comp Neurol 282:555–569

Chun JJM, Nakamura MJ, Shatz CJ 1987 Transient cells of the developing mammalian telencephalon are peptide immunoreactive neurons. Nature 325:617–620

Coyle JT, Bird SJ, Evans RH et al 1981 Excitatory amino acid neurotoxins: selectivity, specificity, and mechanisms of action. Neurosci Res Prog Bull 19:U335–U427

Cragg BG 1975 The development of synapses in the visual system of the cat. J Comp Neurol 160:147–166

Friauf E, McConnell SK, Shatz CJ 1990 Functional synaptic circuits in the subplate during fetal and early postnatal development of cat visual cortex. J Neurosci 10:2601–2613

Ghosh A, Shatz CJ 1992a Pathfinding and target selection by developing geniculocortical axons. J Neurosci 12:39–55

Ghosh A, Shatz CJ 1992b Involvement of subplate neurons in the formation of ocular dominance columns. Science 255:1441–1443

Ghosh A, Shatz CJ 1993 A role for subplate neurons in the patterning of connections from thalamus to neocortex. Development 117:1031–1047

Ghosh A, Shatz CJ 1994 Segregation of geniculocortical afferents during the critical period: a role for subplate neurons. J Neurosci 14:3862–3880

Ghosh A, Antonini A, McConnell SK, Shatz CJ 1990 Requirement for subplate neurons in the formation of thalamocortical connections. Nature 347:179–181

Götz M, Novak N, Bastmeyer M, Bolz J 1992 Membrane-bound molecules in rat cerebral cortex regulate thalamic innervation. Development 116:507–519

Herrmann K, Antonini A, Shatz CJ 1994 Ultrastructural evidence for synaptic interactions between thalamocortical axons and subplate neurons. Eur J Neurosci 6:1729–1742

Kim GJ, Shatz CJ, McConnell SK 1991 Morphology of pioneer and follower growth cones in the developing cerebral cortex. J Neurobiol 22:629–642

Kostovic I, Rakic P 1990 Developmental history of the transient subplate zone in the visual and somatosensory cortex of the macaque monkey and human brain. J Comp Neurol 297:441–470

LeVay S, Stryker MP, Shatz CJ 1978 Ocular dominance columns and their development in layer 4 of cat's visual cortex: quantitative study. J Comp Neurol 179:223–244

Lund RD, Mustari MJ 1977 Development of the geniculocortical pathway in rats. J Comp Neurol 173:289–305

Luskin MB, Shatz CJ 1985a Neurogenesis of the cat's primary visual cortex. J Comp Neurol 242:611–631

Luskin MB, Shatz CJ 1985b Studies of the earliest-generated cells of the cat's visual cortex: cogeneration of subplate and marginal zones. J Neurosci 5:1062–1075

McConnell SK, Ghosh A, Shatz CJ 1989 Subplate neurons pioneer the first axon pathway from the cerebral cortex. Science 245:978–982

McConnell SK, Ghosh A, Shatz CJ 1994 Subplate pioneers and the formation of descending connections from cerebral cortex. J Neurosci 14:1892–1907

Rakic P 1977 Prenatal development of the visual system in rhesus monkey. Philos Trans R Soc Lond B Biol Sci 278:245–260

Shatz CJ, Luskin MB 1986 The relationship between the geniculocortical afferents and their cortical target cells during development of the cat's primary visual cortex. J Neurosci 6:3655–3668

Shatz CJ, Ghosh A, McConnell SK, Allendoerfer KL, Friauf E, Antonini A 1990 Pioneer neurons and target selection in cerebral cortical development. Cold Spring Harbor Symp Quant Biol 55:469–480

Stryker MP, Harris WA 1986 Binocular impulse blockade prevents the formation of ocular dominance columns in cat visual cortex. J Neurosci 6:2117–2133

Valverde F, Facal-Valverde MV 1988 Postnatal development of interstitial (subplate) cells in the white matter of the temporal cortex of kittens: a correlated golgi and electron microscopic study. J Comp Neurol 269:168–192

DISCUSSION

Daw: I would like to talk about the timing of the subplate lesions in relation to ocular dominance segregation and the question of whether subplate neurons have a direct effect or an indirect effect through layer IV. Ocular dominance segregation occurs at around postnatal day (P) 18, and you described an effect at P18 on layer IV when you made a lesion at P4 (Ghosh & Shatz 1994). Consequently, if the effect is direct, then one would expect a lesion of the subplate at P18 to have an effect on ocular dominance segregation. In contrast, if it's an indirect effect through layer IV, then one would expect that injections of kainic acid at P4 would have an effect on the cytology of layer IV at P18, which is what you reported.

Ghosh: I would first like to make a point. We made similar injections at the onset of segregation to see whether this affects ocular segregation (Ghosh &

Shatz 1994). When we ablated subplate neurons at P18, column formation in layer IV was prevented, just as it was after ablations at P4.

Are you referring to a direct effect as being something that might affect synaptic interactions directly between lateral geniculate nucleus (LGN) axons and the neocortex?

Daw: Yes.

Ghosh: Then the evidence is not good enough to distinguish between the two possibilities. If layer IV cells were affected by the injection, it is possible that an effect through layer IV was involved. However, this is unlikely because direct injections into layer IV do not affect ocular segregation. It is interesting to consider why the loss of subplate neurons might have such long-term effects on the cortex. If electrical excitation in the cortex is also affected by subplate ablations, an indirect effect on layer IV may be involved because levels of activity may influence the survival of cells in layer IV. I would like to determine the effect of subplate ablations on synaptic transmission between LGN axons and layer IV cells immediately after the ablations. The cytoarchitectural effect in layer IV is delayed—we did not observe any changes in layer IV for a week after removing the subplate. The layer IV cells are still present, so recording the electrical activity in layer IV should reveal whether there was an effect on synaptic transmission between LGN axons and layer IV that was modulated by subplate neurons. As far as I know, this has not been done.

O'Leary: Your model includes a circuit diagram with projections from the subplate into layer IV, and it predicts that the segregation of geniculocortical afferents into eye-specific stripes occurs not only within layer IV, but also within the subplate. Is there any evidence for this?

Ghosh: No. The model allows for activity-dependent rearrangements to take place in layer IV, but there's no direct evidence for the proposed circuit. According to the model, when there is co-activation of the subplate neuron and the layer IV cell by the LGN axon, the subplate input into layer IV provides the necessary depolarization to allow for activity-dependent synaptic changes.

O'Leary: You presented evidence that subplate cells help geniculocortical axons consolidate their territory in layer IV as these axons segregate into eye-specific stripes. This suggests that a topographic mapping exists in the connections between the subplate and the overlying layer IV, which matches the one between geniculocortical afferents in the subplate and in the overlying layer IV.

Ghosh: No, not necessarily. There might not be any specificity—the subplate axon could depolarize the entire visual cortex, promoting activity-dependent events. This would not require spatial organization.

LaMantia: Generalized activity influences the subplate and layer IV, and the pattern comes from the temporal discontinuity in the activation of the LGN afferents driven by one eye or the other eye. In a sense, this is the same result as a tetrodotoxin experiment or a deprivation experiment in the cortex (reviewed in Goodman & Shatz 1993). However, you are claiming that your results imply that the subplate neurons are the central players in this process of activity-dependent segregation. What is this activity-dependent role, apart from contributing to an increase in general excitation?

Ghosh: One possibility is that, by playing a general role in maintaining excitation, they could regulate the period of plasticity. It may be more difficult for these modifications to occur in the cortex later in development because the subplate cells undergo programmed cell death over time.

LaMantia: Are other activity-dependent patterning events, such as the establishment of the callosal projection at the border between areas 17 and 18, affected by this subplate-mediated mechanism?

Ghosh: We have not yet examined this issue.

Molnár: I would like to mention a few relevant experiments that might help to define these activity-dependent processes. The functional interactions between the ingrowing thalamic fibres and the cortex in the rat has been imaged, at the time of thalamic fibre invasion and synapse formation, using optical recording techniques (see Grinvald 1984) with voltage sensitive dyes (Crair et al 1993, Higashi et al 1993). We examined the cortical activity pattern in thalamocortical slices (prepared according to Agmon & Connors 1991, Bernardo & Woolsey 1987) after thalamic stimulation (M. Crair, S. Higashi, T. Kurotani, N. Yamamoto, K. Toyama & Z. Molnár, unpublished results, 1992, 1993). We found that at P0, when the afferents grow into the cortex and before the majority have arborized in their principal terminal zone in layer IV, they form functional glutaminergic synapses in deep cortical layers. These are possibly transient synapses. We need to examine earlier stages to test these ideas further. However, it is conceivable that there is a platform, just below the cortical plate, where the thalamic fibres line up in a topographically organized fashion immediately after their initial deployment at embryonic day (E) 15–16. The fibres may then start to transmit certain patterns of activity. This initial layout might then be confronted with activity patterns that arrive later from the periphery, such as those produced as the retinal axons reach the thalamus (Lund & Bunt 1976). The activity patterns cannot modify the construction of the initial thalamocortical links, but it may modify the terminals within the subplate. Naegele et al (1988) and Ghosh & Shatz (1991) showed that side branches can develop from thalamic fibres that arrive in certain phases. Anirvan Ghosh, you mentioned in the latter paper that these side branches are not present initially, but that they develop later. It is possible that the initial topographically ordered deployment of thalamic fibres is followed by smaller refinements based on side branch formation. This may cause the fibres to rearrange, cross each other within this region, and perhaps even turn around and change polarity. The evidence to support this speculation may come from the work of Nelson & Le Vay (1985), who demonstrated in the adult cat that thalamic axons cross each other in the mediolateral (but not in the rostrocaudal) axis of the optic radiation only a short distance (200–500 μm) below the visual cortex. It is possible that the decussation pattern seen in the adult represents a modification of the array of thalamocortical fibres that occurred during, or at the end of, the waiting period. In addition, Williams et al (1993) showed that at the side branches, filopodial-like outgrowth can be induced along a developing neurite shaft *in vitro* by an electrical field applied locally. The early activity patterns might play an important role in cortical map formation through these mechanisms.

Bonhoeffer: The subplate neurons may be required for the correct formation of LGN connections to the cortex *in vivo*, but they may not be required *in vitro*. How comparable are these two situations? Because *in vitro* the connections are made over a few hundred micrometres, whereas *in vivo* the connections have to be made over many millimetres. It is possible that the establishment of relatively long range connections requires the subplate cells, whereas the shorter connections do not.

Bolz: I would like to offer an alternative suggestion for the role of subplate cells during the development of thalamocortical projections. In co-cultures, thalamic fibres enter a postnatal cortical slice from the white matter side as well as from the pial side of the cortical slice (Bolz et al 1992). In both cases they establish specific connections with their target cells in layer IV. This experiment shows that thalamic afferents do not have to bypass subplate cells in order to recognize their appropriate target cells, although subplate cells might be required for thalamic axons to invade the developing cortical plate. Some years ago we demonstrated that various growth-promoting factors, which are initially not present in the cortical plate, are up-regulated during later developmental stages (Götz et al 1992). This suggested to us that the invasion of cortical layers by thalamic afferents is controlled by membrane-bound molecules in the cortex. We prepared isolated cortical slice cultures from early developmental stages, and we observed that after a few days *in vitro* the expression of growth-promoting molecules was up-regulated at a similar rate as *in vivo*. Thus, mechanisms intrinsic to the cortex orchestrate the timing of thalamic fibre ingrowth by regulating the expression of growth-promoting molecules. Subplate cells might be part of a coordinated network of cortical elements that is responsible for the differential expression of these molecules. When subplate cells are ablated, the signal that indicates the end of the waiting period might not be up-regulated. Therefore, thalamic fibres would not invade a subplate-deprived cortex (Ghosh & Shatz 1993).

Ghosh: It would be useful to determine whether up-regulation of these molecules occurs in subplate-ablated cats. It would also be useful to ablate the subplate and culture the cortex, so that we could determine whether the subplate-ablated cortex behaves differently in culture than a normal slice when confronted with the thalamus.

Subplate neurons and thalamocortical connections

Blakemore: If it's true that subplate axons play a part in guiding thalamic axons to the subplate compartment, is it possible that local subplate axons within the cortex play a part in guiding thalamic axons when they begin to invade the cortical plate? The local axon collaterals of subplate cells distribute in layer IV, with some extending into layer I—a pattern similar to the final distribution of thalamic axons. There is apparently a preference for thalamic axons to grow on subplate axons, so maybe they grow over the local subplate collaterals to reach their final termination zones.

Ghosh: That's a reasonable suggestion, but there is one reason why it may not be the case. Thalamic axons grow straight towards layer IV and then turn when they reach layer IV. In contrast, the growth of axons from the subplate is not always directed. Their path is more diffuse even though they end up in layer IV.

Molnár: What about the axons that grow towards layer I?

Ghosh: Their growth is also directed.

Molnár: Do you get back-labelled marginal zone cells in the cat if you place carbocyanine dyes into the internal capsule? At early stages we can see some in the rodent (Molnár 1994).

Ghosh: No. There are no marginal zone cells labelled after DiI (1,1′dioctadecyl-3,3,3′,3′-tetramethylindocarbocyanine perchlorate) injections into the internal capsule.

Levitt: Why don't thalamic fibres, which should be able to go to any region of the cortex, grow into the next area of cortex that contains subplate neurons?

Ghosh: We don't know why, in long-term experiments, axons that have grown past their normal target fail to grow into a different target later on.

Levitt: What was the longest number of days that you waited after the lesion?

Ghosh: Thirty-two days, which is long enough for them to have grown into a different target. It is not clear where they could grow in the visual system because area 17 is at the back of the brain and the axons swing around the visual cortex, so there isn't much adjacent cortex to grow into. Although regions of the limbic cortex are nearby, we haven't seen them growing into that part of the cortex. Co-culture experiments have not been done with thalamus and limbic cortex, so it is not known whether it behaves in the same way as neocortex. The auditory system may be a better system to study because medial geniculate nucleus axons that grow past the auditory cortex may have the option of growing into the visual cortex. Our auditory subplate ablation experiments have only gone out to about two weeks, and in that time we did not see them growing into visual cortex.

Blakemore: There may be competition for space in the subplate, so that, if subplate cells are already occupied by their own thalamic axons, there's no room for the others that have abandoned the lesioned region of the subplate.

Bolz: But the first come, first served idea doesn't seem to work because when lesions are made in the subplate zone in a given cortical region, thalamic fibres do not enter this region. They also do not grow into the next available cortex with intact subplate cells, but they grow beyond these regions instead (Ghosh & Shatz 1993).

Molnár: It is important to bear in mind the timing of these events. At E38, when the earliest successful subplate lesions were made, the thalamic fibres have already reached the occipital cortex. Therefore, the issue of long-range guidance cannot be addressed, and we should not dismiss the first come, first served idea completely.

Rakic: What is the evidence that the fibres, even though they're not destroyed, are not affected by kainic acid?

Ghosh: The fibres don't stop or stall at the site of ablation, but they continue to grow.

Rakic: Their growth properties may look unchanged, but they might need specific receptor molecules to make ocular dominance columns.

Ghosh: We cannot rule out the possibility that there may be morphologically undetectable changes in the axons following kainic acid injections. I don't know how good the evidence is for kainic receptors being expressed in presynaptic terminals.

Parnavelas: When I look at electron micrographs of the subplate, I see few synapses. What is the evidence that innervation is occurring?

Ghosh: There are two lines of evidence. Firstly, one can record excitatory postsynaptic potentials in subplate neurons following white matter stimulation in cortical slices (Friauf et al 1990). Secondly, Molnár & Blakemore (1995, this volume) have shown that white matter stimulation leads to increases in intracellular Ca^{2+} in subplate neurons. I don't know why you don't see very many synapses under the electron microscope. It is possible that the synapses are transient because the morphology of axons changes during this period.

Parnavelas: But you should still be able to see transient synapses.

Molnár: There is evidence that in both the dog and the human, synapses are formed in the subplate and the marginal zone (Molliver & Van der Loos 1970, Molliver et al 1973).

Parnavelas: But sometimes what neuroanatomists call the 'subplate' is actually the presumptive layers V and VI.

LaMantia: The subplate is definitely doing something with thalamic axons, but there are also other projections to and from the cortex that are specified regionally. For example, cortical–cortical projections, and subcortical projections to the tectum and to the caudate putamen. Have you looked at any of these other pathways?

Ghosh: We have not looked at whether subplate ablations affect callosal projections, but we have looked at subcortical projections (McConnell et al 1994) because of the pioneer hypothesis and the possibility that subplate neurons might guide the cortical axons down to the thalamus. The results of these experiments are more variable. There is an absence of cortical innervation of the thalamus following subplate ablations in only half of the cases. I'm not sure what it means because it doesn't occur all the time.

O'Leary: We have recently shown that the subplate does not pioneer the corpus callosum, but that it is pioneered instead by a discrete population of cingulate cortical plate cells (Koester & O'Leary 1994). The subplate does not even project through the corpus callosum with the exception of an occasional subplate cell.

Innocenti: We have done experiments similar to those described by Anirvan Ghosh (Innocenti & Berbel 1991a,b) except that our lesions were performed with ibotenic acid and destroyed layers IV, V, VI and part of III, in addition to the subplate. This condition is similar to that of microgyria, a congenital malformation of the human cerebral cortex. We looked at the callosal connections, and we found that their distribution was the same as for intact cortex, but other things were abnormal. The changes immediately after the lesion worried us. There was an abundance of dead neurons and a massive infiltration of microglia in the 'amoeboid' state 24 h after the injection. The astrocytes were wiped out, but regrew in large numbers within one week. One cannot be sure that the deletion of the subplate and the associated cellular changes are not creating obstacles which inhibit the growth of thalamic fibres, rather than actually eliminating guidance cues.

References

Agmon A, Connors BW 1991 Thalamocortical responses of mouse somatosensory (barrel) cortex *in vitro*. Neuroscience 41:365–379

Bernardo KL, Woolsey TA 1987 Axonal trajectories between mouse somatosensory thalamus and cortex. J Comp Neurol 258:542–564

Bolz J, Novak N, Staiger V 1992 Formation of specific afferent connections in organotypic slice cultures from rat visual cortex co-cultured with lateral geniculate nucleus. J Neurosci 12:3054–3070

Crair MC, Molnár Z, Higashi S, Kurotani T, Toyama K 1993 The development of thalamocortical and intracortical connectivity in rat somatosensory 'barrel' cortex imaged by optical recording. Soc Neurosci Abstr 19:702.5

Friauf E, McConnell SK, Shatz CJ 1990 Functional synaptic circuits in the subplate during fetal and early postnatal development of cat visual cortex. J Neurosci 10:2601–2613

Ghosh A, Shatz CJ 1992 Pathfinding and target selection by developing geniculocortical axons. J Neurosci 12:39–55

Ghosh A, Shatz CJ 1993 A role for subplate neurons in the patterning of connections from thalamus to neocortex. Development 117:1031–1047

Ghosh A, Shatz CJ 1994 Segregation of geniculocortical afferents during the critical period: a role for subplate neurons. J Neurosci 14:3862–3880

Goodman CS, Shatz CJ 1993 Developmental mechanisms that generate precise patterns of neuronal connectivity. Cell 72:77–98

Götz M, Novak N, Bastmeyer M, Bolz J 1992 Membrane bound molecules in rat cerebral cortex regulate thalamic innervation. Development 116:507–519

Grinvald A 1984 From growth cones to the intact brain: real-time optical imaging of neuronal activity. Trends Neurosci 16:143–150

Higashi S, Crair MC, Kurotani T, Molnár Z, Toyama K 1993 Imaging neural excitation propagating from thalamus to somatosensory 'barrel' cortex of the rat. Soc Neurosci Abstr 19:49.15

Innocenti GM, Berbel P 1991a Analysis of an experimental cortical network: i) architectonics of visual areas 17 and 18 after neonatal injections of ibotenic acid; similarities with human microgyria. J Neural Transpl Plast 2:1–28

Innocenti GM, Berbel P 1991b Analysis of an experimental cortical network: ii) connections of visual areas 17 and 18 after neonatal injections of ibotenic acid. J Neural Transpl Plast 2:29–54

Koester SE, O'Leary DDM 1994 The corpus callosum is pioneered by early generated neurons of the cingulate cortex. J Neurosci 14:6608–6621

Lund RD, Bunt AH 1976 Prenatal development of central optic pathways in albino rats. J Comp Neurol 165:247–264

McConnell SK, Ghosh A, Shatz CJ 1994 Subplate pioneers and the formation of descending connections from cerebral cortex. J Neurosci 14:1892–1907

Molliver ME, Van der Loos H 1970 The ontogenesis of cortical circuitry: the spatial distribution of synapses in somaesthetic cortex of newborn dog. Ergeb Anat Entwicklungsgesch 42:1–43

Molliver ME, Kostovic I, Van der Loos H 1973 The development of synapses in cerebral cortex of the human fetus. Brain Res 50:403–407

Molnár Z 1994 Multiple mechanisms in the establishment of thalamocortical innervation. DPhil thesis, Oxford University, Oxford, UK

Molnár Z, Blakemore C 1995 Guidance of thalamocortical innervation. In: Development of the cerebral cortex. Wiley, Chichester (Ciba Found Symp 193) p 127–149

Naegele JR, Jhaveri S, Schneider GE 1988 Sharpening of topographical projections ad maturation of geniculocortical axon arbors in the hamster. J Comp Neurol 52:593–607

Nelson SB, Le Vay S 1985 Topographic organization of the optic radiation of the cat. J Comp Neurol 240:322–330

Williams CV, Davenport RW, Dou P, Kater SB 1993 *In vitro* induction of backbranches: implications for an alternate mechanism of target location. Soc Neurosci Abstr 19:449.1

The specificity of interactions between the cortex and the thalamus

Jürgen Bolz, Albrecht Kossel[1] and Dominique Bagnard

INSERM U371 'Cerveau et Vision', 18 avenue du Doyen Lépine, 69500 Bron, France

Abstract. The functioning of the adult mammalian cerebral cortex depends critically upon precise interconnections between specific thalamic nuclei and distinct cortical regions. Therefore, one central issue in understanding cortical development is determining the cellular and molecular strategies underlying the specification of thalamocortical projections. We address the role of axon–axon interactions and membrane-bound guidance molecules in the establishment of connections to and from the cortex. The results of these experiments suggest that the development of layer-specific patterns of afferent and efferent cortical connections does not depend upon neuronal activity. We present evidence that activity conveyed by thalamic afferents is required for the elaboration of the columnar specificity of cortical circuits.

1995 Development of the cerebral cortex. Wiley, Chichester (Ciba Foundation Symposium 193) p173–191

Peripheral influences acting through the thalamus are instrumental in defining the functional architecture of the developing mammalian cerebral cortex. Therefore, an important issue to those interested in the development and plasticity of the cortex is the specification of the projections between the thalamus and the cortex. The synaptic relationships between the cortex and the thalamus are highly ordered: there are reciprocal connections between specific thalamic nuclei and distinct cortical regions, and each projection between a thalamic nucleus and a cortical target area is organized in a topographic fashion within specific cortical layers. To achieve this precise wiring, thalamic axons must grow along specific pathways, select their appropriate cortical target area and then establish synaptic contacts with the correct subset of neurons, their target cells, most of which are located in the cortical layer IV.

---

[1]Present address: Colorado State University, Department of Anatomy and Neurobiology, Fort Collins, CO 80523, USA

Studies of intact rats, cats and primates have revealed the precise sequence of events that leads to the formation of the mature pattern of thalamocortical and corticothalamic connections. For instance, the input from the thalamus reaches the cortex at an early developmental stage, even before the target cells have been generated. Thalamic axons, however, do not innervate the immature cortical plate. They are initially confined to the subplate, the zone beneath the cortical plate, and they invade the cortex only after target layer formation commences (Rakic 1977, Shatz & Luskin 1986, Catalano et al 1991). Neurons that are generated first are contained within the subplate. A subset of these neurons send their axons towards the thalamus long before cells in layer VI of the developing cortical plate, which give rise to the mature corticothalamic projection, have established connections with their thalamic target cells (McConnell et al 1989, 1994, De Carlos & O'Leary 1992). Subplate cells are a transient class of neurons, i.e. many subplate cells disappear after the adult connections have been formed.

These and many other observations have led to a number of proposals on the cellular mechanisms by which specific interconnections between the thalamus and the cortex are achieved during development. For instance, the observation that subplate cells project to the thalamus before cortical plate neurons establish the permanent corticothalamic projection has led to the hypothesis that subplate axons might serve as pioneer fibres for the corticothalamic axons that grow later (McConnell et al 1989, 1994). The outgrowth of thalamocortical and corticothalamic axons occurs in synchrony, and thalamic fibres travelling towards the cortex and subplate fibres travelling towards the thalamus meet in the internal capsule (the gateway between cortical and subcortical structures). These observations suggest that thalamic and cortical fibres grow along each other. The two populations of axons would, thereby, guide each other during the final part of their journey towards their cortical and thalamic targets, respectively (Molnár & Blakemore 1991, Molnár & Blakemore 1995, this volume). However, based on observations made *in vivo*, these and other suggestions are, for the most part, supported only by correlative evidence.

In an attempt to elucidate the cellular and molecular strategies by which specific interconnections between the thalamus and the cortex are established during development, we and other investigators have reconstructed projections between the cortex and the thalamus under *in vitro* conditions (for reviews see Bolz et al 1993, Bolz 1994). These studies have shown that layer-specific and cell-specific thalamocortical and corticothalamic connections are established in organotypic slice cultures that are very similar to those observed *in vivo* (Bolz et al 1990, 1992, Molnár & Blakemore 1991, Yamamoto et al 1989, 1992). By co-culturing cortical and thalamic slices in a plasma gel, which provides a suitable substrate for axonal outgrowth and at the same time stabilizes diffusible substances secreted from the cultured tissue, we have found that the thalamus releases a factor *in vitro* which attracts growing corticothalamic axons at a distance (Bolz et al 1990, Novak & Bolz 1993). Moreover, *in vitro* assays with cortical membranes prepared at different developmental stages or isolated from distinct cortical layers revealed that substrate-bound molecules play an important role in guiding thalamic fibres towards their target and in regulating the innervation of the cortex by thalamic axons at the appropriate time and in the correct cortical layer (Götz et al 1992). In this paper, we give a brief account of our studies on fibre–fibre interactions and other mechanisms that might contribute to specifying the interconnections between the thalamus and the cortex. We shall then consider how, at later developmental stages, interactions between ingrowing thalamic afferents and their cortical target cells lead to the elaboration of distinct cortical circuits.

Specificity of interactions between thalamic and cortical axons

Axonal fasciculation is the growth of axons along the surface of pre-existing fibre tracts (Bray et al 1980). In order to study whether this might play a role in guiding thalamocortical or corticothalamic fibres towards their targets, we are using our co-culture system in combination with time-lapse video microscopy. To test for the specificity of potential interactions between cortical and thalamic axons, we prepared cortex–thalamus and cortex–cortex co-cultures from embryonic rats around the developmental stages when, *in vivo*, the first outgrowing cortical axons meet thalamic axons in the internal capsule. Our time-lapse studies revealed that three different types of reactions occur when thalamic or cortical growth cones encounter another growth cone or an axon: (i) growth cones either continue their growth without changing their morphology or their growth velocity (crossing); (ii) growth cones retract from the axon and collapse (retraction); or (iii) growth cones grow along the axon (fasciculation).

Our preliminary results indicate that the probability of crossing, retraction or fasciculation depends upon the angle at which the growth cone approaches a fibre. Most growth cones cross an axon without any changes in their growth behaviour when they encounter a fibre at an orthogonal angle ($90° \pm 30°$, see Fig. 1). On the other hand, when a thalamic axon grows towards a cortical axon, or vice versa, at an angle of $180° \pm 30°$ (antiparallel, see Fig. 1) or $0° \pm 30°$ (parallel, results not shown), in most cases the growth cone retracts from the fibre. However, if a cortical axon grows towards another cortical axon (either parallel or in the opposite direction), then in the majority of the cases the axons fasciculate (Fig. 1).

The results of these time-lapse studies do not support the idea that thalamic fibres use axons extending from subplate cells as guides towards their cortical targets. Rather, these *in vitro* experiments suggest that growing thalamic and cortical axons tend to segregate from each other. This is consistent with *in vivo*

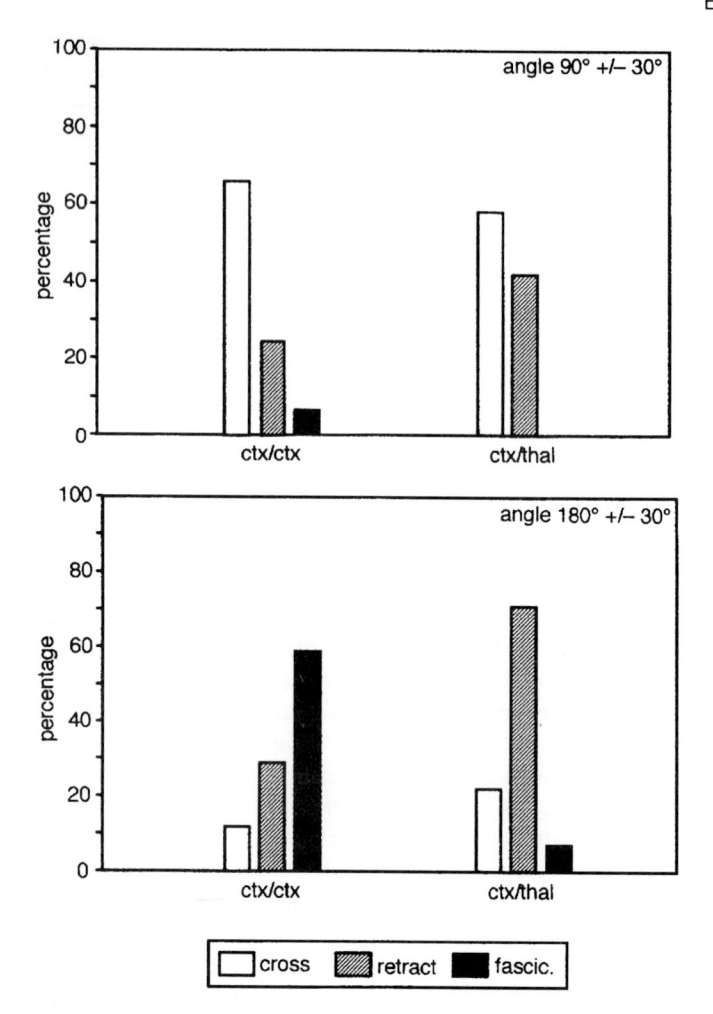

FIG. 1. Reactions of axonal growth cones upon encountering an axon in cortex–cortex (ctx/ctx) and cortex–thalamus (ctx/thal) co-cultures. Top: when growth cones meet an axon at a nearly orthogonal angle, they generally cross the axon without changing their direction or their growth velocity (cross). Bottom: when axons are growing towards each other, the cortical fibres often fasciculate with other cortical fibres (fascic.), whereas most cortical growth cones retract from thalamic axons and most thalamic growth cones retract from cortical axons (retract).

observations which indicate that the trajectories of these two axonal populations are separated within the white matter (Woodward & Coull 1984) and that this segregation is already apparent at early developmental stages (DeCarlos & O'Leary 1992).

On the other hand, our finding that cortical axons tend to fasciculate along each other *in vitro* is consistent with the idea that subplate cells serve as pioneers for the axons of cortical plate neurons that descend later. This notion is further supported by the fact that the fasciculation we observe in culture is selective: cortical axons fasciculate with other cortical axons, but not with thalamic axons. Selective fasciculation has also been observed in other systems where the targeting of growing axons is accomplished with the aid of pioneer cells (Bate 1976, Bastiani et al 1986, Kuwada 1986).

Specificity of interactions between thalamic afferents and cortical target cells

Laminar specificity

The terminal arborizations of thalamic axons within the cortex are highly specialized. The arbors of thalamocortical axons are confined precisely to distinct cortical layers. In higher mammalian species, they may even be confined to sublayers. Within a given layer or sublayer they are often segregated, and thereby they contribute to defining features that characterize a particular cortical area. For instance, in the rodent primary somatosensory cortex, thalamocortical afferents are in register with the cortical barrels (aggregates of layer IV neurons) that form a visible map of the distribution of the sensory hairs on the contralateral body surface (Woolsey & Van der Loos 1970). The primary visual cortex of mammals with frontal vision is another example where thalamocortical afferents are instrumental in defining an important aspect of the functional architecture of the cortex. In these species geniculocortical afferents terminate in layer IV in alternating, eye-specific patches called ocular dominance columns (Hubel & Wiesel 1962).

How is the laminar and columnar patterning of thalamocortical axons achieved during development? The *in vitro* studies mentioned above demonstrate that layer-specific thalamocortical connections can be established in a co-culture system. Moreover, experiments with cortical membranes prepared from different cortical layers provided evidence for a membrane-bound stop signal for thalamic axons in layer IV, their main target layer (Götz et al 1992). These molecules might be located on processes of neurons or glia that pass through this layer, or they might be expressed on the membranes of the target neurons within layer IV. The stop signal for thalamic axons is restricted to layer IV, so it is tempting to speculate that spiny stellate cells (a specific class of neurons characteristic of layer IV that are the major recipients of thalamocortical afferents) carry a surface molecule that causes thalamic axons to cease axonal elongation and to initiate branching and synapse formation. Results obtained in dissociated cell cultures of the cerebellum have already shown that contact-mediated stop signals for afferent mossy fibres are located on the target cells themselves (Baird et al 1992).

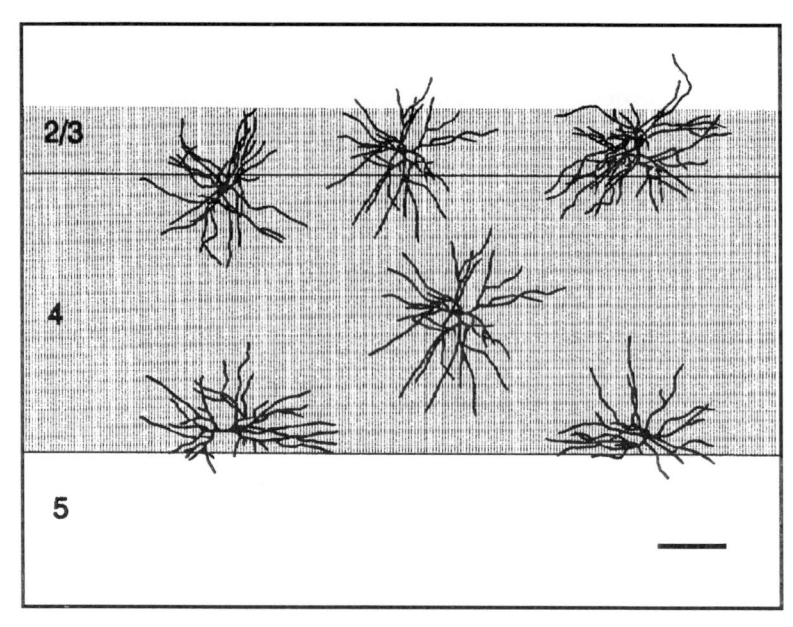

FIG. 2. Dendritic morphology of spiny stellate cells in the primary visual cortex of normal cats. Cells in the middle of layer IV (4) and in the middle of ocular dominance colums have symmetrical dendritic fields. Cells in the vicinity of the upper and lower border of thalamic afferents, however, have asymmetric dendritic fields, and their dendrites are mostly confined to the afferent recipient zone. Note that while the lower border of thalamic afferents corresponds to the border between layer IV and layer V (5), the upper border extends into the deep part of layer III (3) and therefore does not correspond to a laminar border. Bar = 100 μm.

If spiny stellate cells possess a membrane-bound signal that allows growing thalamic axons to recognize their correct target layer, how is the strict laminar stratification of the terminal arborizations of thalamocortical afferents achieved? In the cat primary visual cortex, for instance, the mean dendritic diameter of spiny stellate cells is about 300 μm (Kossel et al 1995). Thus, spiny stellate cells close to the laminar borders could extend their dendrites well beyond 100 μm into the adjacent cortical layer, yet the geniculocortical afferents are sharply confined to their target layer. To tackle this issue, we examined the relationships between the dendritic fields of layer IV cells and the termination pattern of geniculocortical afferents in the primary visual cortex of normal cats and in cats with altered visual experience (Kossel et al 1994, 1995). Thalamic afferents providing the input from the contralateral eye were anterogradely labelled by injecting a fluorescent tracer into lamina A of the lateral geniculate nucleus. Subsequent intracellular staining with Lucifer yellow in slice preparations then allowed simultaneous visualization of the

Thalamocortical interactions

mirror index = n2 - n1 = 7

4

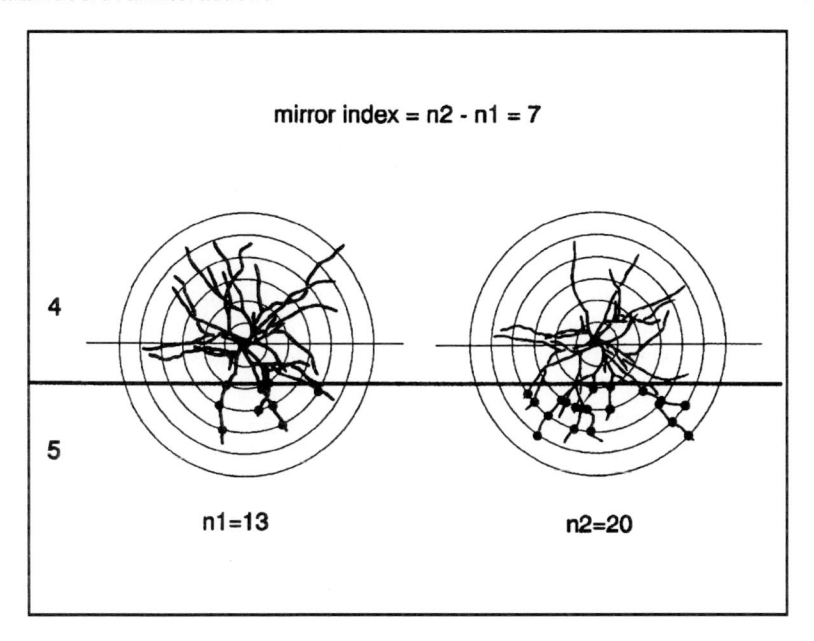

5

n1=13   n2=20

FIG. 3. Quantitative analysis of dendritic asymmetries. Left: the dendritic field was bisected by a line (short thin line) running through the cell body and parallel to the laminar border (long thick line). The number of intersections of all dendrites crossing the border with the concentric circles (spacing 25 μm) was counted (n1). Right: the cell was then mirror imaged at the line running through the cell body and the number of intersections beyond the border was again counted (n2). The difference between this number and the original number was used to calculate a 'mirror index'. A positive mirror index indicates that the cell's dendrites avoided crossing the border. 4, layer IV; 5, layer V.

morphology of individual cells and the patterning of geniculocortical arborizations.

A representative sample of spiny stellate cells in layer IV is illustrated in Fig. 2. All cells in the middle of layer IV had symmetrical dendritic fields relative to the laminar borders. In contrast, cells near the boundary between layer IV and layer V had strongly asymmetric dendritic fields such that most of their dendrites were confined to the thalamic recipient zone. In order to quantify this asymmetry, we drew concentric circles with a spacing of 25 μm around the cell body, and we counted the number of intersections between the circles and the dendrites that crossed the laminar border. The cell was then mirror imaged along a line running through the cell body and parallel to the laminar border, and the number of intersections beyond the border was again counted and compared with the previous value (Fig. 3). The difference between the two values was used to calculate a 'mirror index' (Hübener & Bolz 1992).

A positive mirror index indicates that the cell's dendrites avoided crossing the border between layer IV and layer V, and this was the case for all cells in our sample (Fig. 6). In contrast, the dendrites of spiny stellate cells near the border between layers III and IV crossed this laminar boundary freely (Fig. 2). In the cat primary visual cortex, geniculocortical afferents extend into the deep portion of layer III (LeVay & Gilbert 1976). We visualized the cell's dendritic field and the thalamic afferents simultaneously, and we found that spiny stellate cells located at the top of layer IV and at the bottom of layer III rarely extend their dendrites beyond the termination zone of geniculocortical axons which project into the lower part of layer III (Fig. 2). Thus, spiny stellate cells are a homogeneous class of neurons in that their dendrites are confined to the recipient zone of thalamic afferents.

These results indicate that, in the radial domain, the dendritic fields of spiny stellate cells are restricted regions where geniculocortical axons arborize, and that geniculocortical terminals are strictly confined to the region where spiny cells extend their dendrites. Which is the chicken and which is the egg? Do spiny stellate cells require the presence of thalamic fibres or a certain number of synapses from these afferents in order to extend their dendrites? If, as hypothesized above, molecules bound to dendritic membranes of spiny stellate cells cause the layer-specific termination of thalamic axons, how could thalamic afferents in turn shape the dendritic structure of these neurons? Perhaps dendritic asymmetries of spiny stellate cells near laminar borders are not due to the presence of thalamic axons within layer IV, but are due rather to signals present in the neighbouring layers. There is evidence for laminar differences in neuronal activity between cortical layers during early postnatal development. It has been shown that cells within layer IV of the cat visual cortex become responsive to visual stimuli about two to three weeks before cells in the neighbouring layers III and V exhibit visual responsiveness (Albus & Wolf 1984). These differential activity levels could influence the growth of the dendrites of spiny stellate cells. To test whether different amounts of visually driven activity might have an effect on dendritic fields near laminar borders, we have studied spiny stellate cells in adult cats that had been monocularly deprived by lid suture at the age of three to five weeks. The results obtained from these cats demonstrate that dendritic asymmetries of spiny stellate cells close to the neighbouring layers are similar to those observed in normal cats, irrespective of whether the cells are located in the columns of the open eye or the occluded eye (Fig. 4).

These observations suggest, but do not prove, that the dendritic asymmetries of cells near columnar borders are primarily activity independent. A direct test of this assumption would require an examination of dendritic patterns in a cortex in which activity has been completely blocked. However, the fact that cells in layer IV are the earliest to respond to visual

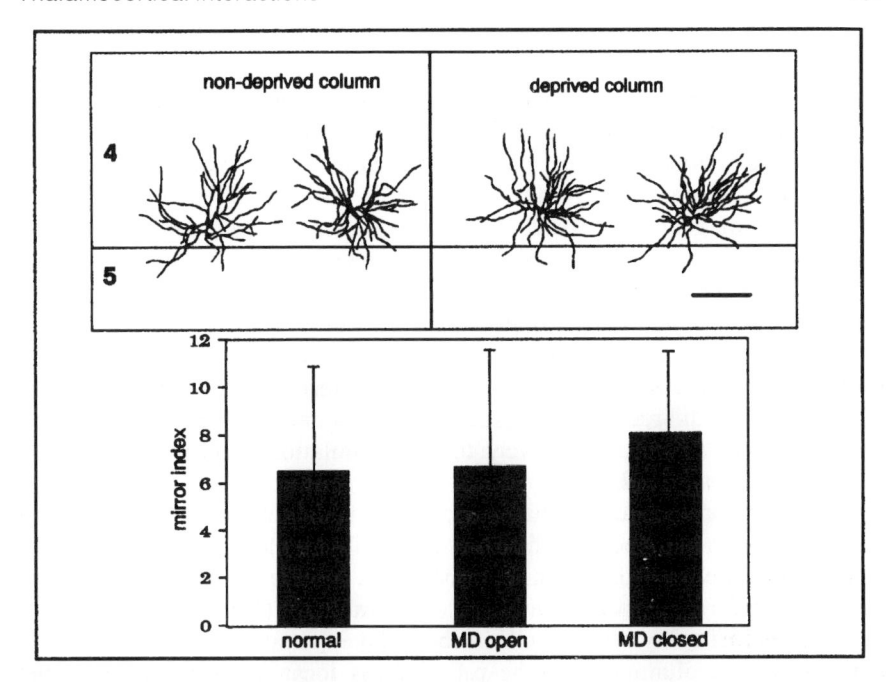

FIG. 4. Top: dendritic fields of spiny stellate cells near the border between layer IV (4) and layer V (5) in monocularly deprived cats. Bottom: quantitative analysis with the mirror index revealed that dendritic asymmetries of spiny stellate cells near laminar borders in monocularly deprived (MD) cats are as strong as in normal animals, independent of whether the cells are located in the column of the deprived (MD closed) or non-deprived (MD open) eye.

stimuli is most likely due to activity relayed by thalamic afferents that have already invaded this layer. Thus, if spiny stellate cells do carry a recognition signal for ingrowing thalamic fibres on the surface of their dendrites, one would not expect that thalamic influences, such as neuronal activity, are the primary cause of dendritic asymmetries near laminar borders. Alternatively, intrinsic cortical cues, independent of neuronal activity, may regulate dendritic growth. For instance, positional cues in distinct cortical layers, located on the surface of cells or within the extracellular matrix, could inhibit and/or stimulate the selective outgrowth of dendrites and thereby shape dendritic fields near laminar borders. Future *in vitro* experiments may clarify the role of layer-specific molecular components in remodelling dendritic fields.

Columnar specificity

Over the last 30 years, the formation of ocular dominance columns has served as a model system for activity-dependent cortical development. During development, the afferents representing the two eyes are initially completely intermixed. Ocular dominance columns emerge gradually as a result of the progressive segregation of geniculocortical axons, which in the cat begins at about two weeks after birth and is completed at about 10 weeks after birth. During this critical period, peripheral manipulations can cause ocular dominance columns to form abnormally. For instance, if one eye is closed, the patches of geniculocortical afferents representing the open eye become much larger than the patches representing the closed eye. As a consequence, most cells in these monocularly deprived cats respond only to visual stimuli presented to the non-deprived eye. On the other hand, if cats are made strabismic, so that the optical axes of the two eyes deviate from each other and the coordinated activity from the two eyes is interrupted, the segregation of geniculocortical afferents into eye-specific patches are more sharply delineated and most cells respond exclusively to the stimulation of only one eye (for reviews see Wiesel 1982, Stryker 1991).

The formation of ocular dominance columns and its perturbation by altered visual experience have been studied in great detail at the level of geniculocortical afferents; however, little is known about the consequences of afferent reorganization on the postsynaptic target cells. Katz et al (1989) have shown that the dendrites of spiny stellate cells in the monkey striate cortex reside preferentially in the column where the cell body is located, suggesting that the dendritic arbors of individual cells in normal monkeys are related to the columnar organization of the cortex. However, they did not determine whether changes in visual experience leading to perturbations in the termination pattern of thalamic afferents also affect the dendritic arborizations of their postsynaptic target cells.

To address this issue, we studied the relationship between ocular dominance columns and the dendritic morphology of spiny stellate cells in the visual cortex of normal, monocularly deprived and strabismic cats (Kossel et al 1994, 1995). In normal cats, the dendrites of spiny stellate cells remained preferentially within one column (Fig. 5). In comparison with dendritic asymmetries observed near laminar borders, the asymmetries near columnar borders were less pronounced (Fig. 6). In strabismic cats, ocular dominance columns were sharper and more precisely demarcated from each other than in normal cats. Spiny stellate cells near these sharp columnar borders had pronounced dendritic asymmetries (Fig. 5) that were as strong as those observed near laminar borders (Fig. 6). On the other hand, monocular deprivation weakened the influence of the columnar borders on the shape of dendritic fields. The dendrites of spiny stellate cells within the columns of the non-deprived eye were confined weakly to these columns. The dendrites of cells in the deprived columns showed the opposite effect: they were extended preferentially into the adjacent columns of the non-deprived eye (Fig. 5). For this group of cells, a quantitative analysis of the

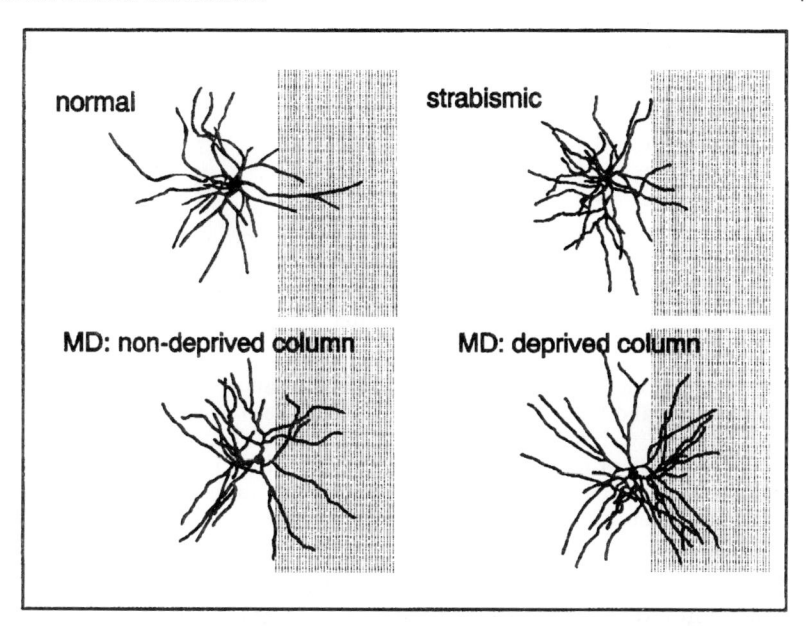

FIG. 5. Spiny stellate cells in normal and visually deprived cats near the border of an ocular dominance column (indicated by shading). Cells in normal cats have fewer dendrites towards the columnar border than away from the border. In strabismic cats these dendritic asymmetries are enhanced. In monocularly deprived (MD) cats the dendrites of cells in the columns of the non-deprived eye tend not to cross into the neighbouring columns of the deprived eye. These asymmetries, however, are not as pronounced as in normal cats. In contrast, cells in the columns of the deprived eye not only freely cross into the neighbouring column, but their dendrites are also biased towards the column of the non-deprived eye.

structure of the dendritic field revealed a negative mirror index (results not shown, see Kossel et al 1995).

The dendritic patterns described above reveal that the segregation of geniculocortical afferents into ocular dominance columns is important in defining the morphology of their cortical target cells. Moreover, different patterns of afferent activity causing different arrangements of thalamocortical axons within layer IV also influences the dendritic morphology of postsynaptic neurons in this layer. A quantitative comparison of dendritic asymmetries in normal and strabismic cats indicated that the degree of correlation of activity relayed by afferents from the left and right eye is an important determinant for the structure of dendritic fields near columnar borders. Finally, our observations of monocularly deprived cats indicated that the dendritic morphology of spiny stellate cells is influenced not only by the relative timing, but also by the relative amount of afferent input.

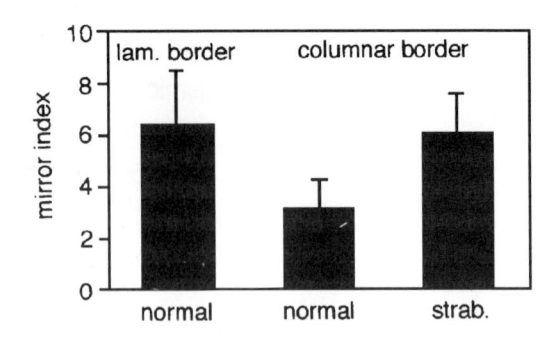

[Bar chart: mirror index for "lam. border" (normal) ≈ 6.5, "columnar border" normal ≈ 3.2, "columnar border" strab. ≈ 6.1]

FIG. 6. Mirror index of spiny stellate cells in normal cats at the laminar (lam.) border between layer IV and layer V (left), at the border between ocular dominance columns (middle), and in strabismic (strab.) cats at the border between ocular dominance columns. Note that in normal cats, dendritic asymmetries near laminar borders are much more pronounced than at columnar borders. In strabismic cats, however, the asymmetries of spiny stellate cells near columnar borders are as strong as asymmetries near laminar borders.

Summary and conclusions

Many lines of evidence suggest that activity-independent mechanisms are important in setting up layer-specific connections between the thalamus and the cortex. In this paper, we emphasize the role of axon–axon interactions and membrane-bound guidance cues in the laminar specification of thalamocortical interconnections. We provide experimental evidence indicating that neuronal activity, acting through thalamic input, is essential for defining the columnar organization of the developing cortex. In line with Hebb's postulate for associative learning (Hebb 1949), physiological studies demonstrated that synaptic contacts in the cortex are enhanced when presynaptic and postsynaptic activity is synchronous, whereas synaptic contacts are weakened when activity is asynchronous (Rauschecker & Singer 1981, Kossel et al 1990, Frégnac et al 1992). These changes in synaptic efficacy have been proposed to lead to a remodelling of the terminal arborizations of thalamocortical axons (LeVay et al 1978, Antonini & Stryker 1993). Our observations suggest that a similar remodelling also occurs for column-specific cortical circuits.

Acknowledgements

We thank Iris Kehrer, Christel Merrouche and Nora Chounlamountri for excellent technical assistance. Thanks also to Siegrid Löwel and Wolf Singer for providing strabismic animals and for stimulating discussions. Mary Behan and Roberto Lent made many helpful comments on the manuscript. This work was supported by the Deutsche Forschungsgemeinschaft.

References

Albus K, Wolf W 1984 Early postnatal development of neuronal function in kitten's visual cortex: a laminar analysis. J Physiol 348:153–185
Antonini A, Stryker MP 1993 Rapid remodeling of axonal arbors in the visual cortex. Science 260:1819–1821
Baird DH, Baptista CA, Wang L-C, Mason CA 1992 Specificity of a target cell-derived stop signal for afferent axonal growth. J Neurobiol 23:579–591
Bastiani MJ, du Lac S, Goodman CS 1986 Guidance of neuronal growth cones in the grasshopper embryo. 1. Recognition of a specific axonal pathway by the pCC neuron. J Neurosci 6:3518–3531
Bate CM 1976 Pioneer neurones in an insect embryo. Nature 260:54–56
Bolz J 1994 Cortical circuitry in a dish. Curr Opin Neurobiol 4:545–549
Bolz J, Novak N, Götz M, Bonhoeffer T 1990 Formation of target-specific neuronal projections in organotypic slice cultures from rat visual cortex. Nature 346:359–362
Bolz J, Novak N, Staiger V 1992 Formation of specific afferent connections in organotypic slice cultures from rat visual cortex cocultured with lateral geniculate nucleus. J Neurosci 12:3054–3070
Bolz J, Götz M, Hübener M, Novak N 1993 Reconstructing cortical connections in a dish. Trends Neurosci 16:310–316
Bray D, Wood P, Bunge RP 1980 Selective fasciculation of nerve fibers in culture. Exp Cell Res 130:241–250
Catalano SM, Robertson RT, Killackey HP 1991 Early ingrowth of thalamocortical afferents to the neocortex of the prenatal rat. Proc Natl Acad Sci USA 88:2999–3003
De Carlos JA, O'Leary DDM 1992 Growth and targeting of subplate axons and establishment of major cortical pathways. J Neurosci 12:1194–1211
Frégnac Y, Shulz D, Thorpe S, Bienenstock E 1992 Cellular analogs of visual cortical epigenesis. I. Plasticity of orientation selectivity. J Neurosci 12:1280–1300
Götz M, Novak N, Bastmeyer M, Bolz J 1992 Membrane bound molecules in rat cerebral cortex regulate thalamic innervation. Development 116:507–519
Hebb DO 1949 The organization of behavior. Wiley, New York
Hubel DH, Wiesel TN 1962 Receptive fields, binocular interaction and functional architecture in the cat's visual cortex. J Physiol 160:106–154
Hübener M, Bolz J 1992 Relationships between dendritic morphology and cytochrome oxidase compartments in monkey striate cortex. J Comp Neurol 324:67–80
Katz LC, Gilbert CD, Wiesel TN 1989 Local circuits and ocular dominance columns in monkey striate cortex. J Neurosci 9:1389–1399
Kossel A, Bonhoeffer T, Bolz J 1990 Non-Hebbian synapses in rat visual cortex. NeuroReport 1:115–118
Kossel A, Löwel S, Bolz J 1994 Plasticity of spiny stellate cells in cat visual cortex of normal and visually deprived animals. Biomed Res 15:11–15
Kossel A, Löwel S, Bolz J 1995 Relationships between dendritic fields and functional architecture in striate cortex of normal and visually deprived cats. J Neurosci 15:3913–3926
Kuwada JY 1986 Cell recognition by neuronal growth cones in a simple vertebrate embryo. Science 233:740–746
LeVay S, Gilbert CD 1976 Laminar patterns of geniculocortical projection in the cat. Brain Res 113:1–19
LeVay S, Stryker MP, Shatz CJ 1978 Ocular dominance columns and their development in layer IV of the cat's visual cortex: a quantitative study. J Comp Neurol 179:223–244

McConnell SK, Ghosh A, Shatz CJ 1989 Subplate neurons pioneer the first axon pathway from the cerebral cortex. Science 245:978–982

McConnell SK, Ghosh A, Shatz CJ 1994 Subplate pioneers and the formation of descending connections from cerebral cortex. J Neurosci 14:1892–1907

Molnár Z, Blakemore C 1991 Lack of regional specificity for connections formed between thalamus and cortex in coculture. Nature 351:475–477

Molnár Z, Blakemore C 1995 Guidance of thalamocortical innervation. In: Development of the cerebral cortex. Wiley, Chichester (Ciba Found Symp 193) p 127–149

Novak N, Bolz J 1993 Formation of specific efferent connections in organotypic slice cultures from rat visual cortex co-cultured with lateral geniculate nucleus and superior colliculus. Eur J Neurosci 5:15–24

Rakic P 1977 Prenatal development of the visual system in rhesus monkey. Philos Trans R Soc Lond Ser B Biol Sci 278:245–260

Rauschecker JP, Singer W 1981 The effects of early visual experience on the cat's visual cortex and their possible explanation by Hebb synapses. J Physiol 310:215–239

Shatz CJ, Luskin MB 1986 The relationship between the geniculocortical afferents and their cortical target cells during development of the cat's primary visual cortex. J Neurosci 6:3655–3668

Stryker MP 1991 Activity-dependent reorganization of afferents in the developing mammalian visual system. In: Lam DMK, Shatz C (eds) Development of the visual system. Massachusetts Institute of Technology, Cambridge, MA, p 267–287

Wiesel TN 1982 Postnatal development of the visual cortex and the influence of environment. Nature 299:583–592

Woodward WR, Coull BM 1984 Localization and organization of geniculocortical and corticofugal fiber tracts within the subcortical white matter. Neuroscience 12:1089–1099

Woolsey TA, Van der Loos H 1970 Structural organization of layer IV in somatosensory region (SI) of mouse cerebral cortex. Description of a cortical field composed of discrete cytoarchitectonic units. Brain Res 17:205–242

Yamamoto N, Kurotani T, Toyama K 1989 Neural connections between the lateral geniculate nucleus and visual cortex *in vitro*. Science 245:192–194

Yamamoto N, Yamada K, Kurotani T, Toyama K 1992 Laminar specificity of extrinsic cortical connections studied in coculture preparations. Neuron 9:217–228

DISCUSSION

Walsh: Do you see different asymmetries depending on whether you look at cells that are innervated by lamina A or lamina A1?

Bolz: We have only injected tracers in lamina A. However, we have found that there are no significant differences in dendritic asymmetries between cells located in the labelled columns and cells located in the unlabelled columns.

Bonhoeffer: You can't take it for granted that unlabelled columns represent the other input. This needs to be demonstrated by injecting both lamina A and A1 in the lateral geniculate nucleus, which would show that the labelling does

Thalamocortical interactions

not overlap and that unlabelled zones do not exist. This experiment has been reported by Hata & Stryker (1994).

Bolz: Yes, to some extent that might be true. The columns from the two eyes might overlap partially. However, although our quantification of dendritic asymmetries with the mirror index does depend on the exact location of the columnar borders, the quantification with the bias index is independent of the exact location of the borders. Therefore, the numbers would not change if there was some overlap between the columns from the two eyes (Kossel et al 1995). This probably does not pose a major problem for the interpretation of our experiments because we get the same results using both types of quantification.

Innocenti: The dendritic polarization results are consistent with what has been seen in the somatosensory system (Steffen & van der Loos 1980), where neurons residing in the barrel walls have their dendrites pointing towards the centre of the barrel. If the vibrissae are removed early, the barrels develop into a different configuration. There is evidence that this occurs simultaneously with the reorganization of thalamocortical input and leads to the reorganization of dendrites of neurons in the wall of the abnormal barrels.

Bolz: Yes, the visual system parallels the somatosensory system in this respect. However, in the visual system one can distinguish between what happens when either the levels of activity or the patterns of activity are changed.

Innocenti: Nevertheless, in both situations one is dealing with a trophic signal for dendrites, which may relate to the presence of the thalamocortical axons, to their activity or to both.

Bolz: In the case of strabismic rats, there are equal amounts of activity in both the left and right eyes. The only difference is that the activities from the two eyes are no longer correlated with each other.

Ghosh: Is your conclusion that the bias of the spiny stellate dendrites towards layers IV and V is not related to activity?

Bolz: This is our hypothesis.

Ghosh: In the monocular deprivation experiments, the axons grow initially into layer IV. At this stage the axons are intermixed, so if one eye is closed, the other eye is effective at driving all of the cells in layer IV. If you showed that an early binocular tetrodotoxin injection did not affect the dendrites, then this would be more compatible with your hypothesis.

Molnár: What is the major initial target of the thalamocortical projections in layer IV? Is it the layer IV segment of layer V cells?

Bolz: It's possible. There is evidence that pyramidal cells are contacted by thalamic afferents and do not respect the thalamic borders, so cell type specificity may be involved, but at the moment these are purely speculations.

Rubenstein: After you ablated the subplate and prevented innervation of the overlying layer IV, did you look at the shape of stellate cells?

Ghosh: Larry Katz (personal communication) has some preliminary results on this. He labelled some cells intracellularly in vibratome slices from subplate-ablated cortex, and he did not observe any major morphological changes in layer IV cells.

Bolz: I predict that spiny stellate cells in layer IV would still exhibit dendritic asymmetries near the border of layer V in subplate-ablated cats because, according to our hypothesis, these asymmetries are due to positional cues in layer V. Actually, this might be a good test for our hypothesis.

O'Leary: Even though some dendrites cross an eye-specific border within layer IV, are they still innervated by inputs driven by the same eye as those that innervate the cell body and the rest of the dendritic arbor? In other words, if you record from layer IV do you get any binocular driven cells, or are they all monocular?

Bolz: There is some overlap between the afferents serving the left and the right eyes. Moreover, in monocularly deprived cats recent studies based on intracellular labelling of individual geniculocortical axons indicated that the terminal arborizations representing the deprived eye are smaller, less dense and have smaller synaptic boutons than those representing the non-deprived eye (Friedlander et al 1991, Antonini & Stryker 1993). Therefore, afferents serving the deprived eye, compared to the non-deprived eye, are less effective in driving cortical target cells. Thus, even though the dendrites from some cells in the columns of the non-deprived eye may cross freely into the columns of the deprived eye, they receive only a weak synaptic input from this eye. The response of these cells is still dominated by the input from the non-deprived eye. The situation is different for cells that are situated in the columns of the deprived eye but are close to a columnar border. Because of the reduced efficacy in synaptic transmission of the afferents serving the deprived eye, the responses of these cells are most likely dominated by the input from the non-deprived eye. Thus, in monocularly deprived cats, ocular dominance columns visualized by anterograde labelling of geniculocortical afferents might not coincide with the physiologically defined grouping of cells according to eye preference. The extent of cortical regions dominated by the input from the deprived eye is smaller than the anatomical distribution of afferents serving the deprived eye. In fact, Shatz & Stryker (1978) have combined anterograde tracing with electrophysiological recordings and they have found that when the electrode entered the columns labelled from the deprived eye, the cells were still dominated by the input from the non-deprived eye.

Blakemore: About 40–50% of layer IV cells in the normal cat are binocularly driven. They may not be spiny stellates, but they may be pyramidal cells with local excitatory interconnections that link neurons in different ocular dominance domains, hence creating binoculariity.

Bolz: We found that 25% of the cells in layer IV have the distinct morphology of pyramidal cells.

LaMantia: The thalamic afferents are confined to the barrel hollow region. Afferents and the dendrites of the cells that are innervated by those afferents are often clustered together in the barrel, for example (Senft & Woolsey 1991). We showed in the olfactory bulb that this also occurs when glomeruli are made (LaMantia & Purves 1989, LaMantia et al 1992). Could you speculate on the relationship between afferent elaboration and dendritic differentiation both in the laminar dimension and in the ocular dominance columns?

Bolz: I imagine that the afferents first generate dendritic asymmetries via an activity-independent mechanism. A substrate-bound factor cannot be involved in the dendritic asymmetries at columnar borders because these borders are dynamic and depend on the setting up of initial differences. Spiny cell dendrites change their shape at columnar borders in response to manipulations of the visual inputs, suggesting that they follow an activity-dependent mechanism.

LaMantia: Therefore, in one case it's a specific substrate-bound constraint, and in the other case it's an activity-dependent event. How is this constraint established? Because in both cases you end up restricting where the dendrites grow.

Bolz: Growing dendrites can respond to substrate interactions at laminar borders or to activities that lead to the sharp segregation of the thalamic afferents into ocular dominance columns.

Blakemore: But neither mechanism can be responsible for the basic fixed periodic pattern in which thalamic afferents segregate. This pattern is aligned with the pattern of cytochrome oxidase blobs (the concentrations of high cytochrome oxidase activity in the superficial layer of the cortex), which can be detected before the segregation in layer II. It seems to me that there must be some inherent property in the cortex that determines the positions of ocular dominance columns in layer IV, the boundaries of which can vary through competitive interaction between right-eye and left-eye axons on the basis of their relative activity. It is possible that the underlying periodicity of the columns is related to some periodically distributed, membrane-bound property of cortical cells in layer IV.

Bolz: I can only say that there are dynamic changes in both dendrites and activity at the border.

Maffei: Do you envisage any physiological significances of this asymmetry?

Bolz: The asymmetry supports the physiological results. If strabismus is induced, almost all the cells are monocular. How do spiny stellate cells in layer IV become monocular if they have dendrites that cross freely into the columns of the left and right eyes? If they were promiscuous, it would be difficult to reconcile this with the strong monocularity of the physiological responses.

Jones: It is possible that the specificity of this effect for spiny stellate cells is not as great as you suggest. One of the populations of pyramidal cells that you injected was the upper row of large cells that ignore the lamina boundary. The lower row of cells are the ones that Lorente de Nó (1949) calls the star pyramids. If you did your analysis on these lower cells, you would probably find that they have more dendritic intersections in the termination zone of thalamic afferents than they have in layer V. In areas of cortex where there are star pyramid cells and no spiny stellate cells, namely the motor cortex and areas 1 and 2 of the somatosensory cortex in primates, the thalamic termination zones extend from the superficial layers to the deep layers. It is possible that the modulation of dendritic form, which extends beyond the spiny stellate cells, affects other cells as well.

Bolz: This raises the issue as to whether pyramidal cells, star pyramids and spiny stellate cells represent a continuum. We defined cells as pyramidal on the basis of having a clear apical dendrite that was twice as long as the basal dendrites. There were cells that had an intermediate morphology, but we did not take these into account because they represented only 12% of the cell population. Large pyramidal cells define the laminar border between layers III and IV, but most pyramidal cells in layer IV are small. The average size of the basal fields of pyramidal cells in layer IV was the same as the average size of the dendritic fields of spiny stellate cells.

Jones: There's no question about that. But I would not feel comfortable in saying that spiny stellate cells, star pyramids and pyramidal cells form a continuum. I believe that these star pyramids are spiny stellate cells that have actually extended their dendritic configurations following the extended distribution of thalamic afferents.

Maffei: What happens to the dendrites that are in the active column, do they withdraw?

Bolz: The dendrites of cells in the non-deprived columns withdraw from the deprived columns, but this withdrawal is not as strong as in normal cats.

Blakemore: You define the boundaries of ocular dominance columns by the distribution of afferent axons. The territory occupied by a deprived eye column becomes much smaller than it normally would be, so its new borders correspond to a region closer to the middle of what would have been the normal full-width column. Therefore, the stellate cells near the boundary of a deprived eye column would originally have been destined for closer to the middle of the column. You concluded that the asymmetrical distribution of dendrites at the border of the deprived eye column is actively formed on the basis of activity in the afferent axons. However, it is conceivable that these cells, which would not normally have been at a border region, have asymmetrical dendrites even in normal cats, and that no active change in dendritic morphology has occurred.

Bolz: But you would then get symmetrical trees, which we do not observe.

Blakemore: Yes, because they would have been near the middle of the original column. Have you looked at the cells near the middle of normal columns?

Bolz: Yes, they are symmetrical.

Blakemore: Do you know anything about the temporal relationship between the redistribution of afferents and the redistribution of dendrites?

Bolz: No, we have not studied this.

LaMantia: Pasko Rakic (1972) showed that the morphology of stellates in the molecular layer of the cerebellum varies, depending on when the axons are added in the pyramidal layer. In addition, Antonini & Stryker (1993) have shown that axon arbors are elaborated within columns as the dendrites are growing. Therefore, it is possible that dendritic morphology is restricted as axons are added.

References

Antonini A, Stryker MP 1993 Development of individual geniculocortical arbors in cat striate cortex and effects of binocular impulse blockade. J Neurosci 13:3549–3573

Antonini A, Stryker MP 1993 Rapid remodeling of axonal arbors in the visual cortex. Science 260:1819–1821

Friedlander MJ, Martin KAC, Wassenhove-McCarthy D 1991 Effects of monocular visual deprivation on geniculocortical innervation of area 18 in cats. J Neurosci 11:3268–3288

Hata Y, Stryker MP 1994 Control of thalamocortical afferent rearrangement by postsynaptic activity in developing visual cortex. Science 265:1732–1735

Kossel A, Lowel S, Bolz J 1995 Relationships between dendritic fields and functional architecture in striate cortex of normal and visually deprived cats. J Neurosci 15:3513–3926

LaMantia A-S, Purves D 1989 Development of glomerular pattern observed in living mice. Nature 341:646–649

LaMantia A-S, Pomeroy SL, Purves D 1992 Vital imaging of glomerular patterns in the olfactory bulbs of developing and adult mice. J Neurosci 12:976–988

Lorente de Nó R 1949 Cerebral cortex: architecture, intracortical connections. In: Fulton JF (ed) Physiology of the nervous system. Oxford University Press, New York p 288–315

Rakic P 1972 Extrinsic cytological determinants of basket and stellate cell dendritic pattern in cerebellar molecular layer. J Comp Neurol 146:335

Shatz CJ, Stryker MP 1978 Ocular dominance in layer IV of the cat's visual cortex and the effects of monocular deprivation. J Physiol 281:267–283

Steffen H, van der Loos H 1980 Early lesions of mouse vibrissal follicles: their influence on dendrite orientation in the cortical barrelfield. Exp Brain Res 40:419–431

General discussion III

The handshake hypothesis

Molnár: It's important to use the appropriate surface for the experiments that Jürgen Bolz described (Bolz et al 1995, this volume). An analogous situation is when two people behave differently in the middle of the desert and at a cocktail party where they have other people to talk to. If two axons are in a hostile environment, for example in the internal capsule or the striatum, they associate closely with each other. However, it is possible that they exhibit a different behavioural pattern on a highly permissive surface (like polylysine and laminin in the culture dish).

Rakic: It depends on whether you think the desert or the cocktail party is the hostile environment!

Bolz: This is a good point. The substrates are important because if a particular substrate is absent, an axon might associate with the next best thing. However, we must not disregard differential effects. When we observe fasciculation, it is mostly between cortical axons, possibly because it is a hostile environment for axonal growth. Thalamic axons do not fasciculate with cortical axons. Sometimes they fasciculate with other thalamic axons, but usually they continue to grow on the substrate that was used in these co-culture experiments, namely laminin with or without postnatal cortical membranes.

Molnár: It is conceivable that different sets of axons respond to different cues. For instance, thalamic axons might find the subplate and their axons more permissive than the corticofugal projections from the cortical plate. We have to distinguish between early corticofugal or subplate projections and true cortical plate corticofugal projections because corticofugal projections from the cortical plate may not mix with the ingrowing thalamic fibres.

Bolz: This point is a valid one. However, thalamic axons sometimes fasciculate with other thalamic axons, whereas cortical and thalamic axons almost never fasciculate with each other, yet both grow on the same substrate.

Blakemore: The age of the cortex is critical. The only postmitotic neurons in cortical explants taken at embryonic day (E) 14 are likely to be preplate cells but those from older fetuses will contain plate neurons of layer VI and even V, whose axons would be expected to behave differently. You have looked at the combined results from all ages.

Bolz: We have combined the results for both E15 and E16. If one analyses the results separately for each age group, one gets basically the same results (we define the day of sperm detection as E 1).

Blakemore: This is important because as soon as layer VI cells arrive in the cortical plate, and perhaps even during migration, they spin out their axons. The axons of layer VI cells don't interact with thalamic axons *in vivo*, so it would not be surprising for them to avoid each other *in vitro*. I have another point to make. Superficially, your results (if they do apply specifically to preplate and thalamic axons) seem to argue against the hypothesis that there is a handshake between the two axon systems *in vivo*. However, there is already evidence that these two axon systems do not interact normally *in vitro*. Nobody has succeeded in replicating the waiting period in co-culture, suggesting that the mechanism of interaction that leads to the waiting period *in vivo* is missing in explant co-culture. It is possible that this missing ingredient is the handshake between the growth cones of thalamic and preplate axons that occurs specifically in the environment of the ventral telencephalon. Co-culture experiments may fail to provide some crucial property for the handshake to occur.

Bolz: Cortex/thalamus co-cultures, after about five days *in vitro*, show some intermixing of outgrowing axons. However, if we look at the fibres with time-lapse video microscopy over time, we do not see a handshake between cortical and thalamic axons.

Blakemore: But we don't know the exact time at which they co-fasciculate *in vivo*. We presume that thalamic axons have actually fasciculated with the preplate scaffold before they reach the cortical plate, but it could occur later, following initial simple intermingling without close contact. Have you looked at axons in co-culture after they've crossed each other to see whether there's any tendency for them to fasciculate some time after they have simply intermingled?

Bolz: Initially, in our *in vitro* experiments, the probability of having two axons growing directly towards each other was relatively low. After two days *in vitro*, there are many axons extending from the explants, and it is difficult to analyse the individual trajectories of hundreds of axons. Therefore, up to now, we do not know what is happening when axons meet a second time after they have crossed each other.

Ghosh: Jürgen Bolz's observations are consistent with the behaviour of axons within the cerebral wall, in that axons of the optic radiations and cortical efferents travel in adjacent zones (Bolz et al 1995, this volume). But what is the situation in the internal capsule? Dennis O'Leary, I believe you have studied this. Would you like to comment?

O'Leary: I agree with Zoltán Molnár and Colin Blakemore. Our observations suggest that the afferents and efferents interact with each other in the internal capsule. However, once they get into the cortical wall, they probably take independent pathways (De Carlos & O'Leary 1992, O'Leary & Koester 1993, Bicknese et al 1994).

Bolz: Were these observations *in vivo* or *in vitro*?

O'Leary: *In vivo*.

Molnár: Dennis O'Leary, the title of one of your papers is 'Thalamic axons preferentially extend along a chondroitin sulfate proteoglycan-enriched pathway coincident with the neocortical subplate and distinct from the efferent path' (Bicknese et al 1994). Where is this pathway?

O'Leary: Early in cortical development, when the preplate sits on top of the neuroepithelium, as preplate axons extend, the majority dive out of the preplate and grow through the upper neuroepithelium, which later becomes the intermediate zone. Later, as the cortical plate develops, cortical plate neurons migrate into the preplate, splitting it into the subplate and marginal zone. Efferents growing out of the cortical plate take up an intracortical path that is deep to the subplate; thus both subplate axons and layer V and VI efferents grow along a path deep to the subplate. In contrast, thalamocortical axons take a path that is centred on the subplate as they grow intracortically back to their appropriate target area.

Molnár: Do subplate projections express chondroitin sulfate proteoglycan?

O'Leary: No. However, both the carbohydrate epitopes and the core proteins of chondroitin sulfate proteoglycans are localized selectively to the subplate layer itself.

Rakic: The internal capsule in man is large enough to distinguish between areas containing thalamocortical projections and corticothalamic projections. If they intermingle as they develop, do they separate later?

Molnár: It is often not made clear that corticofugal projections from layer VI are different from the early corticofugal projections from the preplate. This may lead to misunderstandings when talking about thalamocortical and corticofugal projections. I'm not arguing that thalamocortical projections grow within the bundles of layer VI projections. Unfortunately, most studies published in this field have not distinguished the subplate projections from layer VI projections.

Rakic: What about the areas where the subplate does not project to the thalamus, as reported recently for the ferret by Clasca et al (1994)?

Molnár: According to our results in the rat and mouse (Molnár et al 1993, Molnár 1994), the subplate projections probably do not enter deep into the thalamus even from the putative visual and somatosensory cortices, but stop short and may accumulate just outside in the neighbouring regions. It is not known where and how these early subplate projections terminate.

J. Price: There is a possible problem of selectivity in Jürgen Bolz's culture experiments. He presumably dissociates the entire cortex and plates it, so one could imagine that a minority of these cells are subplate cells. The youngest neurons of an embryo at around E15 would probably survive the best in culture. Jürgen, in terms of the handshake hypothesis, wouldn't you expect the majority of cortical neurons to ignore thalamic afferents?

Bolz: If we do the experiments with cortex/thalamus co-cultures, the fibres retract. If we use the same experimental paradigm with cortex/cortex co-cultures, the fibres fasciculate. Therefore, there is selectivity because the growth cones in this *in vitro* system can distinguish between cortical and thalamic axons.

J. Price: Certainly, some factors will be present that make certain cells fasciculate with others, but I am questioning whether you're testing the specific populations of cells which you would expect to shake hands. If subplate cells are present, they could be influencing your results.

Kennedy: This is an important point because subplate cells are not defined by their laminar distribution, but as the first cells to be generated.

J. Price: Jürgen Bolz hasn't just dissociated subplate cells, he has dissociated the whole of the cortex.

Bolz: We did not dissociate the cortex, we prepared explants which do not undergo neurogenesis in our *in vitro* conditions (Götz & Bolz 1992). The E15 cortical explants therefore consist of subplate cells, whereas the E16 explants consist of subplate cells and layer VI neurons.

Ghosh: I would also like to comment on this point because it also affects the pioneer hypothesis. If fasciculation allows the matching of thalamic axons with the right cortical axons, there must be differences between the regions. If one studies the whole population, one might not be able to see these differences between the thalamus and the cortex. Did you see any differences?

Bolz: There may be regional differences, but we have not checked this so far.

Ghosh: It is possible that all the cortical axons grow towards the internal capsule and continue to fasciculate with each other as they grow towards their targets. However, fasciculation itself may not be important if thalamic axons recognize some other feature of subplate neurons. When an axon starts to invade the telencephalon, the growth cone morphology suggests that it can interact both with the cell bodies of subplate cells as well as with their axons. Although the evidence is not good enough to support one interaction over another, specificity must exist because these axons end up in different regions.

Bolz: According to the original handshake hypothesis (Molnár & Blakemore 1991), the sequence of outgrowth (chronotopy), but not the regional identity (positional information), should be the key for the formation of specific interconnections between cortex and thalamus.

LaMantia: Subplate axons grow towards the thalamus. Axons fasciculate and follow a pioneer, then this pioneer tells the axons to go the other way, so that the subplate pioneer grows away from the cortical plate, whereas the thalamic axons grow towards it. Consequently, another signal is required to establish that directionality. If this signal is absent, the thalamic axons would follow the subplate pioneer, away from the cortical plate. Why don't the thalamic axons follow the pioneer?

O'Leary: Because there are no polarity cues for fasciculation. Once fasciculation is established, axons simply continue to extend along their counterparts.

LaMantia: Then there should be an equal number of thalamic axons growing towards the cortical plate and growing away from it.

Bolz: I showed that cortical axons fasciculate when they are travelling in the same or in opposite directions. Fasciculation is stronger when they are going the same way, but it is also observed when they are going the opposite way.

LaMantia: But why don't you see this *in vivo*?

Bolz: Because they have no reason to do this *in vivo*. Pyramidal cells have a strict polarity. Their apical dendrite goes to the pial surface and their main axon is directed towards the white matter. Consequently, it is almost impossible to get an antiparallel situation *in vivo*, i.e. a situation where two cortical axons grow towards each other in opposite directions.

Ghosh: Polarity is important, but it's not really relevant to the fasciculation problem. As the axons from the cortex dive down, they turn and grow towards the thalamus. We do not understand why they grow in that direction or how the surfaces of the neuroepithelium are involved, but the cues do not have to rely on other axons expressing a polarity signal.

Blakemore: Axons do grow in different directions on pioneer scaffolds in other species; for example, in the grasshopper (Bastiani et al 1986). In mammals both thalamic axons and subplate axons grow unerringly, in opposite directions, towards the internal capsule, so some form of directional cue must be involved. It might depend partly on a simple trophic influence. We don't know if there are constraints on diffusion in the developing brain. It is possible that the internal capsule acts as a diffusion channel between telencephalon and diencephalon, establishing a gradient for trophic influences of thalamus on cortex and cortex on thalamus, and attracting both fibre systems towards the internal capsule.

Molnár: This picture is probably oversimplified because there are other transient cell populations near the thalamus, in the perireticular region and the thalamic reticular nuclei (Mitrofanis 1992), whose function is not understood. However, these early generated cells may all have a role in the guidance of thalamocortical and corticothalamic projections (see Mitrofanis & Guillery 1993).

Walsh: The perireticular cells are interesting because they may represent a 'subplate' for the thalamus (Mitrofanis & Guillery 1993). Do they form a precocious connection to the cortex from the thalamus? And can they be retrogradely labelled from the cortex?

Molnár: Mitrofanis & Baker (1993) discovered that after DiI (1,1'diocta-decyl-3,3,3',3'-tetramethylindocarbocyanine perchlorate) crystals were placed into the rat dorsal thalamus as early as E14 or E15, thalamic reticular cells were back labelled indicating that their projection is the first one to reach the thalamus. Mitrofanis & Guillery (1993) raised the possibility that these cells may guide thalamic fibres from the thalamus, although they were quite cautious in their suggestions for actual mechanisms. It is possible that these perireticular and thalamic reticular cells are involved in the guidance of the early thalamocortical projections. However, I believe that after a localized DiI crystal placement, the labelled cells in the thalamic reticular nucleus lie slightly more anterior to the fibre pathway labelled from the same point. The perireticular cells also project to the cortex (Adams et al 1993) and to the thalamus (Mitrofanis & Guillery 1993, Molnár 1994), so they are in a strategic position to assist in the development of these pathways. It would be interesting to determine whether these cells and the subplate cells have a common evolutionary origin. They are very similar in many ways. They are among the first generated cells in the forebrain, the majority of them are transient and they have similar immunoreactivities for γ-aminobutyric acid, neuropeptide Y and somatostatin (see Mitrofanis & Guillery 1993).

Daw: Between the thalamus and the cortex, there is an inversion of fibres in one dimension, but not in the other dimension (Nelson & LeVay 1985). Do you agree that this cannot be explained by contact guidance? And do you have an alternative explanation?

Blakemore: This could be caused by the migration and rotation of the lateral geniculate nucleus (LGN) after its connections to the occipital cortex are formed. In 'higher' mammals the geniculate moves ventrally from the dorsal thalamus, rotating and presumably dragging the fibres around with it.

Molnár: Conolly and Van Essen (1984), in their analysis of the visual field representation in the dorsal LGN (dLGN) and the visual cortex of the macaque monkey, proposed that there has to be a rotation in the optic radiation along the mediolateral axis, but not along the anteroposterior axis. This work was followed up by Nelson and Le Vay (1985), who injected pairs of different tracers into the dLGN in the adult cat and followed the fibre order. They observed that the differentially labelled optic fibres crossed each other in one dimension (mediolaterally) only about 200–500 μm below the cortical plate and this region could have been the subplate during development. This reversal appears after the thalamic fibres have reached close to the cortex and, therefore, I see no reason why the reversal at the end of the thalamic pathway would challenge the idea of contact guidance for the rest of the pathway. The initial thalamic fibre deployment can be quite different from the topography that is observed later in the adult suggesting that the initial polarity of the representation can change. The side branches developed from the thalamic axons during the waiting period (Naegele et al 1988, Ghosh & Shatz 1992) might form the anatomical substrate for this final reversal. I suspect that the kind of reversal Nigel Daw was referring to is quite general. The whole body surface is represented as a sensory homunculus in the primary somatosensory cortex. In humans this map is fragmented in an interesting way that is different from the motor homunculus (see Szentágothai 1977). It would be interesting to follow the fibre ordering from the somatosensory thalamus, where this fragmentation occurs, and to determine whether the polarity changes are similar to the Nelson and Le Vay (1985) results. It is possible that the findings in rodents can be generalized (see Blakemore & Molnár 1990, Molnár 1994). The initial deployment and layout of thalamic axons is homogeneous and regular in embryonic life but it is then substantially altered and becomes fragmented near the termination site. However, the fibre ordering within the rest of the pathway may preserve the simple juvenile topography.

References

Adams NC, Baker GE, Guillery RW 1993 Perireticular cell bodies are labelled following DiI or HRP injections into the cortex of perinatal rats. Soc Neurosci Abstr 19:29.2
Bastiani MJ, du Lac S, Goodman CS 1986 Guidance of neuronal growth cones in the grasshopper embryo. I. Recognition of a specific axonal pathway by the pCC neuron. J Neurosci 6:3518–3531
Bicknese AR, Sheppard A, O'Leary DM, Pearlman AL 1994 Thalamocortical axons extend along a chondroitin sulfate proteoglycan-enriched pathway coincident with the neocortical subplate and distinct from the efferent path. J Neurosci 14:3500–3510
Blakemore C, Molnár Z 1990 Factors involved in the establishment of specific interactions between thalamus and cerebral cortex. Cold Spring Harbor Symp Quant Biol 55:491–504
Bolz J, Kossel A, Bagnard D 1995 The specificity of interactions between the cortex and the thalamus. In: Development of the cerebral cortex. Wiley, Chichester (Ciba Found Symp 193) p 173–191
Clasca F, Angclucci A, Sur M 1994 Layer V neurons establish the first projection to the dorsal thalamus. Soc Neurosci Abstr 20:214
Conolly M, Van Essen D 1984 The representation of the visual field in parvicellular and magnocellular layers of the lateral geniculate nucleus in the macque monkey. J Comp Neurol 226:544–564
De Carlos JA, O'Leary DDM 1992 Growth and targeting of subplate axons and establishment of major cortical pathways. J Neurosci 12:1194–1211
Ghosh A, Shatz CJ 1992 Pathfinding and target selection by developing geniculocortical axons. J Neurosci 12:39–55
Götz M, Bolz J 1992 Preservation and formation of cortical layers in slice cultures. J Neurobiol 23:783–802
Mitrofanis J 1992 Patterns of antigenic expression in the thalamic reticular nucleus of developing rats. J Comp Neurol 320:161–181
Mitrofanis J, Baker GE 1993 Development of the thalamic reticular and perireticular nuclei in rats and their relationship to the course of growing corticofugal and corticopetal axons. J Comp Neurol 338:575–587
Mitrofanis J, Guillery RW 1993 New views on the thalamic reticular nucleus in the adult and developing brain. Trends Neurosci 16:240–244
Molnár Z 1994 Multiple mechanisms in the establishment of thalamocortical innervation. DPhil thesis, Oxford University, Oxford, UK
Molnár Z, Blakemore C 1991 Lack of regional specificity for connections formed between thalamus and cortex in co-culture. Nature 351:475–477
Molnár Z, Adams R, Blakemore C 1993 3-D confocal microscopic study on the early development of thalamocortical projections in the rat. Soc Neurosci Abstr 19:450.16

Naegele JR, Jhaveri S, Schneider GE 1988 Sharpening of topographical projections and maturation of geniculocortical axon arbors in the hamster. J Comp Neurol 52:593–607

Nelson SB, LeVay S 1985 Topographic organization of the optic radiation of the cat. J Comp Neurol 240:322–330

O'Leary DDM, Koester SE 1993 Development of projection neuron types, axonal pathways and patterned connections of the mammalian cortex. Neuron 10:991–1006

Szentágothai J 1977 Functionalis Anatomia. Az ember anatomiája, fejlödéstana, szövettana és tájanatómiája. III. kötet. Harmadik, javított kiadás. Medicina Könyvkiado, Budapest

Molecular contributions to cerebral cortical specification

Pat Levitt, Raymond Ferri* and Kathie Eagleson

*Department of Neuroscience and Cell Biology, Robert Wood Johnson Medical School–UMDNJ, 675 Hoes Lane, Piscataway, NJ 08854 and *Department of Anatomy and Neurobiology, The Medical College of Pennsylvania, Philadelphia, PA 19129, USA*

Abstract. Evidence is accumulating that decisions of cell fate and commitment to specific regional phenotypes in the cerebral cortex occur through cell interactions that likely begin early in development, perhaps in the proliferative zone. We have focused on the development of the limbic cortex in rats, which includes areas involved in both cognitive and autonomic functions and is marked by expression of the limbic system-associated membrane protein. Transplantation studies show that precursor cells are sensitive to environmental cues which can control expression of area-specific phenotypes, including limbic system-associated protein synthesis and connectivity patterns, but early postmitotic neurons faithfully express traits based on their origin in the donor. We have studied this sensitive period of decision making *in vitro*. Molecules from the epidermal growth factor family influence dramatically the fate of precursor cells, but only in the presence of matrix molecules. *In vivo*, both the epidermal growth factor receptor and collagen type IV are expressed in the progenitor cell pool indicating that they can directly affect the initial decisions in differentiation. We suggest that early patterns of gene expression, influenced by environmental cues, are likely to provide a specific framework for subsequent decisions that lead to establishing cortical areas.

1995 Development of the cerebral cortex. Wiley, Chichester (Ciba Foundation Symposium 193) p 200–213

The areas that comprise the most functionally diverse part of the central nervous system (CNS), the cerebral cortex, are distinguished by unique cytoarchitectural patterns and the specific connections that are made by each area. During development, the cellular constituents that lie in the rostral portion of the neural tube serve as the cortical precursors, and they move through stages of development, eventually producing postmitotic neurons and glia. The neurons move out of the germinal (ventricular) zone, away from the lumen of the tube and, through complex interactions with their environment, take an appropriate route to their final destination in the overlying cortical plate, the forerunner of the cerebral cortex. We hypothesize that, in these early events of corticogenesis, choices and phenotypic commitments are made by precursors or early postmitotic neurons which direct their subsequent differentiation. It is critical to define the factors that direct these early decisions. Recent experiments in our laboratory suggest that the expression of early molecular differences between neurons in separate areas of the cortex may be influenced by local cues in the proliferative zone, prior to neurons settling in the cortex proper.

Molecules highlight early cortical differences

In the brainstem, members of the *Hox* gene family display expression patterns that mark anatomical boundaries of the rhombomeres (Keynes & Krumlauf 1994). These genes encode transcription factors that may control differential gene expression among different cell populations. A number of homologues of invertebrate patterning genes have been discovered that also are expressed in the early developing cerebral hemispheres (Bulfone et al 1993). In the forebrain, however, different groups of patterning genes so far identified allow one to distinguish only between cortical and subcortical regions. Although there are layer-specific patterns in the expression of some genes, for example *Otx1* and *Otx2* (Frantz et al 1994), there are no examples in this gene family that mark the forerunners of discrete cortical domains.

There are, however, at least three clear examples of early molecular differences among neurons situated in different regions of the cerebral cortex. The actual expression or commitment to differential gene expression appears to occur prior to initiation of thalamocortical, corticocortical or other connections. (1) We found that at the onset of neurogenesis, the limbic system-associated membrane protein (LAMP) appears in mesocortical and allocortical areas (Horton & Levitt 1988, Ferri & Levitt 1993). This protein is a 64–68 kDa cell adhesion molecule that is expressed by all neurons in prefrontal, perirhinal, entorhinal and subicular cortex, as well as subcortically in limbic-related nuclei (Levitt 1984, Zacco et al 1990). We recently cloned LAMP and discovered that it is a new member of the Ig superfamily that can selectively augment adhesion and outgrowth of limbic neurons (Pimenta et al 1995). We have suggested that the early expression of LAMP is required for proper targeting of axons to these cortical regions. (2) Latexin (PC3.1) is a protein that was mapped to neurons located in the lateral hemisphere of the rodent cerebral cortex (Arimatsu et al 1994, Hatanaka et al 1994). Latexin crosses defined cytoarchitectonic boundaries. However, within this lateral zone, neurons residing in layers V and VI begin to express the protein at an early stage in postnatal development. This time point is clearly after afferent input to the latexin-positive regions, but an early commitment to the latexin phenotype was shown by explant experiments. Explants of the lateral cerebral wall, obtained from fetuses prior to the invasion of thalamic afferents, express latexin after several days in culture (Arimatsu et al 1992). Explants of cortical regions that are normally latexin negative never express the protein. (3) An enhancer-trap transgenic mouse was created in which the reporter gene *Xgal* is expressed in the layer IV neurons that form the barrels in the primary somatosensory cortex (Cohen-Tannoudji et al 1994). Cells in the dorsal cerebral wall are committed to express this transgene early in neurogenesis, i.e. at embryonic day (E) 13 in the mouse, even when placed in ectopic locations. Other regions of the cerebral wall fail to express the transgene, even when transplanted into the appropriate somatosensory area of the host.

Environment and the commitment to phenotype

Manipulation of the environment in which a cortical cell develops has been used as a strategy to determine the extent to which extrinsic signals drive the differentiation of neurons residing in different cortical areas. In the examples cited above, cells pass through a stage after which they become committed to a particular phenotype, and even significant changes in their environment fail to alter normal expression of latexin or the transgene. Some phenotypes, however, are clearly dependent upon extrinsic signals to drive their expression. Barrel field cytoarchitecture is dependent upon specific thalamic inputs (Woolsey & Wan 1976), and this unique structure can be generated in cortical tissue that normally never expresses such a phenotypic pattern (Schlagger & O'Leary 1991). Patterns of corticobulbar and corticospinal projections also are dependent upon the environment in which the cortical neurons initiate efferents (O'Leary & Stanfield 1989).

We found, through a series of transplantation studies, that cortical cells progress through a period of sensitivity to the environment. Subsequently, the cells become committed to express a panel of limbic or sensorimotor phenotypes. Fetal tissue from E17 limbic cortex, consisting mostly of postmitotic neurons, continued to express LAMP and limbic thalamocortical projection patterns, even when placed in an inappropriate area of the host cortex (Barbe & Levitt 1991, 1992). Younger tissue from E14 limbic cortex, containing both precursor cells and neurons, exhibited phenotypes of both limbic and sensorimotor cortices. LAMP expression was evident in about 50% of the cells (Barbe & Levitt 1991), and the grafts received thalamocortical and corticocortical inputs that reflected both limbic and sensorimotor areal features (Barbe & Levitt 1992, 1995). These mixed projections normally never exist. The grafts of early E12 cerebral wall, containing only precursor cells, exhibited total dependency on the host environment to define specific phenotypes. Thus, we were able to create limbic cortex, complete with LAMP expression and limbic connections, from sensorimotor precursors by placing them into the perirhinal region of the host brain (Barbe & Levitt 1991, 1992, 1995). Limbic precursors responded just as vigorously when placed in a sensorimotor environment, differentiating into neurons that lacked LAMP expression and forming connections with appropriate thalamic and cortical regions.

The time-dependent fashion in which the cortical cells responded to their new environments suggested to us that signals within the early cerebral wall drive the patterning of the cortex (Ferri & Levitt 1993, Levitt et al 1993). There is some evidence that early differences in the wall do exist, but these distinctions are difficult to assign to extrinsic or lineage-based signalling. Dehay et al (1993) describe basic differences in the number of precursors that give rise to area 17 and 18 in the primate visual cortex. We found that the early isolation *in vitro* of precursors from presumptive limbic (lateral) and sensorimotor (dorsal) areas revealed differences in the basic molecular phenotype of the differentiated neurons: lateral precursors produced mostly LAMP-positive neurons, whereas dorsal precursors generated mostly LAMP-negative neurons. Are the progenitor cells indeed fated to produce populations of neurons that exhibit unique, area-specific features? This type of mechanism doesn't necessarily imply a solely lineage-based development. For example, clonally related cells can disperse within the proliferative zone of the cerebral wall (Walsh & Cepko 1993). Thus, two cells produced from the same precursor may become postmitotic in different microenvironments. In addition, our transplant studies suggest that specific cues in the environment of the precursor cells can influence the subsequent phenotypic development of cortical neurons. The studies showing early molecular fating and cell dispersion, although initially appearing to conflict with each other, suggest to us a common mechanism that is reflected in a molecular prepatterning of the ventricular and subventricular zones (Fig. 1). Regionalized cues could underlie the original concept of Rakic (1988), in which a protocortex in the proliferative zone serves as the forerunner of the functionally parcelled cerebral cortex.

Clues to the molecular cues

Studies of neural crest and retina have shown that specific molecular cues can strictly modulate the fate of cells (Anderson 1989, Greenwald & Rubin 1992, Michelsohn & Anderson 1992, Altshuler et al 1993, Fortini & Artavanis-Tsakonas 1993, Kelley et al 1994, Shah et al 1994). We, therefore, designed *in vitro* studies to define the specific signalling molecules that could act on cortical precursors to produce neurons exhibiting a limbic molecular phenotype (Ferri & Levitt 1993). We began by exposing precursors, isolated at E12 from the dorsal cerebral wall, to a variety of growth factors that, in other systems, appear to affect cell differentiation. When cells are grown on polylysine, the addition of epidermal growth factor (EGF), transforming growth factor α (TGF–α), fibroblast growth factor 2 and platelet-derived growth factor failed to induce an increase in the percentage of sensorimotor precursors that differentiated into LAMP-positive neurons. We then began to manipulate the substrate on which the progenitors were grown. Matrigel (Collaborative Biomedical Products, Becton Dickinson Labware, Bedford, MA), a complex of extracellular matrix proteins and growth factors, induced LAMP expression in the sensorimotor progenitors. When reduced matrigel (without growth factors) was used, the precursors failed to differentiate into LAMP-positive neurons. We then added back each of the growth factors listed above separately, and discovered that when EGF or TGF-α was added to the cultures, the sensorimotor precursors once again produced neurons with the limbic phenotype. These results suggested to us that an interaction between growth factor and matrix is required to modulate the fate of the cortical precursors; each alone was not sufficient to induce the LAMP phenotype. Additional experiments narrowed the matrix component to collagen type IV, a protein typically found in epithelial basement membranes. Neither fibronectin nor laminin were able to combine with the exogenous EGF family to induce LAMP expression. We were thus able to reconstitute, to at least some extent, signals perhaps normally found in the microenvironment of the cerebral wall that influence the behaviour of precursors giving rise to neurons destined for different areas of the cerebral cortex.

Expression of signalling molecules in the cerebral wall

What is the likelihood that the components of the signalling system we have identified *in vitro* actually regulate cell commitment in the precursor pool in the cerebral wall? EGF and TGF-α are expressed prenatally and postnatally in the CNS (Lazar & Blum 1992). TGF-α is the most abundant. Neither have been mapped specifically in the developing cerebral wall. In addition, related members of the EGF family, the heregulins, are expressed in unique patterns in the developing cerebral wall (Marchionni et al 1993). The EGF receptor and collagen type IV have not been mapped in the fetal forebrain. Recent immunocytochemical investigations in our laboratory have revealed that both components are present in the developing wall during neurogenesis. Collagen type IV is evident at E12, surrounding precursor cells in the ventricular zone. There is some stratification of collagen type IV immunoreactivity, with the most dense staining at the ventricular surface around dividing cells, and around cells located superficially in the wall, closest to the pial surface. The distribution of collagen type IV staining becomes much less uniform later in gestation, with light staining in the subventricular zone and more dense immunoreactivity in the deep cortical plate/subplate zone. The EGF receptor is also expressed by precursor cells in the cerebral wall, in a more restricted pattern than collagen type IV, with less immunoreactivity of the receptor in ventral and medial regions of the cerebral wall than dorsal and lateral areas (K. Eagleson, M. Rosner & P. Levitt, unpublished observations). The differential distribution of growth factors and receptors, then, in combination with

Cortical specification

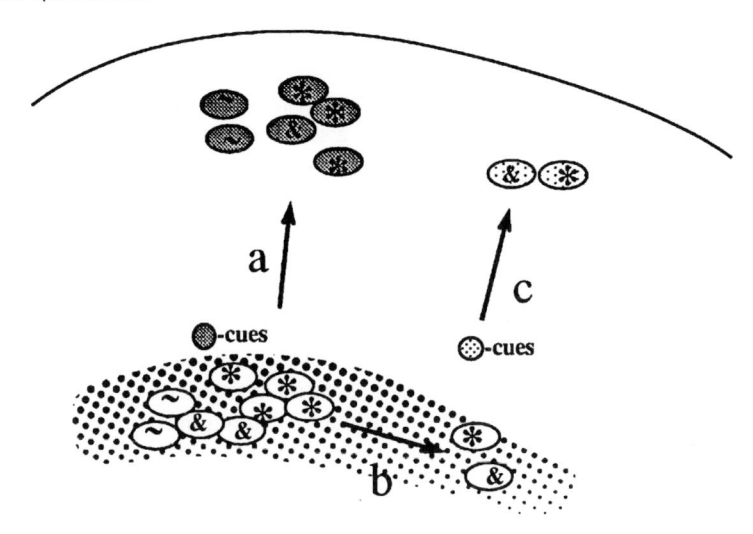

FIG. 1. Schematic diagram depicting the model in which the fate of precursor cells is regulated by different microenvironments in the ventricular zone. Cells related by lineage are marked by different symbols (*, ~, &). Although siblings may end up in the same cortical domain (a) and, therefore, express the same phenotypic features of that area, they may also become dispersed in the ventricular zone (b). When this occurs, the precursors are located in a microenvironment distinct from their original locale. The cells then migrate to a different cortical area (c), expressing phenotypes distinct from their siblings.

appropriate matrix components, could generate significant molecular heterogeneity in the microenvironment of the precursor cells (Fig. 1).

Summary: cell dispersion and prepatterning in the cerebral wall are compatible

Our results suggest that cellular dispersion and early specification of the developing cerebral cortex are compatible events. Specific phenotypic features that are associated with cortical areas are influenced clearly by environmental interactions which occur at different times during development (Rakic 1988, Levitt et al 1993, O'Leary et al 1994). For the precursors that serve as the forerunners of the mature cortex, it is likely that interactions between neuronal progenitors and their environment define the initial commitment to express specific phenotypic traits. Thus, a heterogeneous environmental framework in the ventricular zone would influence the formation of anatomically and functionally discrete areas. For those precursors situated in the zone giving rise to area 17 in the primate, microenvironmental differences could direct the kinetics of the cell cycle differently than neighbouring cells producing neurons for area 18 (see Dehay et al 1993). In the rodent movement of precursors from the dorsal to lateral cerebral wall would also result in different local cues. Here, a descendent of a precursor giving rise to a sensorimotor neuron could attain a final phenotype that is distinctly different, reflecting instead a limbic location in perirhinal cortex. Our model is consistent with findings in other systems, for example the *Drosophila* retina, in which early cues result in defining subsequent steps of differentiation for specific cell populations (Greenwald & Rubin 1992). An early phenotypic commitment does not eliminate the opportunity for a neuron to remain responsive to environmental cues later in development, and thus modify some characteristics that are plastic. Early prepatterning only sets in motion a framework on which later, activity-dependent aspects of cortical development can modulate the establishment of functional circuits (Goodman & Shatz 1993).

References

Altshuler DM, Lo Turco JJ, Rush J, Cepko C 1993 Taurine promotes the differentiation of a vertebrate retinal cell type *in vitro*. Development 119:1317–1328

Anderson DJ 1989 The neural crest cell lineage problem: neuropoiesis? Neuron 3:1–12

Arimatsu Y, Miyamoto M, Nihonmatsu I et al 1992 Early regional specification for a molecular neuronal phenotype in the rat neocortex. Proc Natl Acad Sci USA 89:8879–8883

Arimatsu Y, Nihonmatsu I, Hirata K, Takaguchi-Hayashi K 1994 Cogeneration of neurons with a unique molecular phenotype in layers V and VI of widespread lateral neocortical areas in the rat. J Neurosci 14:2020–2031

Barbe MF, Levitt P 1991 The early commitment of fetal neurons to the limbic cortex. J Neurosci 11:519–533

Barbe MF, Levitt P 1992 Attraction of specific thalamic input by cerebral grafts depends on the molecular identity of the implant. Proc Natl Acad Sci USA 89:3706–3710

Barbe MF, Levitt P 1995 Age-dependent specification of the cortico-cortical connections of cerebral grafts. J Neurosci 15:1819–1834

Bulfone A, Puelles L, Porteus MH, Frohman MA, Martin GR, Rubenstein JLR 1993 Spatially restricted expression of *Dlx-1*, *Dlx-2* (*Tes-1*), *Gbx-2* and *Wnt-3* in the embryonic day 12.5 mouse forebrain defines potential transverse and longitudinal segmental boundaries. J Neurosci 13:3155–3172

Cohen-Tannoudji M, Babinet C, Wassef M 1994 Early determination of a mouse somatosensory cortex marker. Nature 368:460–463

Dehay C, Giroud P, Berland M, Smart I, Kennedy H 1993 Modulation of the cell cycle contributes to the parcellation of the primate visual cortex. Nature 336:464–466

Ferri RT, Levitt P 1993 Cerebral cortical progenitors are fated to produce region-specific neuronal populations. Cereb Cortex 3:187–198

Fortini ME, Artavanis-Tsakonas S 1993 Notch: neurogenesis is only part of the picture. Cell 75:1245–1247

Frantz GD, Weimann JM, Levin ME, McConnell SK 1994 Otx1 and Otx2 define layers and regions in developing cerebral cortex and cerebellum. J Neurosci 14:5725–5740

Goodman CS, Shatz CJ 1993 Developmental mechanisms that generate precise patterns of neuronal connectivity. Cell 72:77–98

Greenwald I, Rubin GM 1992 Making a difference: the role of cell–cell interactions in establishing separate identities for equivalent cells. Cell 68:271–281

Hatanaka Y, Uratani Y, Takaguchi-Hayashi K et al 1994 Intracortical regionality represented by specific transcription for a novel protein, latexin. Eur J Neurosci 6:973–982

Horton HL, Levitt P 1988 A unique membrane protein is expressed on early developing limbic system axons and cortical targets. J Neurosci 8:4653–4661

Kelley MW, Turner JK, Reh TA 1994 Retinoic acid promotes differentiation of photoreceptors *in vitro*. Development 120:2091–2102

Keynes R, Krumlauf R 1994 *Hox* genes and regionalization of the nervous system. Annu Rev Neurosci 17:109–132

Lazar LM, Blum M 1992 Regional distribution and developmental expression of epidermal growth factor and transforming growth factor-α messenger RNA in mouse brain by a quantitative nuclease protection assay. J Neurosci 12:1688–1697

Levitt P 1984 A monoclonal antibody to limbic system neurons. Science 223:229–301

Levitt P, Ferri RT, Barbe MF 1993 Progressive acquisition of cortical phenotypes as a mechanism for specifying the developing cerebral cortex. Perspect Dev Neurobiol 1:65–74

Marchionni MA, Goodearl ADJ, Chen MS et al 1993 Glial growth factors are alternatively spliced erbB2 ligands expressed in the nervous system. Nature 362:312–318

Michelsohn AM, Anderson DJ 1992 Changes in competence determine the timing of 2 sequential glucocorticoid effects on sympathoadrenal progenitors. Neuron 8:589–604

O'Leary DMM, Stanfield BB 1989 Selective elimination of axons extended by developing cortical neurons is dependent on regional locale: experiments utilizing fetal cortical transplants. J Neurosci 9:2230–2246

O'Leary DDM, Schlagger BL, Tuttle R 1994 Specification of neocortical areas and thalamocortical connections. Annu Rev Neurosci 17:419–439

Pimenta A, Zhukareva V, Barbe MF et al 1995 The limbic system-associated membrane protein is an Ig superfamily member that mediates selective neuronal growth and axon targeting. Neuron 14:1–15

Rakic P 1988 Specification of cerebral cortical areas. Science 241:170–176

Schlagger BL, O'Leary DDM 1991 Potential of visual cortex to develop an array of functional units unique to somatosensory cortex. Science 252:1556–1560

Shah N, Marchionni MA, Isaacs I, Stroobant P, Anderson DJ 1994 Glial growth factor restricts mammalian neural crest stem cells to a glial fate. Cell 77:349–360

Walsh C, Cepko CL 1993 Clonal dispersion in proliferative layers of developing cerebral cortex. Nature 362:632–635

Woolsey TA, Wan JR 1976 Areal changes in mouse cortical barrels following vibrissal damage at different postnatal ages. J Comp Neurol 170:53–66

Zacco A, Cooper V, Chantler PD, Fisher-Hyland S, Horton HL, Levitt P 1990 Isolation, biochemical characterization and ultrastructural analysis of the limbic system-associated membrane protein (LAMP), a protein expressed by neurons comprising functional neural circuits. J Neurosci 10:73–90

DISCUSSION

Krubitzer: How did you identify the location of the transplant consistently? You would expect to observe different thalamic and callosal connections for different cortical fields. Some somatosensory regions (e.g. S2) are immediately adjacent to entorhinal cortex; and somatosensory areas S1 and S2, as well as entorhinal cortex, have contralateral connections with both S1 and S2. It's not clear how you would determine that these were entorhinal projections to the opposite hemisphere rather than somatosensory connections. Did you do nissl staining of your transplant to determine which cortical field it was in?

Levitt: Yes. After the DiI (1,1′dioctadecyl-3,3,3′,3′-tetramethylindocarbocyanine perchlorate)-labelled cells were mapped contralaterally, the thick sections were re-sectioned and stained. The situation was complicated by variations among rats in relation to which host cortical fields the transplant was placed and the subsequent DiI placement. For example, in some rats S1 was labelled, and in others S1 and S2 were labelled. The organization of ipsilateral projections to the transplants was also complex because, for example, there were projections from perirhinal cortex to both S1 and S2 in normal rats. The contralateral specificity seemed greater and more strict. Scatter of projection areas around the edge was not observed.

Ghosh: Could you clarify the specificity of corticocortical and thalamocortical connections that you see in the transplants?

Levitt: When we do these transplant experiments, we make a cavity in the cortical grey matter of neonates. The white matter sits underneath the transplant. Thalamic fibres from different nuclei cross beneath the transplant site. We found that if fibres arising from a specific thalamic nucleus did not pass under the transplant of a complementary cortical region, innervation from that particular nucleus did not occur. For example, if we harvested the embryonic day (E) 17 somatosensory cortex and placed it into a perirhinal region, the transplant survived and received somatosensory-like cortical-cortical input, i.e. it received a projection from the contralateral somatosensory region, but it did not receive a ventral basal thalamic input. It is possible that the latter fibres do not innervate the graft because they normally grow dorsally and they don't know that the graft is actually there. Grafts placed dorsally have an advantage in that the fibres running up into the frontal and parietal cortex, for example, run underneath the transplant. If those fibres weren't there, it is possible that they would not innervate the graft.

Ghosh: Are they attracting the axons?

Levitt: No, this is unlikely.

Ghosh: But the growing axons recognize something and stop.

Levitt: That's right. The grafts also grow out rapidly into the host. There is connectivity between the graft and the host, but the nature of the cues that mediate this interaction is not known. There is an alternative model for explaining the observed innervation patterns. Fibres may be partially injured when the lesion cavity is made in the host cortex, inducing sprouting of the neonatal axons. If they interact with a potential target area (i.e. a transplant that they like), then the fibres stay. It is possible that they enter transiently and

Cortical specification

then retract, but we haven't done the timing studies to determine what happens.

Innocenti: The perirhinal region normally projects through the anterior commissure, whereas the somatosensory cortex projects through the corpus callosum. What happens when you switch their positions?

Levitt: The perirhinal projections in the rat two to four weeks after birth actually grow dorsally across the callosum, although some fibres do grow through the anterior commissure. Most of the projections that we see when the somatosensory cortex is placed into the perirhinal region grow up dorsally and then contralaterally, although some do go through the anterior commissure. This is most obvious for the E12 somatosensory transplants which completely transform at the molecular level into limbic system-associated protein (LAMP)-positive tissues and exhibit corticocortical projections, just like perirhinal cortex, when placed in this region in the host.

Bonhoeffer: Does this mean that the final target is correct, but the pathway to the final target is altered?

Levitt: The normal perirhinal cortex has axons going both ways and some pass through the commissure, so it is probably a quantitative issue. It's possible that some of them are growing the wrong way.

Bartlett: Have you actually shown that epidermal growth factor (EGF) switches LAMP-negative cells to LAMP-positive cells, or have you shown that there's an increase in the percentage of cells expressing LAMP due to the preferential survival of LAMP-positive cells?

Levitt: Ray Ferri plated these cells with eight combinations of different matrix and growth factors, and always ended up with about the same number of microtubule-associated protein 2-positive cells. This does not necessarily mean that we're not causing the selective survival of one precursor cell population at the expense of another, although we believe that it is unlikely. We haven't done clonal analysis under these conditions in order to address this directly.

Bartlett: Could you sort these cells on the basis of LAMP expression?

Levitt: Yes, but the cells that express LAMP on their surfaces are already neurons by the time this can be done. Under all the conditions that we've tested, neurons don't respond to the new microenvironment because it's already too late. Only precursor cells respond.

J. Price: Have you tried using bromodeoxyuridine (BrdU) to determine precisely which cells are capable of switching?

Levitt: Mary Barbe and I did some [^{3}H]thymidine studies and obtained a positive correlation, although it was not 100%.

J. Price: Do you mean that the newborn cells were the ones that switched?

Levitt: Yes. A better experiment has been done recently by Mary Barbe in her own laboratory (personal communication). Mary has studied E14–15 perirhinal transplants, and she has found that about half the cells in the graft switch when they are transplanted into the somatosensory region of the host. If this tissue is X-irradiated before transplantation, Mary finds that most of the remaining transplanted cells are LAMP positive. Mary has done [^{3}H]thymidine and BrdU labelling after the irradiation to show that the proliferative cell population is eliminated. In this example, postmitotic perirhinal neurons survive and express LAMP in the heterotopic environment. The non-LAMP-expressing cells fail to appear, as they normally do without X-irradiation. The experiment suggests that if the precursor cell population is eliminated, a group that switches phenotype is not observed. Therefore, in a mixed group of mitotically active and postmitotic cells, only the former will respond to the new environment.

J. Price: It's interesting that all the elements that are plastic are all determined at about the same time.

Ghosh: I would like to address whether or not the newly generated neurons are more plastic. In your E14 transplant, there is a mixed phenotype. All the superficial layer cells are generated later on, so we need to determine whether there is a bias for deeper cells to retain their donor phenotype and the superficial layers to adopt the host phenotype.

Levitt: That's a good point. In order to do that, we would need to both maintain the orientation of our grafts and differentially map which are LAMP positive and which are LAMP negative.

Ghosh: In your previous transplants was the graft just placed in any orientation?

Levitt: We tried to place them into the cavity in one orientation but this is difficult, especially when transplanting the E12 and E14 tissue. We observed a mixed distribution of LAMP-positive and LAMP-negative populations. Occasionally, we saw some lamination, but this tended to be a crude layering. The perirhinal cortex, which is a mesocortical area, is not well laminated in the first place. We took E12 perirhinal tissue and inserted it into somatosensory cortex to see if barrels were formed. We didn't have enough material to do proper staining, but in cresyl violet stains, some barrels seemed to form.

LaMantia: The migrating progenitor cells that Chris Walsh described (1995, this volume) are not regionally committed, whereas the switch that you observe suggests that regional commitment is involved. EGF is a general signalling molecule, so how could it be involved in this regional commitment?

Levitt: This is the reason why we started to look at the receptor and the matrix. If gradients of signal are involved, they will be fairly complex. There are technical problems with determining if gradients exist because we're trying to detect gradients under conditions (tissue preparation) that are unnatural. In the telencephalic vesicle there are segregated populations of EGF receptor-positive cells at E11, which is two days before any of the neurons are generated. It is possible that these receptor-positive cells may correspond to the cells that eventually become LAMP positive. It is also possible that transplanted cells may increase receptor expression if they are placed into the appropriate region of the cerebral wall.

LaMantia: Sue McConnell has observed a dependence of laminar fate on cell cycle time (McConnell & Kaznowski 1991). Is it possible that the cells that are not switching have a further dependence on the cell cycle state?

Levitt: Our experiments have not addressed that. In order to do that, we would have to do sorting and double labelling and we weren't willing to go to those extremes.

Walsh: You showed that anti-LAMP antibody inhibits axon growth. Does it cause axon retraction?

Levitt: No, the antibody does not cause axon retraction. We injected antibody *in vivo* over a seven day period, during the time that the mossy fibre projections are forming. Mary Barbe and I found that the fibre projection loses its proper direction. It grows out into the suprapyramidal region of CA3, a normal zone for targeting, and into the alveus, broad areas of the molecular layer and other subdomains of the hippocampus that it does not normally grow into. We don't impede mossy fibre outgrowth if we block LAMP, but we do cause disorganization. In our *in vitro* studies using the septal and hippocampal explants, we counted the number of cholinergic-positive fibres coming out of the septal explant in the presence of anti-LAMP antibody, and we didn't see a significant decrease. We haven't done the thalamocortical experiments because we would need to label specific subpopulations of thalamic axons coming from LAMP-positive areas. We're not able to do that in those explants.

Blakemore: Is LAMP expressed differentially in distinct regions of the thalamus?

Levitt: Yes. It is expressed in the midline thalamic nuclei, and it is expressed strongly in the anterior nuclear complex. Lower levels of LAMP mRNA are also detected in the ventrobasal complex. We observe two transcripts on our Northern blots and we don't know which one actually encodes the protein. It is possible that they both encode the protein but so far we have evidence for only one form of the protein. The mRNA levels also do not seem to reflect the levels of protein accurately because there is little immunoreactivity in the ventrobasal complex, yet the cells contain LAMP mRNA. We have looked at whether the two transcripts are distributed differentially between regions and we found that the 8 kb mRNA is expressed more heavily in some areas; for example, in most of the classical limbic defined areas. LAMP protein is also present in regions of the cerebellum, although it is not clear why. The levels of the 8 kb mRNA here are low, whereas the levels of the 1.3 kb mRNA are high.

Walsh: Is it possible that the disparity between transcript and protein levels could be due to post-transcriptional events?

Levitt: Yes it is possible, but we haven't looked at that.

J. Price: Does the regulation of expression still occur if you use a larger piece of transplanted tissue? It is possible that regulation is more difficult when a bigger piece of tissue is transplanted because the larger piece of tissue may carry more information with it.

Levitt: We've never addressed this directly. However, if we increase the size of transplanted tissue, we may also enter into other presumptive domains. This would obviously complicate the conclusions if we saw a change.

J. Price: There is obviously a limit to how big a tissue you could transplant. Do you think LAMP is involved in a homophilic interaction?

Levitt: Yes. Our evidence for this is twofold. Firstly, Victoria Zhukareva in the lab coated tissue culture substrates with the affinity-purified, native LAMP protein. She found that only limbic neurons bound to the LAMP substrate and extended axons (Zhukareva & Levitt 1995). Secondly, Vicki and Aurea Pimenta, who cloned the LAMP gene, transfected CHO (Chinese hamster ovary) cells with the LAMP cDNA. They used these cells as a tissue culture substrate and found that although all cell populations from the fetal brain bound to the CHO cells, only the limbic groups extended a large number of axons (unpublished results).

Walsh: When you inject the antibody into the brain *in vivo*, do the axons grow past the normal target and innervate other regions that they don't normally innervate?

Levitt: I don't know whether they actually innervate these regions.

Walsh: But they do grow into them. Are all these regions LAMP positive?

Levitt: The projection forms within the hippocampus, so all the cells there are LAMP positive.

Walsh: How do the antibodies prevent the axons from deciding between several different LAMP-positive regions?

Levitt: Our view is that we're altering axon–axon interactions, which are normally required for tight fasciculation. When these interactions are disturbed, the axons grow and interact with other regions that they might not normally see. We have not followed the antibody injections far enough to determine whether the fibres interact with regions distant from the hippocampus. In order to do these experiments correctly, we need to do genetic manipulations of LAMP expression. We are in the process of doing this now.

Walsh: It is possible that LAMP has another domain, in addition to the one that mediates a homophilic interaction, which is doing something different, for example, interacting with a receptor.

J. Price: How similar is the molecular structure of LAMP to other members of the Ig superfamily?

Levitt: Aurea Pimenta's molecular analysis has revealed some very interesting aspects of LAMP (unpublished results). One interesting feature is that there are only four amino acid substitutions between rat and human

LAMP. Two of these are in the N-terminal signal peptide sequence. Consequently, it is more conserved than the neural cell adhesion molecule, which is only about 85% conserved between rats and humans. LAMP has a cysteine repeat structure, and it has 25–30% homology with other Ig superfamily proteins. Two members of this family, neurotrimmin and opiate-binding cell adhesion molecule, have 50–60% homology to LAMP, but there are no functional data on these proteins. LAMP also maps to the same region on the human chromosome as the genes that encode both the D3 dopamine receptor and GAP-43 (growth-associated protein 43). This is interesting because the gene encoding the D3 receptor is also expressed heavily in the limbic area, and we know that GAP-43 is expressed in the adult limbic forebrain in some areas similar to LAMP. There are no other proteins with which LAMP has a significant level of homology.

References

McConnell SK, Kaznowski CE 1991 Cell cycle dependence of laminar determination in developing cerebral cortex. Science 254:282–285

Walsh C, Reid C 1995 Cell lineage and patterns of migration in the developing cortex. In: Development of the cerebral cortex. Wiley, Chichester (Ciba Found Symp 193) p 21–40

Zhukareva V, Levitt P 1995 The limbic system-associated membrane protein (LAMP) selectively mediates interactions with specific central neuron populations. Development 121:1161–1172

Plasticity in the development of neocortical areas

Dennis D. M. O'Leary, Douglass J. Borngasser, Kevin Fox* and Bradley L. Schlaggar

*Molecular Neurobiology Laboratory, The Salk Institute, 10010 N Torrey Pines Road, La Jolla, CA 92037 and *Department of Physiology, University of Minnesota, Minneapolis, MN 55455, USA*

> *Abstract.* Heterotopic transplantation analysis suggests that individual areas of the developing neocortex have the capacity to differentiate many of the architectural and connectional features normally characteristic of other neocortical areas. Many studies indicate a pivotal role for thalamocortical afferents in the differentiation of the area-specific features that distinguish neocortical areas. Both activity-dependent and activity-independent mechanisms contribute to the patterning of thalamocortical afferent terminations. The available evidence suggests that positional information is established in the cortical subplate and that this information controls the precise targeting of developing thalamocortical axons. In this way appropriate thalamocortical relationships can be established that allow these afferents to promote the differentiation of the functionally specialized and anatomically distinct areas of the adult neocortex.
>
> *1995 Development of the cerebral cortex. Wiley, Chichester (Ciba Foundation Symposium 193) p 214–230*

The mature neocortex is characterized by numerous anatomically and functionally distinct areas that have unique sets of efferent and afferent projections and architecture (Brodmann 1909, Krieg 1946). The developing neocortex lacks many of these area-specific distinctions. We studied the development of area-specific features in normal and experimentally manipulated rat neocortex to define mechanisms that contribute to the differentiation of areas in the developing neocortex.

Plasticity in the development of area-specific properties

We have performed heterotopic transplantations of fetal cortex to determine whether areas of the immature cortex are competent to develop the connectional and architectural features normally associated with other cortical areas. The organization of the adult neocortex into functionally specialized areas requires that each area has efferent projections to appropriate subcortical structures. Layer V is the only source of cortical projections to the midbrain, hindbrain and spinal cord. In the adult layer V neurons in each neocortical area only project to a subset of these targets. This projection differs from area to area. However, during development, layer V neurons in each area first extend a spinally directed primary axon that forms collateral projections to the complete set of brainstem targets (O'Leary et al 1990). Thus, the functionally appropriate, area-specific layer V projections characteristic of the adult do not develop from the outset, but emerge through the selective loss of different subsets of the initial set of projections. For example, layer V neurons in the motor cortex lose their axon collateral to the superior colliculus, whereas those in the visual cortex lose their collateral projections to the inferior olive and dorsal column nuclei, as well as the entire primary axon caudal to the basilar pons (Stanfield et al 1982, O'Leary & Stanfield 1985, Stanfield & O'Leary 1985a, Thong & Dreher 1986, O'Leary & Terashima 1988, O'Leary et al 1990).

By heterotopically transplanting pieces of late embryonic neocortex, we have found that the cortical location at which the transplant develops has a prominent influence on the development of its area-specific layer V projections. For example, layer V neurons in the visual cortex transplanted to the sensorimotor cortex extend and permanently retain axons to the spinal cord, a subcortical target of the sensorimotor cortex, whereas layer V neurons in the sensorimotor cortex transplanted to the visual cortex initially extend and later lose spinal axons, but they retain a projection to the superior colliculus, a subcortical target of the visual cortex (Stanfield & O'Leary 1985b, O'Leary & Stanfield 1989). Heterotopic transplants also establish callosal and thalamic connections, both efferent and afferent, appropriate for their new location (Chang et al 1986, O'Leary & Stanfield 1989, O'Leary et al 1992). These experiments indicate that different areas of the developing neocortex are sufficiently alike such that if they are heterotopically transplanted in the developing neocortex, they can develop the area-specific layer V projections appropriate for their new location.

Similar heterotopic transplantations have addressed plasticity in the development of area-specific architecture. Architectural differences between neocortical areas are due, in part, to unique functional groupings; for example, the vibrissae-related barrels of rodent primary somatosensory cortex (S1) (Woolsey & Van der Loos 1970). Barrels are discrete aggregates of layer IV neurons innervated by clusters of ventroposterior thalamic afferents, and are arranged in the same pattern as the vibrissae on the face. Barrels are also characterized by coincident or complementary distributions of certain enzymes and extracellular matrix (ECM) molecules (reviewed in Schlaggar & O'Leary 1993). Sensory information is relayed from the vibrissae to S1 through a series of synaptic connections: the peripheral processes of trigeminal sensory neurons innervate vibrissae follicles and their central processes connect to brainstem trigeminal nuclei, which in turn project to the ventroposterior thalamus. In both the brainstem and the ventroposterior thalamus, as well as in S1, discrete groupings of neurons are innervated by clustered afferents distributed in a vibrissae-related pattern (reviewed in Woolsey 1990). The differentiation of a normal barrel pattern is dependent upon information relayed from an intact sensory periphery via ventroposterior thalamocortical afferents during the first several days of postnatal development, termed the critical period (reviewed in Woolsey 1990). For example, destruction of a row of vibrissae follicles in neonatal rodents results in a loss of barrel patterning in the corresponding row in S1 cortex.

We wanted to address whether ventroposterior thalamocortical afferents impart barrel patterning information to S1 or whether they respond to patterning information intrinsic to S1. If ventroposterior thalamocortical afferents impart barrel patterning information to S1 then the first barrel-related patterning in S1 must be apparent in their distribution. An evaluation of the development of ventroposterior thalamocortical afferents using sensitive techniques, such as anterograde DiI (1,1'dioctadecyl-3,3,3',3'-tetramethylindocarbocyanine perchlorate) axon labelling or acetylcholine esterase histochemistry which transiently mark developing principal sensory thalamic axons, has revealed that ventroposterior afferents are organized in a vibrissae-related pattern before any of the known cortical components are organized (Erzurumlu & Jhaveri 1990, Jhaveri et al 1991, Schlaggar & O'Leary 1994). Although these observations do not rule out that an undiscovered molecular framework may direct barrel differentiation, they are consistent with the hypothesis that ventroposterior thalamocortical afferents carry barrel patterning information to S1.

To address more definitively whether ventroposterior afferents direct barrel differentiation, we transplanted heterotopically late embryonic visual cortex (V1) into S1 in newborn rats. This tests whether S1 is uniquely specified to differentiate barrels and their characteristic array or if other areas of the developing neocortex are competent to form these functional groupings (Schlaggar & O'Leary 1991). V1 has some characteristics in common with S1. They both have a layer IV that contains a high density of small stellate neurons and receives a dense innervation from a principal sensory thalamic nucleus. In the case of V1, this is the dorsal lateral geniculate nucleus (dLGN). However, in contrast to the clustered distribution of ventroposterior afferents and layer IV cells seen in the barrelfield of S1, stellate neurons and dLGN thalamocortical afferents are distributed uniformly in layer IV of V1. We found that transplanted V1 develops the distinct features of normal S1 barrels, including clustered distributions of ventroposterior afferents coincident with aggregates of layer IV neurons and complemented by certain ECM molecules

(Schlaggar & O'Leary 1991). Thus, ventroposterior afferents can cluster appropriately in layer IV of a foreign piece of neocortex, and layer IV neurons aggregate in response to these afferents. Late embryonic V1 has the capacity to develop this functional grouping normally unique to S1. Therefore, it is not committed to develop the area-specific architecture and thalamocortical afferent distribution characteristic of mature V1. Instead, it is competent to form features characteristic of other neocortical areas. These findings also indicate that barrel architecture is not a prespecified feature of S1.

In summary, these results demonstrate that different parts of the developing neocortex have similar capacities to develop some of the unique anatomical and functional organizations which distinguish neocortical areas in the adult. Further, these findings indicate that thalamocortical afferents play a predominant and even primary role in cortical area differentiation.

Role of neural activity in the patterning of ventroposterior thalamocortical afferents

The segregation of geniculocortical afferents into eye-specific stripes in developing V1 is driven by an activity-dependent mechanism. Therefore, it has been suggested that the segregation of ventroposterior thalamic afferents into a vibrissae-related pattern in developing S1 is also driven by an activity-dependent mechanism, which is based on the postsynaptic detection of synchronously active afferents. The N-methyl-D-aspartate (NMDA) type of glutamate receptor has been implicated as a postsynaptic detector of synchronously active afferents, partly because of its voltage-gated properties. For this reason, we set out to test the hypothesis that peripheral activity drives the patterning of ventroposterior afferents in S1, and that NMDA receptors regulate this process. This hypothesis is plausible because peripherally evoked responses can be recorded in rat S1 on the day of birth (McCandlish et al 1993). Also, both NMDA and non-NMDA types of glutamate receptors are functional in S1 (Kim et al 1995) at thalamocortical synapses (Agmon & O'Dowd 1992) before ventroposterior afferents, which use glutamate as their neurotransmitter, cluster into a vibrissae-related pattern.

To test this hypothesis, we implanted pieces of Elvax (DuPont, Wilmington, DE; a polymer used for the prolonged and controlled release of drugs) loaded with the NMDA receptor antagonist, D-2-amino-5-phosphovaleric acid (D-APV), subdurally over S1 in newborn rats before ventroposterior afferents cluster into a vibrissae-related pattern (Schlaggar et al 1993). Elvax loaded with a vehicle solution or L-APV (the inactive isomer of APV) were used as controls. The patterning of ventroposterior afferents was assessed using acetylcholine esterase histochemistry at postnatal day 8, several days after patterning emerges. Ventroposterior afferents developed a vibrissae-related pattern that appeared to be normal, even in the D-APV treated rats. This suggests that this process is not dependent on neural activity. We assessed the effectiveness of the D-APV treatment by employing single unit recording coupled with iontophoresis of NMDA and 2-(aminomethyl)phenylacetic acid (AMPA), which is an agonist for non-NMDA glutamate receptors. We demonstrated that the concentration of D-APV in S1 was high enough to block not only NMDA receptors, but also non-NMDA glutamate receptors, over the entire critical period for barrel development (Schlaggar et al 1993). We also found that the D-APV treatment blocked all peripherally evoked activity in S1, which was expected because the treatment effectively blocked glutamatergic transmission in S1. Thus, ventroposterior afferents still clustered into a vibrissae-related pattern even though we completely blocked their ability to activate their postsynaptic target cells in S1.

Two other groups, using distinct manipulations to block peripherally evoked activity at birth in rats, have also been unable to demonstrate a role for this activity in the segregation of ventroposterior afferents into vibrissae-related clusters in S1 (Chiaia et al 1992, Henderson et al 1992). These investigators used chronic application of the sodium channel blocker tetrodotoxin either to the infraorbital nerve, the branch of the trigeminal nerve innervating the mystacial vibrissae (Henderson et al 1992), or to S1 (Chiaia et al 1992). Therefore, these experiments indicate major differences between the roles of activity in the development of eye-specific stripes in V1 and of barrels in S1.

In contrast, in a subsequent set of experiments using electrophysiological analysis of the functional organization of S1 treated with D-APV, we found that layer IV cells are driven by inappropriate whiskers at short latencies (Fox et al 1993). To determine the distribution of functional thalamic inputs in the barrel field of rats treated with D-APV, we analysed the spatial distribution of short-latency responses of individual layer IV neurons at three to six weeks of age to stimulation of each vibrissa composing their receptive field. Short-latency responses are believed to reflect monosynaptic input from the thalamus. In normal adult rodents short-latency responses for each vibrissa and ventroposterior thalamic afferents corresponding to individual vibrissa are restricted to a single barrel and the zone immediately surrounding it (Armstrong-James & Fox 1987, Armstrong-James et al 1992, Jensen & Killackey 1987, Senft & Woolsey 1991). However, in the rats treated with D-APV, short-latency responses to individual vibrissa were not restricted to the appropriate barrel, but were also found in surrounding barrels. These results suggest that ventroposterior afferents formed synaptic connections to inappropriate barrels in D-APV-treated S1. Thus, the clustering of ventroposterior afferents into a somatotopic pattern does not require neural activity, whereas the topographic refinement of this projection resulting in the one-to-one functional relationship between a vibrissa and a barrel depends upon functional glutamate receptors and postsynaptic activity. These findings suggest that there are both similarities and differences in the role of activity between the development of eye-specific stripes in visual cortex and the development of barrels in S1.

Development of area-specific thalamocortical projections

Some of the findings presented above, and other evidence from several groups including ours, indicate that thalamocortical afferents influence the developmental differentiation of the functionally specialized, architectonic areas of the neocortex (for reviews see Rakic 1988, O'Leary 1989, O'Leary et al 1994). Therefore, we have been interested in defining mechanisms that control the development of area-specific thalamocortical projections. In the adult thalamocortical projections, which terminate principally in layers IV and VI, are organized in an area-specific manner, i.e. certain thalamic nuclei project to specific neocortical areas. For example, dLGN afferents project to V1 and ventroposterior afferents project to S1 (Hohl-Abrahao & Creutzfeldt 1991). We used DiI as a retrograde tracer to investigate the targeting specificity of thalamocortical axons from the principal sensory thalamic nuclei to the primary sensory areas of the neocortex during the two distinct stages of their development (De Carlos et al 1992). Two types of DiI injections were made into various sites in the embryonic and early postnatal cortex. (1) Injections involving the cortical subplate (which is formed by the earliest generated cortical neurons and is positioned just deep to the cortical plate, from which cortical layers II–VI will differentiate, Luskin & Shatz 1985). These were used to assess the targeting specificity of thalamic axons as they grow tangentially through their intracortical pathway, centred on the subplate, to their appropriate target regions. (2) Injections restricted to the cortical plate. These were used to assess the specificity with which thalamic axons extend collateral branches from the subplate into the cortical plate overlying their intracortical pathway. Regardless of the age or injection type, the distributions of retrogradely labelled thalamic neurons were restricted to thalamic nuclei appropriate for the areal location of the injection site. Other investigators have reported findings consistent with ours (Crandall & Caviness 1984, Miller et al 1993). Thus, developing thalamocortical axons show remarkable specificity, both in their initial targeting and in their subsequent invasion of the cortical plate. This finding favours the hypothesis that position-dependent information in the developing cortex controls the establishment of area-specific thalamocortical connections.

It is likely that thalamocortical axons use information available in the subplate layer to target their appropriate cortical areas. The intracortical pathway of thalamocortical axons is centred on the subplate (Miller et al 1993, Bicknese et al 1994). Therefore, they are in a position to be influenced by potential targeting cues found in the subplate. Indeed, they appear to make their targeting decisions within it (Reinoso & O'Leary 1990, Ghosh & Shatz 1992). In addition, thalamocortical axons fail to invade regions of the cortical plate that overlie pharmacologically neuron-depleted regions of the subplate. Instead, these axons extend aberrantly beyond their appropriate sites (Ghosh et al 1990). Thus, the information that governs the development of area-specific thalamocortical projections likely operates in the subplate. A straightforward way to establish this targeting information would be for progenitor cells in the neocortical neuroepithelium to have positional identity and to impart this identity to their progeny.

This mechanism requires that gene expression in the developing subplate is regulated in a position-dependent manner. At present, *Emx2* is the best candidate for providing position-dependent gene regulation involved in the area-specific targeting of thalamocortical projections. *Emx2* encodes a putative homeodomain transcription factor and is a homologue of *Drosophila empty spiracles* (Simeone et al 1992). *Emx2* has a caudal to rostral graded distribution in the developing neocortex (Simeone et al 1992), and it is expressed by neuroepithelial cells but not by postmitotic neurons (Gulisano et al 1994). Although the graded expression of *Emx2* neither delineates nor predicts neocortical areas, the early expression of *Emx2* could initiate a genetic cascade that regulates the position-dependent expression of genes encoding molecules that control the area-specific targeting of thalamocortical axons directly. If *Emx2* operates in such a fashion, the progeny of neighbouring neuroepithelial cells would have to maintain neighbour relationships in the subplate. If the dispersion of these cells is substantial, targeting information would probably have a source other than subplate cells or they would be specified by interactions that occur after they finish their migration.

Several studies have shown that cortical plate neurons and neuroepithelial cells are considerably dispersed during the assembly of the cortical plate, which occurs at a later stage of cortical development. For example, a battery of recombinant retroviruses with unique tags to mark clonally related cells has been used to show that cortical plate neurons disperse widely across diverse areas of the neocortex (Walsh & Cepko 1992, 1993). This dispersion could occur in a number of ways during the generation and migration of cortical plate neurons. For example, Fishell et al (1993) have shown that cells in the neocortical neuroepithelium move tangentially in a random direction. In addition, a proportion of cells migrating radially from the neuroepithelium change course abruptly in the intermediate zone and migrate tangentially on a course orthogonal to radial glia (O'Rourke et al 1992), whereas other migrating cells shift from one radial glial fascicle to another (Austin & Cepko 1990, Misson et al 1991).

The recombinant retroviral method has not been applied successfully to study neuronal dispersion at the earliest stages of corticogenesis, during the time that subplate cells are generated because it is difficult to inject agents into the ventricular space prior to embryonic day 14 in rats. However, time-lapse videomicroscopy of fluorescently labelled neuroepithelial cells and their progeny in fetal rat cortex maintained *in vitro* indicates that subplate progenitors do maintain neighbour relationships in the neuroepithelium during the period that subplate cells are generated. This is also true of the postmitotic subplate cells that migrate and establish the nascent subplate (O'Leary & Borngasser 1992). Subplate cells rapidly form an intricate network of processes (De Carlos & O'Leary 1992) that would presumably restrict their movements relative to one another. This would preserve their neighbour relationships as the cortex expands and as the subplate is displaced deeper by the subsequent formation of the cortical plate. Therefore, positional information imparted to subplate cells around the time they are generated could regulate the later deployment of guidance molecules that control the area-specific targeting of thalamocortical axons. In this way, molecular information laid down in the subplate could influence the subsequent differentiation of arca-specific features within the overlying cortical plate.

Acknowledgements

This work was supported in part by NIH grants NS 31558 (D. D. M. O.) and NS 27759 (K. F.).

References

Agmon A, O'Dowd DK 1992 NMDA receptor-mediated currents are prominent in the thalamocortical synaptic response before maturation of inhibition. J Neurophysiol 68:345–348

Armstrong-James M, Fox K 1987 Spatiotemporal convergence and divergence in the rat S1 'barrel' cortex. J Comp Neurol 263:264–281

Armstrong-James M, Fox K, Das Gupta A 1992 Flow of excitation within rat barrel cortex. J Neurophysiol 68:1345–1358

Austin CP, Cepko CL 1990 Migration patterns in the developing mouse cortex. Development 110:713–732

Bicknese AR, Sheppard AM, O'Leary DDM, Pearlman AL 1994 Thalamocortical axons extend along a chondroitin sulfate proteoglycan-enriched pathway coincident with the neocortical subplate and distinct from the efferent path. J Neurosci 14:3500–3510

Brodmann K 1909 Vergleichende Lokalisationslehre der Groshirnrinde in ihren Prinsipien dargestellt auf Grund des Zellenbaues. J. A. Barth, Leipzig

Chang F, Steedman JG, Lund RD 1986 The lamination and connectivity of embryonic cerebral cortex transplanted into newborn rat cortex. J Comp Neurol 244:401–411

Chiaia NL, Fish SE, Bauer WR, Bennett-Clarke CA, Rhoades RW 1992 Postnatal blockade of cortical activity by tetrodotoxin does not disrupt the formation of vibrissa-related patterns in the rat's somatosensory cortex. Dev Brain Res 66:244–250

Crandall JE, Caviness VS 1984 Thalamocortical connections in newborn mice. J Comp Neurol 228:542–546

De Carlos JA, O'Leary DDM 1992 Growth and targeting of subplate axons and establishment of major cortical pathways. J Neurosci 12:1194–1211

De Carlos JA, Schlaggar BL, O'Leary DDM 1992 Targeting specificity of primary sensory thalamocortical axons in developing rat neocortex. Soc Neurosci Abstr 18:57

Erzurumlu RS, Jhaveri S 1990 Thalamic axons confer a blueprint of the sensory periphery onto developing rat somatosensory cortex. Dev Brain Res 56:229–234

Fishell G, Mason CA, Hatten ME 1993 Dispersion of neural progenitors within the germinal zones of the forebrain. Nature 362:636–638

Fox K, Schlaggar BL, O'Leary DDM 1993 The effect of postsynaptic activity blockade on development of barrel cortex. Soc Neurosci Abstr 19:616

Ghosh A, Shatz C 1992 Pathfinding and target selection by developing geniculocortical axons. J Neurosci 12:39–55

Ghosh A, Antonini A, McConnell SK, Shatz CJ 1990 Requirement for subplate neurons in the formation of thalamocortical connections. Nature 347:179–181

Gulisano M, Broccoli V, Boncinelli E 1994 Emx2 is selectively expressed in germinal neuroepithelium of developing dorsal telencephalon of the mouse. Soc Neurosci Abstr 20:1079

Henderson TA, Woolsey TA, Jacquin MF 1992 Infraorbital nerve blockade from birth does not disrupt central trigeminal pattern formation in the rat. Dev Brain Res 66:146–152

Hohl-Abrahao JC, Creutzfeldt OD 1991 Topographic mapping of the thalamocortical projections in rodents and comparison with that in primates. Exp Brain Res 87:283–294

Jensen KF, Killackey HP 1987 Terminal arbors of axons projecting to the somatosensory cortex of the adult rat. I. The normal morphology of specific thalamocortical afferents. J Neurosci 7:3529–3543

Jhaveri S, Erzurumlu RS, Crossin K 1991 Barrel construction in rodent neocortex: role of thalamocortical afferents versus extracellular matrix molecules. Proc Natl Acad Sci USA 88:4489–4493

Kim HG, Fox K, Connors BW 1995 Properties of excitatory synaptic events in neurons of primary somatosensory cortex of neonatal rats. Cereb Cortex 5:148–157

Krieg WJS 1946 Connections of the cerebral cortex. I. The albino rat. A. Topography of the cortical areas. J Comp Neurol 84:221–275

Luskin MB, Shatz CJ 1985 Studies of the earliest-generated cells of the cat's visual cortex: cogeneration of subplate and marginal zones. J Neurosci 5:1062–1075

McCandlish CA, Li CX, Waters RS 1993 Early development of the Si cortical barrel field representation in neonatal rats follows a lateral-to-medial gradient: an electrophysiological study. Exp Brain Res 92:369–374

Miller B, Chou L, Finlay BL 1993 The early development of thalamocortical and corticothalamic projections. J Comp Neurol 335:16–41

Misson J-P, Austin CP, Takahashi T, Cepko CL, Caviness VS Jr 1991 The alignment of migrating neural cells in relation to the murine neopallial radial glial fiber system. Cereb Cortex 1:221–229

O'Leary DDM 1989 Do cortical areas emerge from a protocortex? Trends Neurosci 12:400–406

O'Leary DDM, Borngasser D 1992 Minimal dispersion of neuroepithelial cells and their progeny during generation of the cortical preplate. Soc Neurosci Abstr 18:925

O'Leary DDM, Stanfield BB 1985 Occipital cortical neurons with transient pyramidal tract axons extend and maintain collaterals to subcortical but not intracortical targets. Brain Res 336:326–333

O'Leary DDM, Stanfield BB 1989 Selective elimination of axons extended by developing cortical neurons is dependent on regional locale: experiments utilizing fetal cortical transplants. J Neurosci 9:2230–2246

O'Leary DDM, Terashima T 1988 Cortical axons branch to multiple subcortical targets by interstitial axon budding: implications for target recognition and 'waiting periods'. Neuron 1:901–910

O'Leary DDM, Bicknese AR, De Carlos JA et al 1990 Target selection by cortical axons: alternative mechanisms to establish axonal connections in the developing brain. Cold Spring Harbor Symp Quant Biol 55:453–468
O'Leary DDM, Schlaggar BL, Stanfield BB 1992 The specification of sensory cortex: lessons from cortical transplantation. Exp Neurol 115:121–126
O'Leary DDM, Schlaggar BL, Tuttle R 1994 Specification of neocortical areas and thalamocortical connections. Annu Rev Neurosci 17:419–439
O'Rourke NA, Dailey ME, Smith SJ, McConnell SK 1992 Diverse migratory pathways in the developing cerebral cortex. Science 258:299–302
Rakic P 1988 Specification of cerebral cortical areas. Science 241:170–176
Reinoso BS, O'Leary DDM 1990 Correlation of geniculocortical growth into the cortical plate with the migration of their layer 4 and 6 target cells. Soc Neurosci Abstr 16:439
Schlaggar BL, O'Leary DDM 1991 Potential of visual cortex to develop arrays of functional units unique to somatosensory cortex. Science 252:1556–1560
Schlaggar BL, O'Leary DDM 1993 Patterning of the barrelfield in somatosensory cortex with implications for the specification of neocortical areas. Perspect Dev Neurobiol 1:81–92
Schlaggar BL, O'Leary DDM 1994 Early development of the somatotopic map and barrel patterning in rat somatosensory cortex. J Comp Neurol 346:80–96
Schlaggar BL, Fox K, O'Leary DDM 1993 Postsynaptic control of plasticity in developing somatosensory cortex. Nature 364:623–626
Senft SL, Woolsey TA 1991 Growth of thalamic afferents into mouse barrel cortex. Cereb Cortex 1:308–335
Simeone A, Gulisano M, Acampora D, Stornaiuolo A, Rambaldi M, Boncinelli E 1992 Two vertebrate homeobox genes related to the *Drosophila empty spiracles* gene are expressed in the embryonic cerebral cortex. EMBO J 11:2541–2550
Stanfield BB, O'Leary DDM 1985a The transient corticospinal projection from the visual cortex during the postnatal development of the rat. J Comp Neurol 238:236–248
Stanfield BB, O'Leary DDM 1985b Fetal occipital cortical neurons transplanted to the rostral cortex can extend and maintain a pyramidal tract axon. Nature 313:135–137
Stanfield BB, O'Leary DDM, Fricks C 1982 Selective collateral elimination in early postnatal development restricts cortical distribution of rat pyramidal tract neurones. Nature 298:371–373
Thong IG, Dreher B 1986 The development of the corticotectal pathway in the albino rat. Dev Brain Res 25:227–238
Walsh C, Cepko CL 1992 Widespread dispersion of neuronal clones across functional regions of the cerebral cortex. 255:434–440
Walsh C, Cepko CL 1993 Clonal dispersion in proliferative layers of developing cerebral cortex. Nature 362:632–635
Woolsey TA 1990 Peripheral alteration and somatosensory development. In: Coleman JR (ed) Development of sensory systems in mammals. Wiley, New York, p 461–516
Woolsey TA, Van Der Loos H 1970 The structural organization of layer IV in the somatosensory regions (S1) of mouse cerebral cortex: the description of a cortical field composed of discrete cytoarchitectonic units. Brain Res 17:205–242

DISCUSSION

Daw: What is the difference between the colliculus and the cortex in terms of axon targeting?

O'Leary: In the superior colliculus of the rat, there is initially a diffuse connectivity which is then remodelled into precise connections (Simon & O'Leary 1992a,b). This doesn't occur in the thalamocortical system. In terms of establishing positional information, however, the two systems could be similar.

Daw: Why should there be such a large difference in the two mechanisms?

O'Leary: I don't know the answer to that.

Kennedy: Perhaps the answer lies in the evolutionary history of the two structures. During development in lower vertebrates, it is necessary to accommodate the increase in the size of the retina functionally by shifting retinal terminals in the colliculus. Consequently, the initial and final position of a retinal cell can be different during active life (Gaze et al 1979). In contrast, the more recently evolved cortex does not have to accommodate changes in the sensory periphery during active life.

Molnár: I am surprised that we use the terms somatosensory thalamus, visual thalamus and visual cortex. During the early stages of development, there is no peripheral input to these regions. Your labelling studies at embryonic day (E) 17, and ours at E16, showed that there is no discontinuum between thalamic nuclei. This labelling does not respect any boundaries because there are no boundaries in the thalamus at this early stage. It is possible that these thalamic cells are not as smart as we think they are—they just occupy these places, and the nuclear boundaries are carved out later within the thalamus. There is a quotation from Cogeshall (1964) in Ted Jones' book on the thalamus (1985) about the possibility that the nuclear differentiation in the dorsal thalamus can be correlated with the arrival of subcortical afferents. This transforms the embryonic dorsal thalamus, a unit that is forming connections with the telencephalon alone, into the adult dorsal thalamus, a mosaic of units some of which connect the telencephalon synaptically with lower centres. It seems likely that optic, tactile and other incoming pathways plug into circuits that are, to some extent, already formed.

O'Leary: I disagree. Thalamic nuclei can be defined accurately at these stages. For example, the principal thalamic nuclei are sharply demarcated, as revealed with a variety of histochemical stains (Schlaggar et al 1993).

I use the terminology of somatosensory thalamus and visual thalamus prior to these cells processing sensory information because it's a simpler terminology for more general audiences and easier for me too.

Molnár: Do you agree that these areas are not necessarily processing information at these early stages?

O'Leary: Yes.

Levitt: But that doesn't mean that they are not organized.

Molnár: They are not organized according to the future nuclei.

O'Leary: Yes they are.

Levitt: I also disagree with Zoltán Molnár. They have not received peripheral sensory information, but they are generated and then they migrate and settle in specific patterns.

O'Leary: When they are in place can you define their boundaries?

Molnár: I placed the different carbocyanine dyes at E16 in different sequences at points in a coronal or parasagittal row along the hemisphere and I have not seen any distortions in the pattern of back-labelled thalamic cells in these early stages. Coronal placement corresponded to anteroposterior and parasagittal to mediolateral sequence of back-labelled groups of cells without obvious sharp boundaries.

O'Leary: I showed you localization of the staining at E17. If I showed you the counterstaining and the histochemical sections you could delineate the various nuclei perfectly.

Bolz: You describe more widespread short-latency responses to whisker stimulation as being due to a lack of topographic refinement in ventroposterior thalamocortical connections to layer IV of barrel cortex. Could these findings be accounted for by changes in corticocortical connections without changes in thalamocortical connections?

O'Leary: We're almost positive that these are direct thalamocortical inputs. We are now studying the anatomy to confirm this point. The reports of topographic inaccuracies of ventroposterior thalamocortical afferents to the developing barrel field support this. Senft & Woolsey (1991) have reported that these afferents arborize broadly across layer IV, and gradually restrict their arbors to appropriate barrels. Agmon and colleagues (1993) report that topographic inaccuracies occur by the misdirection of afferent axons to inappropriate locations within layer IV.

In our physiology experiments, we divided the responses of layer IV cells to the stimulation of an individual vibrissa into three latency groups: short, intermediate and long. All three latency groups have a more diffuse distribution pattern in the 2-amino-5-phosphovaleric acid-treated rats than in the control rats. This finding is consistent with inappropriate refinement of thalamocortical as well as intracortical connections.

Bolz: So changes do occur in thalamocortical connections?

O'Leary: Yes. Short latency responses are thought to represent direct monosynaptic excitation from the thalamus because they are the fastest inputs to the cortex from the vibrissae. In our study, we categorized responses with a latency of 10 ms or less as short latency. We are fairly confident that these represent direct thalamocortical inputs.

Bolz: How do you relate the distribution of the ventroposterior thalamocortical connections in terms of whisker-related clusters and inappropriate topographic distribution?

O'Leary: We know, by studying the anatomy, that they are segregated to these clusters. The physiology suggests that underlying this clustered segregation, there is an inappropriate precision in terms of topology; that is, some afferents terminate in an inappropriate cluster.

Bolz: Do they cross into the cortical regions?

O'Leary: This is a broader issue, and one that we haven't looked at.

Bonhoeffer: You showed the anatomy of the transplantion of the visual cortex into the barrel cortex. Have you studied the physiology?

O'Leary: We have not studied the physiology of those transplants extensively. Brent Stanfield and I did some microstimulation experiments in collaboration with Asanuma and colleagues in rats in which we transplanted visual cortex into the motor cortex and allowed it to develop (Porter et al 1987). We had previously shown that visual cortex transplanted to motor cortex not only extends an axon to the spinal cord, but also permanently retains this connection characteristic of motor cortex (Stanfield & O'Leary 1985, O'Leary & Stanfield 1989). We then showed, in the microstimulation experiments, that forelimb movements could be evoked from the transplant, but we were cautious in our interpretation because of the possibility that current had spread to host motor cortex.

Rakic: This is a remarkable example of specificity in the cortex. The reason why it was not possible to show this before was that we didn't have such good markers and it was difficult to know exactly where the dye was placed. We did some double labelling in the auditory and visual cortex of the ferret, which has a larger brain (M. Sheng & P. Rakic, unpublished results). We found that the fibres bypass unattended areas and just go precisely where they're supposed to go. We did not find any aberrant contacts.

Jones: Dennis O'Leary alluded to the prevailing dogma in the somatosensory cortex which says that there is initially an extreme divergence and overlap of thalamocortical connections which are then pruned back in order to generate this highly specific concentrated barrel-like formation. Evidence for this is based upon preparations in which large injections or placements of tracer have been made. This means that it is extremely difficult to dissect out individual fibres, except perhaps at their tips. We are fortunate (Amil Agmon, Diane O'Dowd, Lee Yang and myself) in that we have access to a reduced preparation of thalamocortical connectivity—a thalamocortical slice which retains connectivity between the ventrobasal thalamus and the somatosensory cortex. It is 400 μm thick and was prepared primarily for physiological studies (Agmon et al 1993, 1995). By placing deposits of DiI (1,1'dioctadecyl-3,3,3',3'-tetramethylindocarbocyanine perchlorate) in the thalamus, we showed that fibres in the subplate and subsequently in layer VI of the cortex were organized irregularly when they directed themselves upwards into the presumptive barrel field. They grew up vertically without any branching. The branching occurred subsequent to their arrival and termination within a single presumptive barrel. Errors that we saw were those in which a fibre came up radially in conjunction with a number of other fibres directed to a particular 'proto-barrel', but then suddenly realized it had made a mistake and crossed, without making any terminations, into the correct adjacent barrel. We did not see any branching in the underlying plexus of fibres within the layer V–VI border, although the density and meandering configuration of the fibres prevented us from saying absolutely that thalamocortical fibres do not branch in the deep strata and go straight across the barrel. We have done some retrograde labelling, which enabled us to confirm that observation. We used DiI 3 and DiI 5 which fluoresce at different wavelengths. Lee Yang became skilled at placing tiny crystals of dye into individual barrels where they could be visualized in mice older than four days or at equivalent interbarrel distances in mice younger than four days. He tried to place the different crystals closer and closer together to determine the extent of double labelling of cells in the underlying thalamus. We studied those preparations at postnatal day (P) 0 and P4 in which the dyes either did not overlap or overlapped minimally. The retrograde labelling of the whole pathway was outlined. The two sets of fibres come together in the vicinity of the reticular nucleus and there are two populations of labelled cells in the ventroposterior nucleus. There are few if any densely labelled cells at P0, whereas the organization becomes tighter at P4, when there are two populations of cells, none of which are double labelled. In this kind of preparation at any age, the numbers of double-labelled cells were never greater than about 5% (Agmon et al 1995). The clusters tighten up between P0 and P4, but there are some outlying cells of both types that are probably lost over time. The topography is set up at the earliest age (P0), and there is no refinement or loss of connections in the cortex, the cells just become more compacted in the thalamus. In addition, axons arising from these cells take a varied course to get out of the ventroposterior nucleus, but once they get into the region of the reticular nucleus, they become more compact. I originally believed that initial diffuseness, with a refinement of projections by the dying back of exuberant branches, prevailed in this system. However, I have now come to the conclusion, like Dennis O'Leary, that it has a high degree of precision from the beginning.

Molnár: You mentioned that there is no branching in the underlying white matter of the cortex at P4, but this does not exclude the possibility that, during the embryonic stages, there were transient side branches, one of which was strengthened whereas the others disappeared, so that at the later stages only the final, meticulous result is observed.

Jones: I can't answer that because we have only made preparations at P0, P4 or older.

Rakic: There is a beautiful study by Steve Easter and his colleagues, who looked at the first connections in the forebrain. They found that these connections are precise and do not make mistakes (Burrill & Easter 1994).

LaMantia: Can Dennis O'Leary interpret his transplant results in the light of Ted Jones' observations? Thalamic afferents recognize a specific region of the cortex. However, on the other hand, cortical cells, once innervated by a distinct thalamic input, give rise to a projection that is consistent with new innervation.

O'Leary: It is important that I describe some of our methodology to answer this question. In order to transplant E17 or E18 cortex into a newborn host, we make a precise hole in the cortical plate and leave the underlying subplate intact. It is important to leave this intact because the thalamocortical afferents are in the subplate and if we remove them, the transplant will not be innervated. The donor cortex is the full thickness of the cortex from the neuroepithelium out to the pial surface. This is the same paradigm that Lund and his colleagues pioneered and published (Jaeger & Lund 1981). The reason why the transplanted cortex is innervated by a foreign thalamic nucleus is probably because the transplant is performed after thalamocortical axons have completed their targeting phase. Consequently, they've now switched over to an innervation phase, or an invasion of the cortical plate phase. So we're basically challenging thalamic axons to re-extend into a foreign piece cortex and observing what they do within that foreign piece of cortex and how the transplanted cortical cells respond.

LaMantia: Is your argument that once the regional positional information is established, the cortical plate itself becomes less specified regionally so that it's receptive to colonization and organization by foreign thalamic innervation?

O'Leary: The cortical plate is probably receptive to foreign innervation throughout the period when thalamic afferents normally grow into it. The available evidence suggests that there is targeting information in the subplate during the target phase (about E16–20 in the rat), but we don't know the time course of expression of that information.

Rakic: Transplants show remarkable plasticity, but one cannot necessarily assume that transplantation does not change some aspects of their development in the new environment. There are aspects of cortical organization, like ocular dominance columns, that depend entirely on input from the periphery. Other aspects may be less malleable. Now that we have more sophisticated knowledge of events *in vivo* we can interpret the results differently.

Levitt: Have all the layer IV neurons been generated at the time that you do your transplants?

O'Leary: Essentially all of the layer IV neurons have been generated.

Levitt: Miller (1986) has reported that only about 50–60% have been generated, so there's some discrepancy. Is there any variation in the extent of barrel formation between a piece of tissue taken out at E17 and E18? Because tissue taken at E17 or E18 from the occipital portion of a wall is different in time than tissue taken at E17 or E18 from the rostral part of the wall. Younger embryonic tissues may have a more pronounced formation of barrels, and this capacity may shift. The barrel architecture might not be as dramatic with tissue that is a day and a half older.

O'Leary: We haven't done a systematic study comparing the capacity to develop barrels with different ages of transplants. This is partly because at later ages, the viability of the transplant diminishes rapidly. We have done [^{3}H]thymidine and bromodeoxyuridine birth-dating studies of cortical neurogenesis and, in our hands, the peak generation day of layer IV is E16, and the generation of layer IV neurons is essentially complete by E17.

Innocenti: Dennis O'Leary, would it be fair to say that the heterotopically transplanted cortex does not become identical to the cortex it replaces? Because Barth & Stanfield (1994) showed that grafts of occipital cortex are not as good as grafts of frontal cortex for behavioural recovery. Also, Ebrahimi-Gaillard et al (1994) mapped the efferent projections originating from homotopic and heterotopic transplants systematically and found that many of them are appropriate for the site of origin of the graft rather than for its new locale.

Kennedy: Ebrahimi-Gaillard et al (1994) grafted visual cortex into the somatosensory field at the same age as in Dennis O'Leary's experiments. They injected *Phaseolus vulgaris* leucoagglutinin into these grafts and showed that the visual cortex that had been placed in the somatosensory field formed connections with the appropriate thalamic nuclei, including the lateral geniculate nucleus.

O'Leary: We've never seen any evidence of input coming in from the thalamus that's appropriate for the origin rather than the location. It is possible that this happens if you choose cortical donor tissue that's spatially close to the site where you put it in the host. We take pieces of cortex from distantly located visual cortex and sensory motor cortex, so it is possible that if we used pieces that were closer together, we might see a mixing of afferent input. Other explanations could be that it is related to the size of the transplant, the integrity of a transplant or how well the injection site was restricted to the transplant and did not involve underlying axonal pathways.

I find the Barth & Stanfield (1991) experiments intriguing. They did homotopic and heterotopic transplants into the cortex, and they found that homotopic transplants reduce a behavioural deficit caused by a lesion cavity, whereas heterotopic transplants exacerbated the deficit. On the other hand, Stein and colleagues (1991) have shown in adult rats that if they lesion the visual cortex and insert an embryonic cortical graft, regardless of the cortical source of the graft, it restores certain visually guided behaviours.

Levitt: Did they analyse the anatomical connections?

O'Leary: No. One problem with these experiments is that you have to be careful when you make the injection because if you involve the underlying subplate, you're going to label nuclei that have axons passing underneath that location. Consequently, if you put a visual cortical transplant in the sensorimotor cortex and the injection site involves the underlying subplate, you're going to label retrogradely lateral geniculate nucleus axons and cells.

References

Agmon A, Yang LT, O'Dowd DK, Jones EG 1993 Organized growth of thalamocortical axons from the deep tier of terminations into layer IV of developing mouse barrel cortex. J Neurosci 13:5365–5382

Agmon A, Yang LT, Jones EG, O'Dowd DK 1995 Topological precision in the thalamic projection to neonatal mouse barrel cortex. J Neurosci 15:549–561

Barth TM, Stanfield BB 1991 Heterotopic and homotopic fetal cortical transplants to the developing rostral cortex of rats produces different behavioral effects. Soc Neurosci Abstr 17:51

Barth TM, Stanfield BB 1994 Homotopic, but not heterotopic, fetal cortical transplants can result in functional sparing following neonatal damage to the frontal cortex in rats. Cerebral Cortex 4:271–278

Burrill JD, Easter SS 1994 Development of the retinofugal projections in the embryonic and larval zebrafish (*Brachydanio rerio*). J Comp Neurol 346:583–600

Cogeshall RE 1964 A study of diencephalic development in the albino rat. J Comp Neurol 122:241–269

Ebrahimi-Gaillard A, Guitet J, Garnier C, Roger M 1994 Topographic distribution of efferent fibers originating from homotopic or heterotopic transplants: heterotopically transplanted neurons retain some of the developmental characteristics corresponding to their site of origin. Dev Brain Res 77:271–283

Gaze RM, Keating MJ, Ostberg A, Chung SH 1979 The relationship between retinal and tectal growth in larval *Xenopus*. Implications for the development of the retinaltectal projection. J Embryol Exp Morphol 53:103–143

Jaeger CB, Lund RD 1981 Transplantation of embryonic occipital cortex to the brain of newborn rats. An autographic study of transplant histogenesis. Exp Brain Res 40:265–272

Jones EG 1985 The Thalamus. Plenum Press, New York

Miller MW 1986 The migration and neurochemical differentiation of gamma-aminobutyric acid (GABA)-immunoreactive neurons in rat visual cortex as demonstrated by a combined immunocytochemical-autoradiographic technique. Dev Brain Res 28:41–46

O'Leary DDM, Stanfield BB 1989 Selective elimination of axons extended by developing cortical neurons is dependent on regional locale. Experiments utilizing fetal cortical transplants. J Neurosci 9:2230–2246

Porter LL, Cedarbaum JM, O'Leary DDM, Stanfield BB, Asanuma H 1987 The physiological identification of pyramidal tract neurons within transplants in the rostral cortex taken from the occipital cortex during development. Brain Res 436:136–142

Schlaggar BL, De Carlos JA, O'Leary DDM 1993 Acetylcholinesterase as an early marker of the differentiation of dorsal thalamus in embryonic rats. Dev Brain Res 75:19–30

Senft SL, Woolsey TA 1991 Growth of thalamic afferents into mouse barrel cortex. Cereb Cortex 1:308–335

Simon DK, O'Leary DDM 1992a Development of topographic order in the mammalian retinocollicular projection. J Neurosci 12:1212–1232

Simon DK, O'Leary DDM 1992b Responses of retinal axons in vivo and in vitro to molecules encoding position in the embryonic superior colliculus. Neuron 9:977–989

Stanfield BB, O'Leary DDM 1985 Fetal occipital cortical neurons transplanted to rostral cortex develop and maintain a pyramidal tract axon. Nature 313:135–137

Stein DG 1991 Fetal brain tissue grafting as therapy for brain dysfunctions: unanswered questions, unknown factors and practical concerns. J Neurosurg Anesthes 3:170–192

The roles of growth factors and neural activity in the development of the neocortex

David J. Price, R. Beau Lotto, Natasha Warren, Gillian Magowan and Julia Clausen

Department of Physiology, The University of Edinburgh Medical School, Teviot Place, Edinburgh EH8 9AG, UK

Abstract. Previous research on primarily the peripheral nervous system has shown that soluble growth factors help control key developmental events by contributing to dynamic autocrine and paracrine signalling systems. Much less is known about the roles of these substances in neocortical development. Using cell and tissue culture paradigms, we have demonstrated that soluble growth factors are produced by the neocortex and its subcortical targets, and that these tissues can respond to them. There are several possible functions for these factors in neocortical development *in vivo*: they may initiate axonal growth from neocortical neurons and/or their afferents; accelerate or guide that growth; and/or play a role in the later refinement of connections. Although none of these possibilities can be excluded, the existing evidence strengthens the hypothesis that soluble growth factors are important for the early postnatal growth and refinement of neocortical connections, when their levels of release may be regulated by neocortical activity. At present we do not know which growth factors are involved in these processes, but the results of preliminary experiments indicate that neurotrophins and fibroblast growth factor are prime candidates.

1995 Development of the cerebral cortex. Wiley, Chichester (Ciba Foundation Symposium 193) p 231–250

Several lines of evidence, from both *in vitro* and *in vivo* studies, indicate the importance of growth factor-mediated intercellular interactions in neocortical development. Taking the *in vitro* results first, we have shown that developing neocortical explants promote the growth of neurites from embryonic thalamic explants cultured in serum-free medium (Rennie et al 1994, Lotto & Price 1994, 1995). Explants from other regions of the brain, such as the cerebellum, also have a growth-promoting effect on thalamic neurites, but other control tissues, such as additional thalamic or non-neural tissues such as liver, do not (Rennie et al 1994, Lotto & Price 1995). Conditioned medium and conditioned substrate experiments have shown that these interactions are mediated by diffusible growth factors (Rennie et al 1994, Lotto & Price 1994). We have not yet identified the molecules involved, although cross-species co-cultures have indicated that they are highly conserved in divergent mammalian species (Lotto & Price 1994).

Further indications of the importance of growth factor-mediated intercellular interactions in the development of the cerebral neocortex have come from the *in vivo* experiments of Cunningham et al (1987) and Haun & Cunningham (1993). These authors have described how thalamocortical and corticocortical neurons can be rescued from axotomy-induced death by the addition of a macromolecular fraction of medium preconditioned with cultured neocortical slices to the sites of the lesions. These findings are in agreement with those of Hisanaga & Sharp (1990), who demonstrated a trophic dependence of at least some dissociated thalamic neurons on diffusible factors released by the cortex *in vitro*.

Although there is now compelling evidence from our work and that of others that diffusible growth factors are produced by the neocortex and its subcortical targets, and that these tissues can respond to these substances, two major questions remain: (1) what are the roles of these factors in normal neocortical development; and (2) what are the identities of the relevant molecules? The aim of this work is to address these two issues.

Methods

To address the first question, we used three types of culture. These were: (1) explant cultures in liquid medium (Yamamoto et al 1989, Molnár & Blakemore 1991, Rennie et al 1994, Lotto & Price 1994, 1995); (2) explant cultures in 3D collagen gels (Lumsden & Davies 1983); and (3) dissociated cultures (e.g. Hisanaga & Sharp 1990). Each method had its own advantages. Explant cultures in liquid medium enabled us to study soluble growth factor-mediated interactions between tissues that retain many of their organotypic features. Similar cultures in 3D collagen gels allowed us to examine axonal growth through

---

FIG. 1. (a) A camera lucida drawing of a slice through the embryonic day (E) 15 mouse brain, hybridized with a probe to the *Dlx2* gene. To study the effects of the cortex on the embryonic thalamus, we cut coronal slices through the brain at the level indicated: DT, dorsal thalamus; VT, ventral thalamus; LGE, lateral ganglionic eminence. The asterisks indicate lines along which the slices were cut so that they included mainly dorsal and ventral thalamus. The stippled area was expressing the *Dlx2* gene; this gene is expressed in the ventral but not the dorsal thalamus (Bulfone et al 1993). (b) A drawing of a slice of the right thalamus, dissected as shown in (a), after three days in culture on a collagen-coated membrane in liquid medium. Crystals of DiI (1,1'dioctadecyl-3,3,3',3'-tetramethyl indocarbocyanine perchlorate) and DiA (4-[p-dihexadecylaminostyryl]-N-methylpyridinium iodide) were placed in the dorsal and ventral regions of the thalamus to label the growth that occurred *in vitro*. This growth had a strong tendency to follow

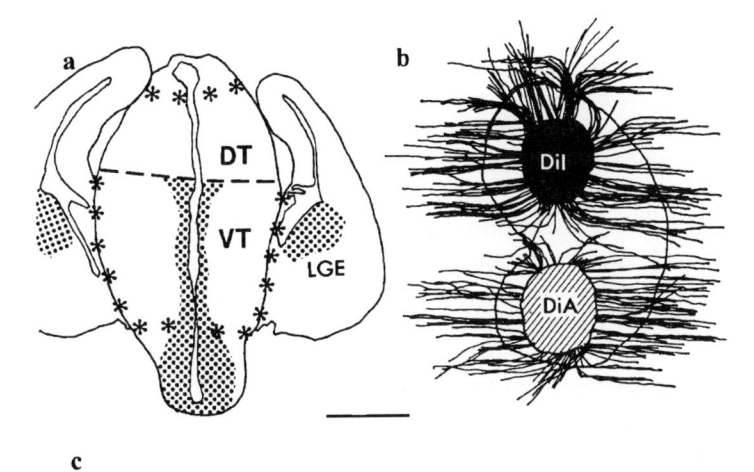

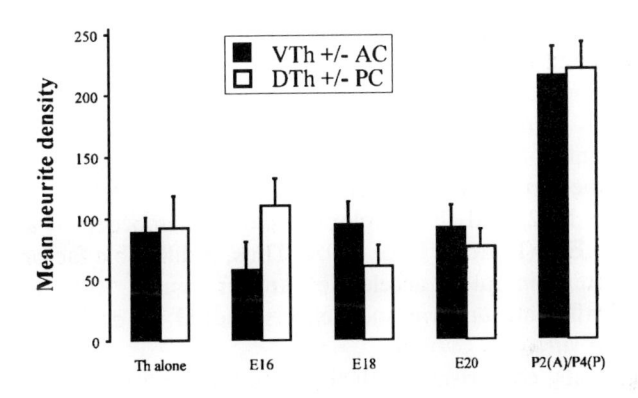

Growth factors and neural activity

the grooves on the collagen-coated membrane (as described in Rennie et al 1994) and to emerge from the explants, where its density was measured to generate the graph in (c). Bar = 0.5 mm. (c) Histograms show the density of outgrowth from E15 thalamic slices cultured either alone (Th alone) or with embryonic or postnatal cortical slices. Embryonic cortical slices were obtained at E16, E18 and E20. Anterior (A) postnatal cortical slices were obtained at postnatal day 2 (P2) and were cultured with ventral thalamus (Vth). Posterior (P) postnatal cortical slices were obtained at postnatal day 4 (P4) and were cultured with dorsal thalamus (DTh). Filled bars are data on outgrowth from the VTh, cultured with or without the anterior cortex (AC); open bars are data on outgrowth from the DTh, cultured with or without the posterior cortex (PC). The only significant effects were with postnatal cortex; interestingly, the anterior cortex stimulated significant outgrowth from the ventral thalamus a day or two earlier than the posterior cortex stimulated significant outgrowth from the dorsal thalamus. Values are mean ± SEM; $n = 4$.

a medium that, unlike liquid, can retain diffusion gradients of soluble factors. Dissociated cultures permitted greater access of diffusible factors to individual neurons and were easier to quantify than explant cultures. All our experiments were carried out in defined serum-free culture medium (Romijn et al 1984). This was essential since serum contains many growth factors in variable amounts that may prevent the detection of factors released by the cultured tissues.

We have also begun to address the second question by testing the effects of known growth factors on the neocortex and its afferents *in vitro*, and by assaying these tissues for expression of the receptors for such candidate molecules both *in vivo* and *in vitro* (using reverse transcription PCR).

All of our experiments were performed in the mouse. An isolated breeding colony was used so that the time of mating was known.

A role for diffusible cortex-derived growth factors in prenatal thalamocortical development?

The first hypothesis that we tested was that diffusible factors released from the developing neocortex initiate, accelerate and/or guide the prenatal growth of axons between the thalamus and the neocortex. Embryonic day (E) 15 thalamic explants were cultured in liquid medium, either alone or with neocortical explants of various prenatal ages. E15 thalamic explants were chosen because the birth of the posterior thalamus is completed between E14 and E15 (Rennie 1992) and the development of thalamocortical afferents is well advanced by E16 (Lotto & Price 1995). Thus, if diffusible factors released by the neocortex initiate, accelerate and/or guide the growth of thalamocortical afferents *in vivo*, such effects should be detectable on E15 thalamic tissue *in vitro*.

Thalamic explants, comprising both the dorsal and ventral thalamus, were prepared by sectioning the brain in the coronal plane (Fig. 1a). We verified that both of these structures were included by studying: (1) the anatomical features of the slices and comparing them to an atlas of the embryonic mouse brain; and (2) the expression patterns of three homeobox-containing genes, *Dlx2*, *Gbx2* and *Wnt3*, that are expressed in either the dorsal but not the ventral thalamus (*Wnt3* and *Gbx2*) or the ventral but not the dorsal thalamus (*Dlx2*) (Fig. 1a, Bulfone et al 1993). After three days *in vitro*, outgrowth from the explants was measured, as described in Rennie et al (1994) and Lotto & Price (1994, 1995), and results on neurites emerging from the ventral and dorsal regions were kept separate. Crystals of the carbocyanine dyes, 1,1′dioctadecyl-3,3,3′,3′-tetramethylindocarbocyanine perchlorate (DiI) and 4-(p-dihexadecylaminostyryl)-N-methylpyridinium iodide (DiA), were placed at each end of every explant, and the labelling that they produced confirmed that the outgrowth emerging from the dorsal thalamus originated in that area, and that the same was true

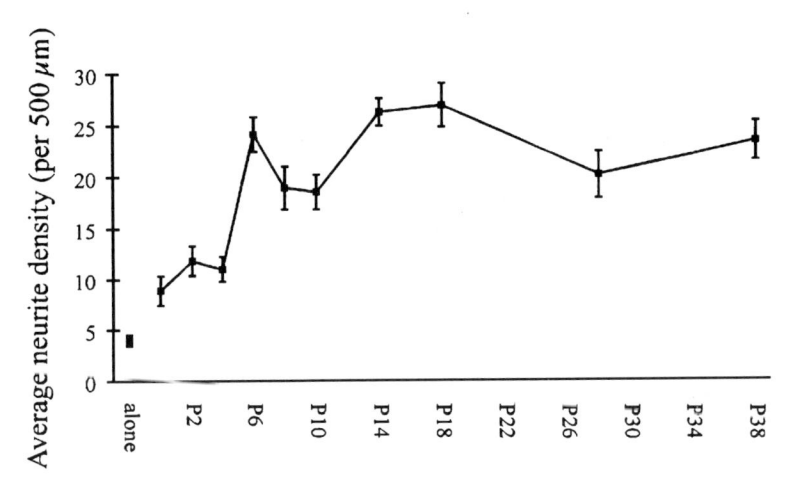

FIG. 2. Increase in the mean density of outgrowth from explants of the embryonic day (E) 15 dorsal thalamus with posterior cortical slices of increasing age. The outgrowth from thalamic explants cultured alone is shown on the left. P, postnatal day. Values are mean ± SEM; $n = 8-12$.

for the ventral thalamus (Fig. 1b). The amounts of outgrowth observed when these explants were cultured either alone or with embryonic or postnatal slices of the anterior or posterior cortex are shown in Fig. 1c. The results indicate that the anterior prenatal cortex does not stimulate increased outgrowth from the ventral thalamus, nor does the posterior prenatal cortex stimulate increased outgrowth from the dorsal thalamus. On the other hand, postnatal cortex does stimulate significantly more outgrowth than that which emerges from thalamic explants cultured alone (final two columns in Fig. 1c, Rennie et al 1994, Lotto & Price 1995).

Overall, these findings argue against a role for diffusible cortex-derived growth factors in the very early prenatal stages of thalamocortical development, e.g. in the acceleration or guidance of advancing thalamocortical axons.

A role for diffusible cortex-derived growth factors in early postnatal thalamocortical development?

One indication that cortex-derived growth factors are more likely to play an important role in the early postnatal development of thalamocortical projections has come from our study of the effects of cortical slices of increasing postnatal age on the outgrowth of neurites from thalamic explants (Fig. 2, Lotto & Price 1995). There was an overall increase in the degree to which cortical slices stimulated outgrowth from the thalamus between birth and postnatal day (P) 38. In these experiments, posterior cortical explants (all

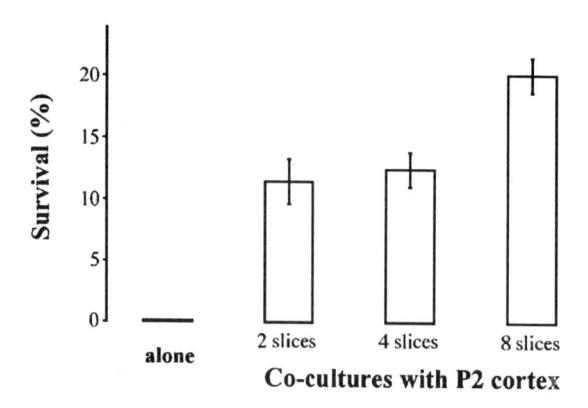

FIG. 3. Mean percentages of postnatal day (P) 2 thalamic neurons surviving after three days in dissociated culture. The cells were either cultured alone (none survived) or with 2, 4 or 8 slices of P2 cortex. Values are mean ± SEM; $n=3$.

of similar volumes) were combined with dorsal thalamic explants and neither the explants nor the neurites growing from them made direct contact. Within this general trend, the results indicated two peaks in the growth-promoting activity of the cortical slices, a first sharp peak around P6 and a second broader peak around P14–18. The first of these rises is likely to follow closely the termination of thalamocortical fibres in layer IV (Lund & Mustari 1977, Molnár & Blakemore 1991). The increased release of growth factors may result from enhanced production stimulated by thalamocortical synaptogenesis with layer IV cells, perhaps reflecting the arrival of spontaneous afferent activity driven from the retina (Galli & Maffei 1988). Consistent with this idea, increased neurotrophin release occurs coincident with afferent arrival at targets in the peripheral nervous system (Davies et al 1987, Rohrer et al 1988). The second rise between P10 and P14 coincides with the opening of the eyes. This temporal relationship suggested to us that the rise may be caused by increased neural activity in the cortex, a possibility that is discussed below.

Further evidence for the importance of diffusible cortex-derived growth factors in the early postnatal development of thalamocortical connections has come from our experiments on the survival-promoting effects of these factors on thalamic cells grown in dissociated culture. We have found that when dissociated thalamic cells from E15 mice are cultured alone, most of these cells survive for at least three days (in agreement with results from E15 thalamic explants cultured alone, Rennie et al 1994). When P2 thalamic cells are cultured under identical conditions, they all die by apoptosis. Many of these dissociated P2 neurons can be rescued from death by culturing them with slices of P2 cortex (Fig. 3). The survival promoting effect is increased by increasing the numbers of cortical slices in each well. In these experiments, the cortical explants are separated by a microporous membrane from the dissociated cells and interactions between the two are mediated by diffusible factors released into the medium that bathes the whole culture. The number of neurons (i.e. cells with neuronal morphology and immunoreactivity for microtubule-associated protein) in each culture is estimated after three days and expressed as a percentage of the numbers of neurons placed in culture.

To date, we have been able to rescue a sizeable proportion (up to 25%) of thalamic cells by including cortical slices in these cultures. It is probable that more would be rescued if we could achieve a higher concentration of growth factor(s) in the medium (more than eight cortical slices cannot be used because there is a limitation of space in the culture wells, and we have not yet carried out experiments with concentrated conditioned medium). Our results indicate that embryonic thalamic neurons are able to survive well in isolation, whereas early postnatal thalamic neurons are not. Many of them may require growth factors that are derived from their major target.

Our overall conclusion that diffusible growth factors may play their most significant role in cortical development after birth is in agreement with the observations of others on the levels of several known growth factors in the developing brain. Levels of mRNA for nerve growth factor (NGF), brain-derived neurotrophic factor (BDNF) and fibroblast growth factor (FGF) all increase in the postnatal brain (Large et al 1986, Thomas et al 1991, Maisonpierre et al 1990, Castren et al 1992). Thus, our hypothesis is that these and/or other diffusible growth factors stimulate the growth of axons that have already reached the cortex and promote the viability of these afferents.

Evidence for a similar role of diffusible growth factors in the development of corticofugal and corticocortical connections

We have also shown that dissociated neurons from the embryonic cortex can survive when they are cultured alone, whereas postnatal cortical neurons die under identical conditions (Lotto 1994). Many of these postnatal cortical neurons are rescued by co-culture with either thalamic or cortical explants (separated from the dissociated cells by a microporous membrane, as above). Our findings on the effects of cortical explants are in agreement with *in vivo* results of Haun & Cunningham (1993), who demonstrated that corticocortical neurons can be rescued from axotomy-induced death by the addition of a macromolecular fraction of medium preconditioned with cultured neocortical slices to the sites of the lesions. Thus, it appears that the postnatal requirement for soluble target-derived growth factors is not confined to the thalamocortical projections, but may be general to other cortical pathways.

238   D. Price et al

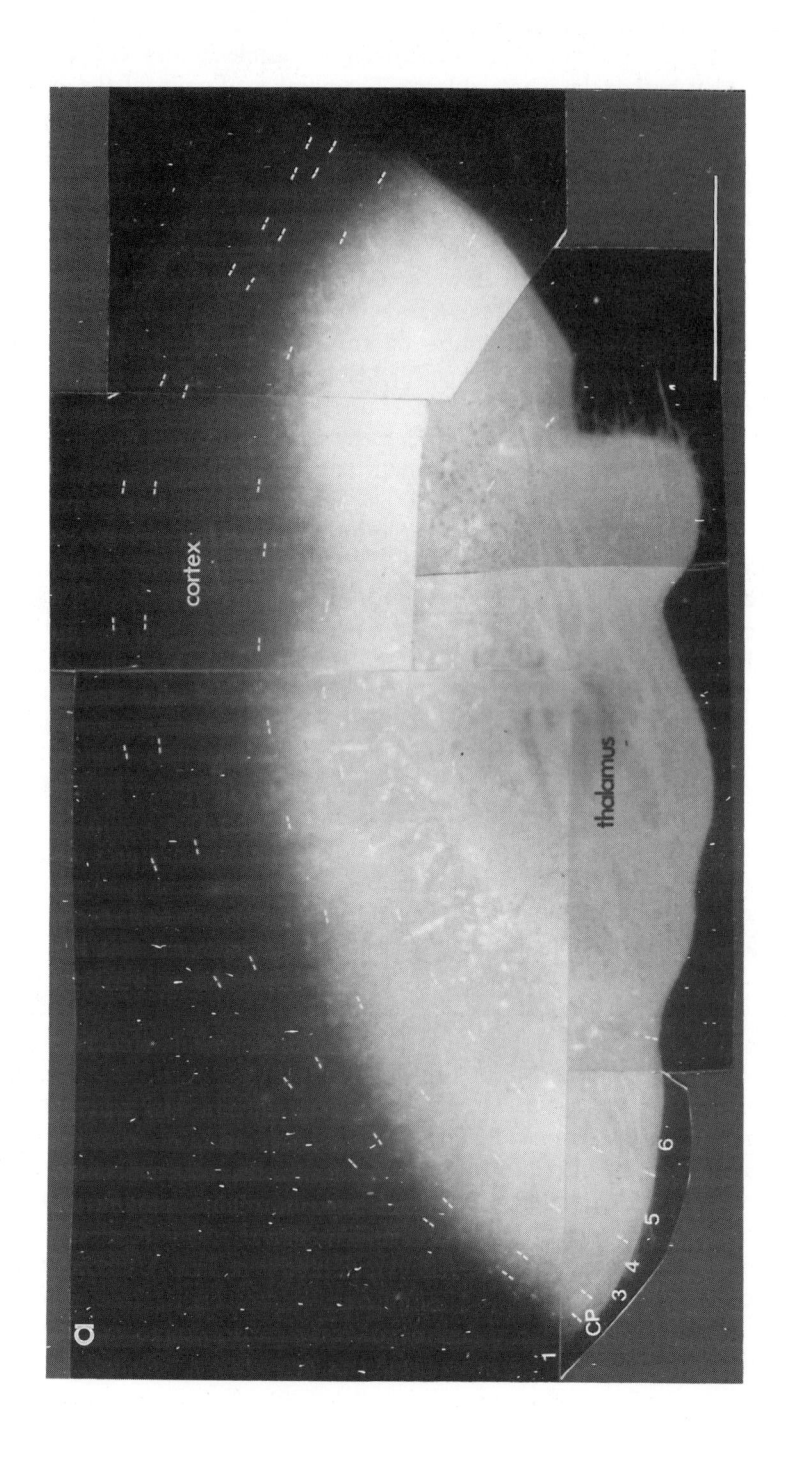

Growth factors and neural activity 239

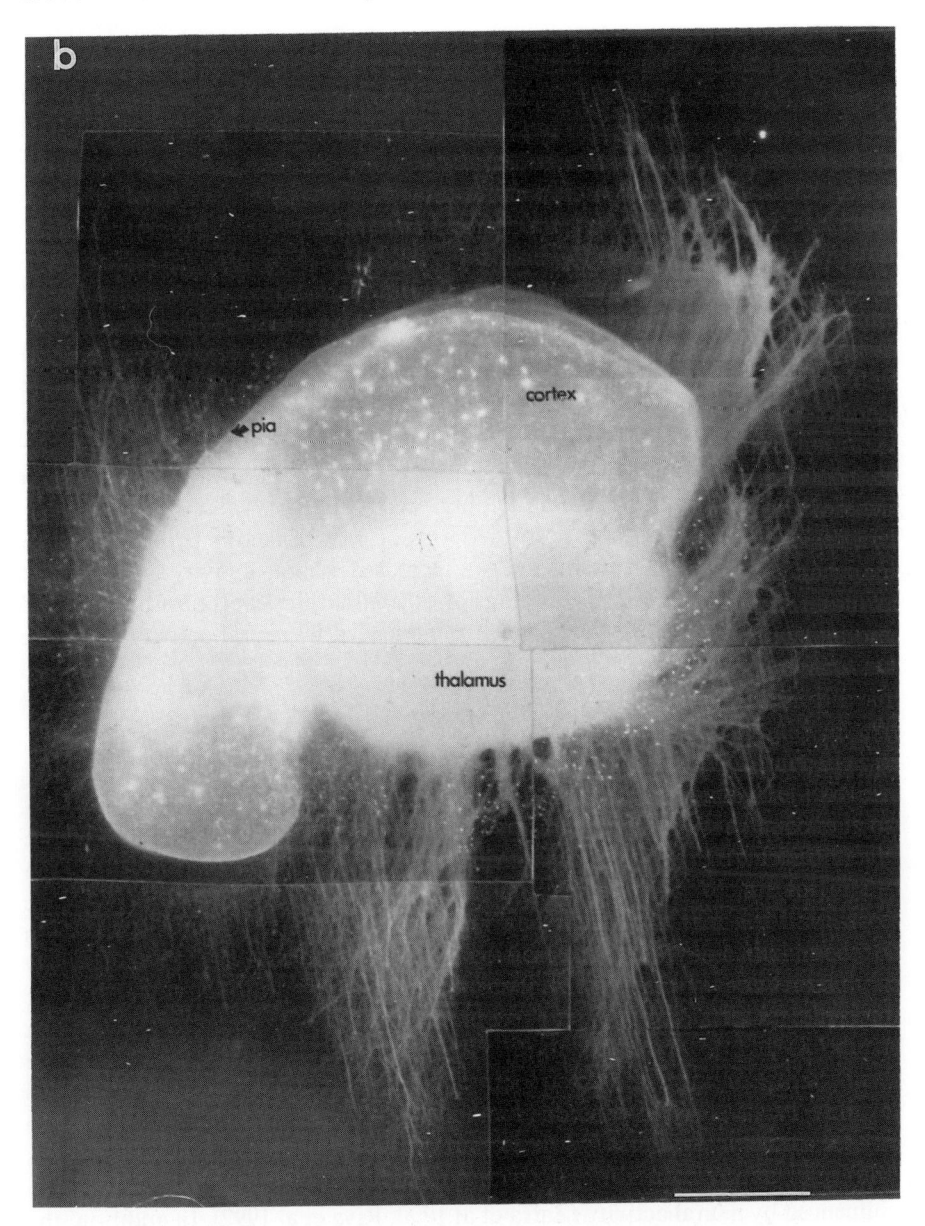

FIG. 4. (a) An organotypic co-culture of the postnatal day (P) 6 occipital cortex with the embryonic day (E) 16 thalamus grown for six days. The thalamus has been labelled with DiI (1,1'dioctadecyl-3,3,3'3'-tetramethylindocarbocyanine perchlorate) and thalamic fibres have innervated the cortex and stopped in layer IV (as described previously by Yamamoto et al 1989, Molnár & Blakemore 1991, Bolz et al 1992). (b) A similar organotypic co-culture to that in (a) but with 50 mmol KCl added to the medium. Bar = 0.5 mm.

Neural activity and growth factor production

The results illustrated in Fig. 2 suggest that increases in the level of growth factor production by the developing neocortex result from heightened neocortical activity. We obtained further evidence for this from two series of co-culture experiments: in one, we cultured E15 posterior thalamic explants (centred on the lateral geniculate nucleus; Rennie et al 1994, Lotto & Price 1995) with occipital cortical slices from P22 mice that had been reared in the dark from birth; in the other, we enhanced activity in the cultured explants by adding KCl to the culture medium. Cortical slices obtained from mice reared in the dark reduced the amount of outgrowth to the levels seen after culture with cortical slices from mice aged P2–4 (Lotto & Price 1995). The effect of adding 50 mmol KCl is shown in Fig. 4. In these experiments, E15 thalamic explants were co-cultured with P6 cortical explants. A co-culture in normal culture medium is shown in Fig. 4a; after addition of KCl, there was an enormous increase in the amount of thalamic outgrowth (Fig. 4b).

Thus, the postnatal increase in the levels of growth factors produced in the neocortex may be a consequence of increased neural activity. Again, this notion is compatible with the findings of others that the levels of production of known growth factors in the cortex are influenced by neural activity (Zafra et al 1990, Riva et al 1992, Castren et al 1992).

The chemical nature of the growth factors

At present, our search for the diffusible molecules that are important in mediating growth-promoting and survival-enhancing interactions during the early postnatal period is focused on the thalamocortical pathway. We have found that medium conditioned with cortex can rescue P2 dissociated thalamic cells, and that heating this medium to 60 °C reduces the effect. This suggests that the trophic interaction described above is mediated at least in part by proteins. To date, we have considered three candidate molecules: FGF; NGF; and BDNF. We have found that if FGF and NGF are added (in nanogram quantities) to thalamic explants cultured alone, they stimulate neurite outgrowth (Lotto & Price 1995). Furthermore, both molecules are synthesized in the cortex (Large et al 1986, Thomas et al 1991) and are released from neurons (although the exact mechanism of FGF release is unclear, Stock et al 1992). FGF mRNA production in the cortex is also influenced by neural activity (Zafra et al 1990, Riva et al 1992). In addition, the tissues that are known to produce FGF and NGF in the developing brain are those that have stimulatory effects in our co-cultures, whereas those that produce them at very low levels do not (Large et al 1986, Shelton & Reichardt 1986, Whittemore et al 1986, Gonzalez et al 1990, Matsuyama et al 1992, Rennie et al 1994, Lotto & Price 1995).

Growth factors and neural activity

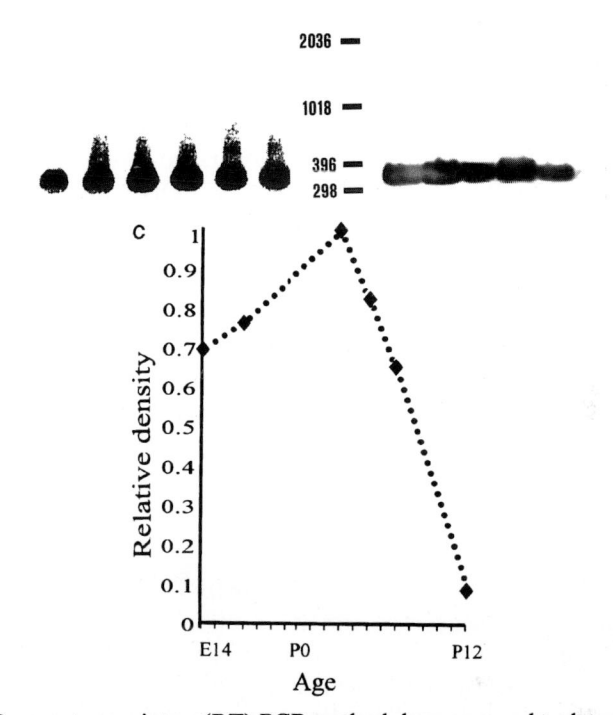

FIG. 5. (a) Reverse transcriptase (RT)-PCR methodology was used to detect expression of the *trkB* gene in explant cultures of the embryonic day (E) 15 thalamus. The primers were both 19 base sequences (bases 2040–2059 and 1706–1725 of the cDNA described by Klein et al 1989). mRNA was extracted, cDNA was synthesized and the PCR was carried out according to standard protocols. PCR products were subjected to electrophoresis on an agarose gel, transferred to nitrocellulose and probed with a labelled oligonucleotide complementary to the sequence from bases 1775–1835 of the cDNA (Klein et al 1989). Lanes: 1 & 2, E15 thalamus immediately after dissection, not cultured; 3, E15 thalamus cultured for two days alone; 4, E15 thalamus cultured for two days with postnatal day (P) 3 cortex; 5, E15 thalamus cultured for four days alone; 6, E15 thalamus cultured for four days with P3 cortex. (b) The RT-PCR carried out as described in (a) on E14 thalamus (lane 1), E17 thalamus (lane 2), P3 thalamus (lane 3), P5 thalamus (lane 4), P7 thalamus (lane 5) and P12 thalamus (lane 6). (c) A quantitative analysis of the Southern blot in (b). The relative density on a scale of 0 (background) to 1.0 (most intense band in b) reflects the amount of *trkB* expressed, as assayed by RT-PCR methodology. These results suggest that the level of *trkB* expression in the thalamus falls rapidly at the same time as the amount of growth factor released from the cortex increases.

BDNF also has a distribution that is compatible with a possible role in neocortical development—its levels in the cortex increase during postnatal development and they are influenced by neural activity (Maisonpierre et al 1990, Isackson et al 1991, Castren et al 1992). In other species (Allendoerfer et al 1994) receptors for BNDF (full-length TrkB) are present in the developing thalamus. This is also likely in the mouse, since full-length mRNA from the *trkB* gene is produced in the embryonic murine thalamus (Fig. 5a). Other results argue against the involvement of BDNF in the postnatal development of thalamocortical connections. First, our preliminary work has indicated that the expression of *trkB* decreases in the thalamus during early postnatal life, at the same time as the level of growth factor production in the cortex is increasing (Figs. 5b and c). Second, we have failed to demonstrate a growth-promoting or a survival-promoting effect of BDNF on either prenatal or postnatal thalamic explants and cells. This result needs to be resolved with the results indicating high levels of *trkB* expression in the prenatal thalamus.

Overall, it is still possible that BDNF has an important autocrine function in the early thalamus, as it may also do in the cortex (Ghosh et al 1994). FGF and NGF may be involved in mediating postnatal interactions between the developing cortex and thalamus. In support of a role for NGF, Maffei et al (1992) have shown that infusions of NGF prevent the shift in ocular dominance distribution of visual cortical neurons that occurs in monocularly deprived rats.

Conclusion

We have provided evidence that diffusible growth factors are important in the development of the neocortex and its connections. This conclusion is in agreement with results obtained with different approaches from several other laboratories. We suggest that these factors play a major role during the later, postnatal stages of neocortical development. One attractive idea is that growth factors produced by the developing neocortex provide a resource for which developing neurons compete. These neurons may require such factors not only for their growth but also for their survival. The rise in the concentrations of growth factors in the cortex during postnatal life (probably driven by increased neural activity) may stabilize patterns of connections and cell numbers as competition is gradually abolished, due to the attainment of sufficient quantities of a vital resource. This may contribute to the ending of the critical period in early postnatal development.

Acknowledgements

Our research is funded by the Medical Research Council, The Royal Society and The Wellcome Trust.

References

Allendoerfer KL, Cabelli RJ, Escandon E, Kaplan DR, Nikolics K, Shatz CJ 1994 Regulation of neurotrophin receptors during the maturation of the mammalian visual system. J Neurosci 14:1795–1811

Bolz J, Novak N, Staiger V 1992 Formation of specific afferent connections in organotypic slice cultures from rat visual cortex cocultured with lateral geniculate nucleus. J Neurosci 12:3054–3070

Bulfone A, Puelles L, Porteus MH, Frohman MA, Martin GR, Rubenstein JLR 1993 Spatially restricted expression of *Dlx-1*, *Dlx-2* (*Tes-1*), *Gbx-2*, and *Wnt-3* in the embryonic day 12.5 mouse forebrain defines potential transverse and longitudinal segmental boundaries. J Neurosci 13:3155–3172

Castren E, Zafra R, Thoenen H, Lindholm D 1992 Light regulates expression of brain-derived neurotrophic factor mRNA in rat visual cortex. Proc Natl Acad Sci USA 89:9444–9448

Cunningham TJ, Haun F, Chantler PD 1987 Diffusible proteins prolong survival of dorsal lateral geniculate neurons following occipital lesions in newborn rats. Dev Brain Res 37:133–141

Davies AM, Bandtlow C, Heumann R, Korsching S, Rohrer H, Thoenen H 1987 Timing and site of nerve growth factor synthesis in developing skin in relation to innervation and expression of the receptor. Nature 326:353–363

Galli L, Maffei L 1988 Spontaneous impulse activity of rat retinal ganglion cells in prenatal life. Science 242:90–91

Ghosh A, Carnahan J, Greenberg ME 1994 Requirement for BDNF in activity-dependent survival of cortical neurons. Science 263:1618–1623

Gonzalez A-M, Buscaglia M, Ong M, Baird A 1990 Distribution of basic fibroblast growth-factor in the 18-day rat fetus: localization in the basement membranes of diverse tissues. J Cell Biol 110:753–765

Haun F, Cunningham TJ 1993 Recovery of frontal cortex-mediated visual behaviours following rescue of axotomized neurons in frontal cortex. J Neurosci 13:614–622

Hisanaga K, Sharp FR 1990 Marked neurotrophic effects of diffusible substances released from non-target cerebellar cells on thalamic neurons in culture. Dev Brain Res 54:151–160

Isackson PJ, Huntsman MM, Murray KD, Gall CM 1991 BDNF mRNA expression is increased in adult rat forebrain after limbic seizures: temporal patterns of induction distinct from NGF. Neuron 6:937–948

Klein R, Parada LF, Coulier F, Barbacid M 1989 *trkB*, a novel tyrosine protein kinase receptor expressed during mouse neural development. EMBO J 8:3701–3709

Large TH, Bodary SC, Clegg DO, Weskamp G, Otten U, Reichardt LF 1986 Nerve growth factor gene expression in the developing rat brain. Science 234:352–355

Lotto RB 1994 A search for factors controlling the formation and maintenance of connections between the thalamus and cortex *in vitro*. PhD thesis, University of Edinburgh, UK

Lotto RB, Price DJ 1994 Evidence that molecules influencing growth and termination in the developing geniculocortical pathway are conserved between divergent mammalian species. Dev Brain Res 81:17–25

Lotto RB, Price DJ 1995 The stimulation of thalamic neurite outgrowth by cortical-derived growth factors *in vitro*: the influence of cortical age and activity. Eur J Neurosci 7:318–328

Lumsden AGS, Davies AM 1983 Earliest sensory nerve fibres are guided to peripheral targets by attractants other than nerve growth factor. Nature 306:786–788

Lund RD, Mustari MJ 1977 Development of the geniculocortical pathway in rats. J Comp Neurol 173:289–305

Maffei L, Berardi N, Domenici L, Paris V, Pizzorusso T 1992 Nerve growth factor (NGF) prevents the shift in ocular dominance distribution of visual cortical neurons in monocularly deprived rats. J Neurosci 12:4651–4662

Maisonpierre PC, Belluscio L, Friedman B et al 1990 NT-3, BDNF, and NGF in the developing rat nervous system: parallel as well as reciprocal patterns of expression. Neuron 5:501–509

Matsuyama A, Iwata H, Okumura N et al 1992 Localization of basic fibroblast growth factor-like immunoreactivity in the rat brain. Brain Res 587:49–65

Molnár Z, Blakemore C 1991 Lack of regional specificity for connections formed between thalamus and cortex in coculture. Nature 351:475–477

Rennie S 1992 *In vivo* and *in vitro* development of the geniculocortical pathway in mice. PhD thesis, University of Edinburgh, UK

Rennie S, Lotto RB, Price DJ 1994 Growth-promoting interactions between the murine neocortex and thalamus in organotypic co-cultures. Neuroscience 61:547–564

Riva MA, Gale K, Mocchetti I 1992 Basic fibroblast growth factor mRNA increases in specific brain regions following convulsive seizures. Mol Brain Res 15:311–318

Rohrer J, Heumann R, Thoenen J 1988 The synthesis of nerve growth factor (NGF) in developing skin is independent of innervation. Dev Biol 128:240–244

Romijn HJ, van Huizen F, Wolters PS 1984 Towards an improved serum-free, chemically defined medium for long-term culturing of cerebral cortex tissue. Neurosci Biobehav Rev 8:301–334

Shelton DL, Reichardt LF 1986 Studies on the expression of the β nerve growth factor (NGF) gene in the central nervous system: level and regional distribution of NGF mRNA suggest that NGF functions as a trophic factor for several distinct populations of neurons. Proc Natl Acad Sci USA 83:2714–2718

Stock A, Kuzis K, Woodward WR, Nishi R, Eckenstein FP 1992 Localization of acidic fibroblast growth factor in specific subcortical neuronal populations. J Neurosci 12:4688–4700

Thomas D, Groux-Mascatelli B, Raes M-B et al 1991 Developmental changes of acidic fibroblast growth factor (aFGF) transcription and expression in mouse brain. Dev Brain Res 59:117–122

Whittemore SR, Ebendal T, Larkfors L et al 1986 Developmental and regional expression of β nerve growth factor messenger RNA and protein in the rat central nervous system. Proc Natl Acad Sci USA 83:817–821

Yamamoto N, Kurotani T, Toyama K 1989 Neural connections between the lateral geniculate nucleus and visual cortex *in vitro*. Science 245:192–194

Zafra R, Hengerer B, Leibrock J, Thoenen H, Lindholm D 1990 Activity dependent regulation of BDNF and NGF mRNAs in the rat hippocampus is mediated by non-NMDA glutamate receptors. EMBO J 9:3545–3550

DISCUSSION

Maffei: I would like to mention some of our results on the role of nerve growth factor (NGF) in the light of David Price's talk. We found that in postnatal rats, if we implant hybridoma cells producing antibodies against NGF into the lateral ventricles during the critical period, i.e. from postnatal day (P) 14 to P15, the development of the visual cortex is disturbed—visual acuity is reduced, binocularity of the cortical neurons remains at an embryonic level and there is a shrinkage of the lateral geniculate nucleus (LGN) neurons (Berardi et al 1994). In addition, treatment with anti-NGF antibodies prolongs the critical period. If these rats are deprived at P45, after the end of the critical period, there is a clear shift in the ocular dominance columns of cortical neurons (Domenici et al 1994). In another series of experiments, we have shown that all the effects of monocular deprivation during the critical period are prevented by injection of NGF in the lateral ventricle both in rats (Maffei et al 1992) and in cats (Carmignoto et al 1993). This suggests that NGF or an NGF-like factor has a role in the development of the visual cortex.

D. Price: I would also like to add that Cabelli et al (1994) have shown that if brain-derived neurotrophic factor (BDNF) or neurotrophin 4/5 (NT-4/5) are injected into the cortex of the ferret before the time of ocular dominance segregation, that segregation doesn't occur.

We found some growth-promoting effects of NGF on thalamic explants, although they're relatively small (Lotto & Price 1995). One problem with NGF is that, unlike BDNF, its level of expression is not regulated by activity (Castren et al 1992). For this reason, NGF may not be the molecule that explains our *in vitro* findings.

J. Price: We made a retrovirus that expressed NGF and β-galactosidase and injected it into cortex. These experiments are analogous to the experiments I described previously (Price et al 1995, this volume). The results were indistinguishable from the controls that expressed only β-galactosidase. This is what we expected because no one has found TrkA receptors in the cerebral cortex. These experiments also have an inbuilt control, which proves that the virus is expressing NGF, because when we inject into the lateral ventricle we occasionally infect cells in the septum/striatal regions which contain TrkA receptors. These clones are quite different from the control clones—they are large and contain many processes which travel all over the forebrain.

Ghosh: We also tried to narrow down the factors that might influence the cortex and the thalamus (A. Ghosh & M. E. Greenberg, unpublished results). We dissociated cortical and thalamic cells and stimulated them with various neurotrophins, and we then looked at the phosphorylation of the TrkA receptors using immunoprecipitation followed by Western blotting with anti-phosphotyrosine antibodies. We found that both in the thalamus and the cortex, TrkA receptors were phosphorylated in response to BDNF and NT-4/5, but not in response to NGF. It is possible that the basal forebrain afferents (which are NGF responsive) in the cortex might play an important role in plasticity.

Rubenstein: I don't believe that the various peptide knockouts have phenotypes which are readily discernible on the cerebral cortex suggesting that they don't act alone. Serafini et al (1994) purified two factors (netrin-1 and netrin-2) that are involved with chemotropism. This suggests that we should be thinking along the lines of developing *in vitro* functional assays rather than trying known molecules. Does the adult cortex also produce factors that support thalamic growth?

D. Price: I have not studied the adult cortex. Presumably, you're asking that because a large source of tissue is required. For example, in experiments on the floor-plate, Serafini et al (1994) isolated the netrins from the brain rather than the smaller spinal cord. However, the cortex is still a substantial size, even at these early times, so there would be enough starting material.

Regarding the knockout experiments, it is not clear to me whether they showed abnormal cortical and thalamic phenotypes because Ernfors et al (1994) did not address this issue in any detail.

Blakemore: Do you control for the volume of tissue as a function of age? Could the lack of prenatal effects be due to the fact that the cortex is thin?

D. Price: No, we always control for volume. At younger ages, we insert more explants deliberately. There is a progressive reduction in cortical neuronal density with age, so these results may be an underestimate of the changes that are really occurring.

Levitt: Tim Cunningham's laboratory has done some work relevant to this (Cunningham et al 1987, Eagleson et al 1992). They have shown that embryonic visual cortex can promote the survival of both neonatal and adult visual thalamus, and the dorsal LGN (dLGN), and they are now trying to isolate the factor that is responsible. In their hands, BDNF, fibroblast growth factor (FGF) and NGF don't promote the survival of dLGN thalamic neurons in transplant experiments, whereas conditioned media from the cortex does (although conditioned media from other controlled tissues does not). They have narrowed it down to a particular molecular weight that is responsible, although it's not clear whether one or two peptides are responsible.

Blakemore: David Price, what was the age of the thalamic explants on which you tested the effect of trophic influence from the cortex?

D. Price: Embryonic day (E) 15.

Molnár: Yes, but this was in the mouse.

Blakemore: So that is equivalent to about E17 in the rat, which means that axon outgrowth will already have occurred from the thalamus before the explant is taken. Therefore, the relay neurons in the explant will have to be axotomized. Any neurite outgrowth that you see can, therefore, be interpreted as being due to the promotion of growth from axotomized cells rather than the stimulation of initial neurite outgrowth. This might explain why the maximum effect of the cortex on axon outgrowth from the thalamus in your experiments is at a stage, namely P12, after all axon growth and synapse formation have stopped *in vivo*. Do you not find that result surprising?

D. Price: No. It depends on what you think the factors are doing. I believe that they have a more important role in promoting survival. They may also be involved in competition and it's possible that they induce arborization. For example, it is possible that the degree to which segregation occurs is dependent upon the levels of growth factors.

Kennedy: Do you have any evidence that the influence of the thalamus on the cortex is mediated by the neurons in the thalamic culture? It could be mediated by glia that are proliferating in the thalamus. Have you tried to block that proliferation to see if you still get the effects?

D. Price: We have only looked at an age where there are few cells other than neurons in the thalamus.

Parnavelas: How does the decrease in density and length of neurites occur after rearing in the dark? It is possible that it is activity dependent, but the retina is present, so the whole pathway is active.

D. Price: I anticipate that in adult mice reared in the dark, the overall level of activity would be reduced within the occipital cortex.

Blakemore: Is cell survival also influenced by rearing in the dark?

D. Price: We haven't looked at that.

Blakemore: Would you expect that rearing in the dark would cause cell death in the LGN?

D. Price: In this situation in culture my prediction would be that dark-reared cortex would be less effective in promoting thalamic survival.

Daw: It has not been resolved whether rearing in the dark has a specific effect on the visual system or a general effect on the whole mouse. Did you look at the effects on neurite length and neurite density outside the cerebral cortex?

D. Price: No, we have not looked at that.

Daw: How do you expect rearing in the dark to have an effect on growth factors?

D. Price: In the hands of Castren et al (1992), the rate of transcription of the genes is reduced, but I don't know what the steps are that are involved in influencing transcription.

Daw: In many ways one might expect there to be an effect on the whole mouse rather than an effect specifically on the visual system.

D. Price: Do you mean that the activity patterns of the mouse are going to change generally?

Daw: Yes, and day–night cycles will be removed.

Maffei: If NGF is injected into mice reared in the dark, most of the effects of rearing in the dark are avoided and vision is almost entirely normal.

Daw: But one wants to know if both the somatosensory cortex and the visual cortex are affected.

Levitt: If activity changes are induced in other pathways, such as in the hippocampus or in the amygdala, by kindling experiments, in which defined pathways are hyperexcited by electrical activity, the induction of growth factor gene expression follows activity changes in the neurons that are activated and doesn't spread. Therefore, it's possible that you have seen activity-driven changes in gene expression that remain within that pathway. This is certainly true for non-visual areas.

Daw: One aspect that's not clear about rearing in the dark is whether it has a significant effect on the overall level of activity in the visual system, because there are 'on' and 'off' cells in the retina. Their spontaneous activity probably hasn't changed all that much, and when light is shined onto the retina, the activity of some cells increases and in others it decreases (Czepita et al 1994).

LaMantia: There are two phases of trophic interactions: one in which cell numbers are shaped and cells die; and the other in which dendritic morphology and synaptic connections are sorted out. Can you work out from your results whether these two phases exist for thalamocortical interactions? Williams & Rakic (1988) showed that there was an early phase of LGN cell death, and that the pattern of connections was established later. Do these represent two similar phases?

D. Price: They could be mediated by the same events. Growth factors cannot explain everything, but cell death continues during the postnatal period in rodents and ends around the time when growth factor levels are reaching high levels. The sorting out of connections is occurring at the same time, and it's possible that growth factors play a role in this. Presynaptic cells may have a requirement for a certain amount of growth factor to support them. The amount required may depend on their own level of activity, and postsynaptic cells may be capable of producing growth factors in quantities that depend on their level of activity. Consequently, a balance between these two could explain some of the changes we're seeing during the refining process.

LaMantia: You're proposing that in the rodent, cell survival and sorting are temporally coincident. However, in other situations, such as in the primate visual system or the chick neurotransmitter system (Williams & Rakic 1988), they're not.

Rakic: We examined cell death in the primate brain, in which the protracted developmental period provides greater temporal resolution. In particular, we wanted to determine whether segregation of ocular dominance columns in the visual cortex is associated with differential cell death in the LGN. We found that about 40% of the cells in this nucleus were eliminated; however, they die before onset of ocular dominance segregation in the cortex (Williams & Rakic 1988).

LaMantia: This is also true for motor neuron death and neuromuscular synapse elimination in the chick and rodent (Purves 1988). Motor neuron cell death occurs at about five to ten days before segregation.

Blakemore: Is it possible that cell death in the thalamus occurs during the period of innervation of the subplate? It is possible that competition for synaptic space in the subplate leads to death of a fraction of neurons in the thalamus.

Rakic: It is possible that these two events coincide.

Molnár: Could you take me through your protocol again? You injected bromodeoxyuridine at E12 and cultured the cortical slices at E20 with E16 thalamus. How long did you follow this co-culture?

D. Price: For nine days.

Molnár: The subplate cells seemed to survive in the solitary cortical slices but die in the co-cultured ones.

D. Price: These cells did not die when the thalamus was present.

Molnár: Why is there a discrepancy between the *in vivo* and *in vitro* results? Why don't they die in culture?

D. Price: I don't know. Presumably, whatever it is that kills them, or the events that lead to their death *in vivo*, are not being mimicked in this situation. You would expect that during that time *in vivo* they would die.

Molnár: Do the subplate cells commit suicide or are they murdered? Is the onset of the subplate death independent from the various interactions during cortical circuit formation or a natural consequence of them? If these interactions are prevented, subplate death will not occur. My suspicion is that the death of the subplate cells might be initiated as early as the waiting period, by exposure to glutamate released from the accumulating thalamic fibres. This 'excitotoxic' idea is just speculation, nevertheless there is evidence that subplate neurons are sensitive to kainate (Ghosh et al 1990) and their intracellular $[Ca^{2+}]$ responds to glutamate agonists (Herrmann and Shatz 1992).

D. Price: A certain level of activity within the culture itself may be required to release the glutamate that kills the cells. If that level is not achieved, then the killing process may not occur.

Molnár: Alternatively, an insufficient reproduction of the waiting period in culture may kill the cells.

References

Berardi N, Cellerino A, Domenici L, Fagiolini M, Pizzorusso T, Cattaneo A 1994 Monoclonal antibodies to nerve growth factor affect the postnatal development of the visual system. Proc Natl Acad Sci USA 91:684–688

Cabelli RJ, Radeke MJ, Wright A, Allendoerfer KL, Feinstein SC, Shatz CJ 1994 Developmental patterns of localization of full-length and truncated TrkB proteins in the mammalian visual system. Soc Neurosci Abstr 20:24

Carmignoto G, Canella R, Candeo P, Comelli MC, Maffei L 1993 Effects of nerve growth factor on neuronal plasticity of the kitten visual cortex. J Physiol 464:343–360

Castren E, Zafra R, Thoenen H, Lindholm D 1992 Light regulates expression of brain-derived neurotrophic factor mRNA in rat visual cortex. Proc Natl Acad Sci USA 89:9444–9448

Cunningham TJ, Haun F, Chantler PD 1987 Diffusible proteins prolong survival of dorsal lateral geniculate neurons following occipital cortex lesions in newborn rats. Dev Brain Res 37:133–141

Czepita D, Reid SNM, Daw NW 1994 Effect of longer periods of dark-rearing on NMDA receptors in cat visual cortex. J Neurophysiol 72:1220–1226

Domenici L, Cellerino A, Berardi N, Cattaneo A, Maffei L 1994 Antibodies to nerve growth factor (NGF) prolong the sensitive period for monocular deprivation in the rat. Neuroreport 5:2041–2044

Eagleson KL, Cunningham TJ, Haun F 1992 Rescue of both rapidly and slowly degenerating neurons in the dorsal lateral geniculate nucleus of adult rats by a cortically derived neuron survival factor. Exp Neurol 116:156–162

Ernfors P, Lee K-F, Jaenisch R 1994 Mice lacking brain-derived neurotrophic factor develop with sensory deficits. Nature 368:147–150

Ghosh A, Antonini A, McConnell SK, Shatz CJ 1990 Requirement for subplate neurons in the formation of thalamocortical connections. Nature 347:179–181

Herrmann K, Shatz CJ 1992 Glutamate-induced calcium responses of developing subplate cells. Soc Neurosci Abstr 18:389.9

Lotto RB, Price DJ 1995 The stimulation of thalamic neurite outgrowth by cortical-derived growth factors *in vitro*: the influence of cortical age and activity. Eur J Neurosci 7:318–328

Maffei L, Berardi L, Domenici L, Parisi V, Pizzorusso T 1992 Nerve Growth Factor (NGF) prevents the shift in ocular dominance distribution of visual cortical neurons in monocularly deprived rats. J Neurosci 12:4651–4662

Purves D 1988 Body and brain: a trophic theory of neural connections. Harvard University press, Harvard, MA

Serafini T, Kennedy TE, Galko MJ, Mirzayan C, Jessell TM, Tessier-Lavigne M 1994 The netrins define a family of axon outgrowth-promoting proteins homologous to C. elegans UNC-6. Cell 78:409–424

Williams RW, Rakic P 1988 Elimination of neurons in the rhesus monkey's lateral geniculate nucleus during development. J Comp Neurol 272:424–436

General discussion IV

In vitro **and** *in vivo* **evidence for areal commitment of neuron precursors in the ventricular zone of the neocortex**

Kennedy: I would like to present some evidence, obtained by Colette Dehay and myself, for the early specification of the development of the neocortex. Our results argue that: (1) the proliferative characteristics of the ventricular zone corresponding to individual neocortical areas are specialized; (2) in some instances, these specializations are susceptible to influences from the thalamus; (3) the ventricular zone exhibits a degree of commitment in its proliferative activity; and (4) these specializations of the ventricular zone are not restricted to primates but are likely to be a universal feature in the generation of the neocortex.

The first point relates to the specialization of the proliferative activity of the ventricular zone. A characteristic feature of the primate striate cortex is that it possesses approximately double the number of neurons between pia and white matter compared with the rest of the neocortex (Rockel et al 1980). Using pulse injections of [^{3}H]thymidine, we have shown that the ventricular zone giving rise to striate cortex has cell cycle parameters which distinguishes it from the ventricular zone giving rise to the adjacent area, i.e. area 18 (Dehay et al 1993). These results also show that the high rates of cell production which characterize the ventricular zone of striate cortex are limited to the last stages of corticogenesis when supragranular layers are being generated. Interestingly, these higher rates of neuron production occur when there is a morphological specialization of the germinal zones (H. Kennedy, I. Smart & C. Dehay, unpublished observations).

The second point addresses the susceptibility of the ventricular zone to thalamic influences. The removal of both eyes early in fetal development leads to a drastic reduction in the dimensions of striate cortex and lateral geniculate nucleus (LGN) (Rakic 1988, Dehay et al 1989). Pasko Rakic has suggested that the reduction in the size of the striate area is the consequence of cortex which was destined to become area 17 failing to acquire the normal cytoarchitectonics of that area. We have tested this idea in the primate by measuring the dimensions of striate, extrastriate and total neocortex following enucleation at different ages (Dehay et al 1995). These results are quite clear: early enucleation leads to a reduction in the size of the striate cortex but does not influence the overall size of the neocortex. This is because the dimensions of the extrastriate visual cortex are actually larger in the experimental primates and that cortex which was destined to become striate cortex has developed the cytoarchitectonics of normal area 18. Cell counts from pia to white matter in striate and peristriate cortex in experimental primates give a similar ratio as in normal primates, i.e. that the reduced striate cortex has a larger cell number than the expanded extrastriate cortex. Although these results need to be confirmed using pulse injections of [^{3}H]thymidine, they suggest that the greater proliferative activity of the germinal zone giving rise to striate cortex is under the control of the early thalamic input to the cortex. This would fit with the developmental timetable of the thalamocortical pathway in primates (Rakic 1976).

The third point concerns the early commitment of the ventricular zone. So far, our results are compatible with peripheral control of cortical development as proposed by Van der Loos & Woolsey (1973). The question is whether the development of the striate cortex is determined fully by the cellular environment. To address this issue, we have examined rates of proliferation in dissociated cultures prepared from presumptive striate and extrastriate cortex of the primate at embryonic day (E) 78. After three days in culture, the density of dissociated striate cortex cells is higher than that in cultures prepared from extrastriate cortex. Anti-PCNA (proliferating cell nuclear antigen) antibody recognizes the pool of proliferating cells. A pulse of BrdU (bromodeoxyuridine) shows that the labelling index of the PCNA-positive cells is higher in cultures from presumptive striate cortex compared to presumptive area 18 (Kennedy et al 1995). The precursors generating striate cortex are characterized by a faster cell cycle time and these *in vitro* experiments suggest that this feature reflects an early commitment.

The fourth point establishes that the early specializations of the ventricular zone are not a unique primate feature. In rodents, differences in the thickness of infragranular layers, in terms of cell numbers, distinguish the motor and the somatosensory cortex (Beaulieu 1993). Using pulse injections of [^{3}H]thymidine, we have compared the cell cycle parameters during the generation of the infragranular layers. The results show that the ventricular zone generating the infragranular layers of the somatosensory cortex shows higher rates of neuron production than the ventricular zone generating the adjacent motor cortex (F. Polleux, C. Dehay & H. Kennedy, unpublished results). Quantitative analysis of the intensity of labelling shows important differences between the two areas suggesting that differences in the mode of division (symmetrical versus asymmetrical) contribute to differences in the growth fraction (i.e. the fraction of cycling cells). Interestingly, these proliferative differences are largely established prior to the arrival of thalamic afferents in the vicinity of the cortical plate, so that thalamic input could only modulate proliferation during the generation of layer IV.

Rakic: We have carried out a similar series of experiments on monkey fetuses with long survival times. When binocular enucleation was performed during the middle of gestation, area 17 of monkeys with a survival time of up to three years retained its proper thickness and cytoarchitectonic properties, but was considerably reduced in total surface area (Rakic et al 1991). However, most of the adjacent cortex anterior to area 17 did not look like normal area 18, but displayed some properties that are characteristic of both areas. We have used the term 'area X' to designate this modified cortex until we discover more about it. As a working hypothesis, we interpreted this finding as an indication that when cells destined to become area 17 receive input appropriate for area 18 they respond in a different way and form a 'hybrid' area that is not found in the normal monkey cerebrum. This may also be a useful model for understanding the function of the occipital cortex in cases of secondary congenital anophthalmia in humans.

I would also like to comment on Henry Kennedy's suggestion that differential cell proliferation in the ventricular zone subjacent to areas 17 and 18 are controlled by thalamic afferents. There is no evidence for this as yet. However, even if the areal differences in cell proliferation are indeed induced by thalamic afferents then one still has to consider how each afferent system innervates specific areas of the ventricular surface.

Kennedy: At first, we were not sure about the identity of the cortex next to area 17. It didn't seem likely that it was cortex which had been dragged back onto the occipital lobe. However, we looked at numbers of neurons and the thickness of area X and area 17, and we found that the ratio is the same as in the normal cortex. There are numerous neurons in the upper layers of the reduced striate cortex that are not present in the adjacent cortex. This suggests that there are fewer neurons in the monkey enucleated earlier than there are in the monkey enucleated later or in the normal monkey. Consequently, the reduction of input from the periphery results in a decreased number of neurons.

O'Leary: Do you think this decrease is at the level of proliferation or cell death?

Kennedy: We're doing [^{3}H]thymidine injections and looking at the labelling index. If the labelling index difference between area 17 and the peristriate cortex is the same as in the normal cortex, we can conclude that the effect is on proliferation.

Daw: Did these monkeys have blobs and stripes?

Kennedy: Yes. They were present at the same frequency, and they were the same size.

Rakic: Area 17 in anophthalmic monkeys has the same pattern and spacing of cytochrome oxidase-positive blobs, but they are stained more lightly than in normal monkeys (Kuljis & Rakic 1991). Subsequently, we found that area X also has blobs but, paradoxically, they are stained more darkly than in area 17. Our interpretation of this new finding is that area X may be functionally more active than area 17 in anophthalmic monkeys. We also examined the distribution of major neurotransmitter receptors in area 17 of prenatally enucleated monkeys. We found that most of the receptor distribution is remarkably similar to that of age-matched controls (Rakic & Lidow 1995). For example, α_2-adrenergic receptors are distributed in sublayers IVA and IVCβ, which are normally associated with colour in this species, even though they have been deprived of any information from the retina.

Krubitzer: It's interesting that area 17 is present at all. A number of species have reduced visual capacity, microchiropteran bats and the platypus for example (Krubitzer et al 1995), and they still retain portions of area 17, even though it's not apparently functional. This suggests that there are regions which are maintained regardless of use and upon which the complexity is built.

Rakic: We have quantitative results from enucleation experiments performed at E60, E70, E80 and E90 in rhesus monkeys. There are gradients of size and cell numbers in the LGN. The size and number of cells in area 17 is proportional to this gradient. If we are precise with the timing of enucleation, we can predict the size of area 17.

Krubitzer: Can you eliminate area 17 completely?

Rakic: No. However, we could not enucleate embryonic retina before E60 because the uterus is too small in relation to the placenta.

Blakemore: Why do you think that early enucleation results in an apparently normal, but smaller, area 17? At the time of enucleation, the LGN has presumably already distributed its axons to area 17, or at least to the subplate of area 17. Two thirds of the cells in the LGN subsequently die as a result of the enucleation, so why is there not just patchy loss of individual afferent axons within a normal sized area 17, rather than shrinkage of the whole territory of innervation?

Kennedy: I don't know. There are holes in the cortex, but they are minor. No one has done DiI (1,1′dioctadecyl-3,3,3′,3′-tetramethylindocarbocyanine perchlorate) labelling yet, so we do not know much about the input.

Molnár: It is possible that area 17 cannot be entirely eliminated by binocular enucleation at any age, no matter how early the experiment is performed, because of the relatively autonomous nature of thalamocortical organization. Perhaps a more direct approach would be to cut the entire internal capsule before thalamic fibre outgrowth and prevent the fibres from reaching the cortex at an earlier stage.

Rakic: But we would not be able to interpret the results because the corticothalamic fibres would also be ablated. I predict that if the whole of the geniculate could be removed, area 17, as we know it, would not be present. My reasoning is that both cortical cells and the thalamic input play a synergistic role in forming area 17. It is not enough just to have cortical cells to form area 17, their differentiation is also dependent on input from the thalamus. If a portion of the cortex is predetermined to be responsive to geniculate input, area

17 is formed. However, if the cortex receives a different input, it may respond in a different way and make a new area. This new area would not be normal area 18, and it wouldn't be area 17 either because it has not received input from the geniculate.

Molnár: Perhaps the hybrid area was exposed to all these thalamic factors, but only at the beginning.

O'Leary: Windrem & Finlay (1991) have published the result of making lesions in the LGN early in hamster life. They showed that the presumptive visual cortex lost its granularity and looked like motor cortex. There's also a natural case of this in the primate (Huntley & Jones 1991). They showed that area 4 is granular at birth, and many weeks later it takes on the characteristic agranular appearance of the motor cortex.

Jones: Area 4 is not highly granular. In the fetus there is a layer IV that continues from the parietal cortex through the postcentral gyrus into the precentral gyrus. These cells disperse as area 4 thickens and as the dendrites of layer III and IV pyramidal cells expand and spread. It looks like a developing area 4 with an added granular layer, rather than agranular cortex.

I would like to mention one other experiment. Steve Wise and I removed the thalamus of newborn rats and looked at the barrel field in the adults (Wise & Jones 1978). We found that there was a homogeneous granular cell layer without barrels. This indicated that there is normally a degree of specification of cortex to become granular cortex, but imposed upon that are the elements of fine-grain architecture, namely the barrels.

LaMantia: In visual system experiments, where both eyes are removed, much of the visual system remains; for instance, area 17, blobs and a geniculocortical map are present. This was also true in the experiments of Schlaggar & O'Leary (1991) where they didn't take out the whiskers, but they removed activity in the somatosensory cortex. This suggests that activity from the periphery doesn't have an organizing effect on regional and modular circuitry. Therefore, what is the role of the periphery in this?

Kennedy: It probably has a role in connectivity because there are numerous callosal connections in the cortex and area X. These connections are not present in area 17 throughout any normal developmental period.

LaMantia: Can peripherally driven activity functionally identify and modify the individual synapses and fine-grain synaptic architecture, but not large-scale patterns of connections?

Rakic: Periphery-driven activity can certainly modify synaptic pattern, but because synapses are distributed differently in areas 17 and 18, the effect on the layers of the cortex would be different in each area. Accumulation of a large number of synapses forms neuropil, which separates cells into layers. Therefore, thalamic fibres can have an influence on the layering pattern of the cortex, but in an area-specific manner.

Levitt: I initially proposed (Levitt et al 1993) that there are some phenotypic features of cortex that are acquired early and are not modified by environmental perturbation. However, the phenotypic properties of cells in the cortex are complex and are likely to be acquired over a period of time, and each of these may be specified in a slightly different manner. This concept of a gradual acquisition of phenotypic traits is similar to that which has been described in invertebrate tissues (see Greenwald & Rubin 1992).

LaMantia: What is the range of choices? And what determines whether the decision is made through cell–cell interactions, activity or some other earlier effect?

Levitt: One of the problems we have in answering those questions is that it is difficult to mark these areas of cortex.

References

Beaulieu C 1993 Numerical data on neocortical neurons in adult rat, with special reference to the GABA population. Brain Res. 609:284–292

Dehay C, Horsburgh G, Berland M, Killackey H, Kennedy H 1989 Maturation and connectivity of the visual cortex in monkey is altered by prenatal removal of the retinal input. Nature 337:265–267

Dehay C, Giroud P, Berland M, Smart I, Kennedy H 1993 Modulation of the cell cycle contributes to the parcelation of the primate visual cortex. Nature 366:464–466

Dehay C, Giroud P, Berland M, Killackey H, Kennedy H 1995 The contribution of thalamic input to the specification of cytoarchitectonic cortical fields in the primate: effects of bilateral enucleation in the fetal monkey on the boundaries, dimensions and gyrification of striate and extrastriate cortex. J Comp Neurol, in press

Greenwald I, Rubin GM 1992 Making a difference: the role of cell–cell interactions in establishing separate identities for equivalent cells. Cell 68:271–282

Huntley GW, Jones EG 1991 The emergence of architectonic field structure and areal borders in developing monkey somatosensory cortex. Neuroscience 44:287–310

Kennedy H, Dehay C, Berland M, Savatier P, Cortay V 1995 In vitro proliferation of primate cortical precursors and specification of cortical areas. Neurosci Abstr 21:593.1

Krubitzer L, Manger P, Pettigrew J, Calford M 1995 Organization of somatosensory cortex in monotremes: in search of the prototypical plan. J Comp Neurol 351:261–306

Kuljis RO, Rakic P 1990 Hypercolumns in primate visual cortex develop in the absence of cues from photoreceptors. Proc Natl Acad Sci USA 87:5303–5306

Levitt P, Ferri RT, Barbe MF 1993 Progressive acquisition of cortical phenotypes as a mechanism for specifying the developing cerebral cortex. Perspect Dev Neurobiol 1:65–74

Rakic P 1976 Prenatal genesis of connections subserving ocular dominance in the rhesus monkey. Nature 261:467–471

Rakic P 1988 Specification of cerebral cortical areas. Science 241:170–176

Rakic P, Lidow MS 1995 Distribution and density of neurotransmitter receptors in the absence of retinal input from early embryonic stages. J Neurosci 15:2561–2574

Rakic P, Suner I, Williams RW 1991 A novel cytoarchitectonic area induced experimentally within the primate visual cortex. Proc Natl Acad Sci USA 88:2083–2087

Rockel AJ, Hiorns RW, Powell TPS 1980 The basic uniformity in structure of the neocortex. Brain 103:221–244

Schlaggar BL, O'Leary DDM 1991 Potential of visual cortex to develop an array of functional units unique to somatosensory cortex. Science 252:1556-1560

Van der Loos H, Woolsey TA 1973 Somatosensory cortex: structural alteration following early injury to sense organs. Science 179:395–398

Windrem MS, Finlay BL 1991 Thalamic ablations and neocortical development: effects on cortical cytoarchitecture and cell number. Cereb Cortex 1:220–240

Wise SP, Jones EG 1978 Developmental studies of thalamocortical and commissural connections in rat somatic sensory cortex. J Comp Neurol 178:187–208

Factors that are critical for plasticity in the visual cortex

N. W. Daw, S. N. M. Reid, X.-F. Wang and H. J. Flavin

Department of Ophthalmology and Visual Science, Yale University School of Medicine, 330 Cedar Street, PO Box 208061, New Haven, CT 06520-8061, USA

> *Abstract.* Factors that may be critical for plasticity in the visual cortex are evaluated according to three criteria. (1) Do antagonists to the factor abolish plasticity? (2) Does the concentration or activity of the factor peak with the critical period for plasticity? (3) Does rearing in the dark, which postpones the critical period, affect the factor in a similar fashion? N-methyl-D-aspartate receptors fulfil all three criteria. Metabotropic glutamate receptors fulfil two of them. Most other putative factors do not fulfil more than one.
>
> *1995 Development of the cerebral cortex. Wiley, Chichester (Ciba Foundation Symposium 193) p 258–276*

The visual cortex is mutable or plastic early in life. Any deficits in the image that reaches the retina, or in the coordination of the image on the right retina with the image on the left retina, can lead to permanent irreversible alterations in the synaptic connections in the visual cortex, if the deficit is not treated early. There is a critical period for these effects, which starts soon after the eyes open and ends around puberty. The system is most sensitive early in this period— between four and six weeks of age in the cat, between six months and one year of age in the human and between three and six weeks of age in the rat.

The paradigm used most often to study plasticity in the visual cortex is monocular deprivation. The eyelids of one eye are sutured shut for a period of time. The eye is then opened again, a sample of cells in the visual cortex is assessed for eye dominance and an ocular dominance histogram is constructed. In the cat after one eye is closed for several days at the peak of the critical period, nearly all cells are driven exclusively by the open eye. Very few cells can be driven from the eye that had been closed (Wiesel & Hubel 1963). This is due partly to the shrinkage of axon terminals that project from the lateral geniculate nucleus to the visual cortex (LeVay et al 1980), but there are also changes in the connections within the visual cortex.

A number of treatments have been shown to reduce or abolish these ocular dominance changes:

Factors critical for plasticity

(1) tetrodotoxin (TTX) (Stryker & Harris 1986);
(2) infusion of antagonists to the N-methyl-D-aspartate (NMDA) receptor (Kleinschmidt et al 1987);
(3) reduction in the inputs of noradrenaline and acetylcholine to the cortex (Bear & Singer 1986);
(4) injection of cortisol (Daw et al 1991);
(5) infusion of nerve growth factor (NGF) (Carmignoto et al 1993);
(6) lesions of the internal medullary lamina and medial dorsal nucleus in the thalamus (Singer 1982);
(7) lesions of the cingulum (Gordon et al 1990);
(8) infusion of glutamate into the cortex (Shaw & Cynader 1984);
(9) anaesthesia and paralysis (Freeman & Bonds 1979).

However, this list includes some treatments that are general non-specific depressants of neural activity, such as anaesthesia, and others that are hard to interpret, such as lesions of the medial dorsal nucleus of the thalamus. The latter affects both attention and eye movement. Which effect is relevant in this situation has not been determined. Thus, the list needs to be evaluated by additional criteria.

One criterion for a factor or substance that is critical for plasticity is that the factor should be most abundant, or have highest activity, at the peak of the critical period. Such a factor would be present in a young animal, thus allowing plasticity, and relatively absent in the adult. There are numerous changes in both the levels and the activities of various transmitters, second messengers and proteins during development in the visual cortex, but there are only three cases where quantitative measurements show that the activity follows the time-course of the critical period accucately: binding of [^{3}H]MK 801 to NMDA receptors (Gordon et al 1991); stimulation of phosphoinositide turnover by ibotenate through metabotropic glutamate receptors (Dudek & Bear 1989); and activity of protein kinase C in the cytosol (Sheu et al 1990). As a counter-example, consider the growth-associated protein, GAP-43, which has been implicated in long-term potentiation in the hippocampus. In the visual cortex its concentration peaks before the critical period starts (McIntosh et al 1990). GAP-43 is concentrated in growth cones; consequently, its prime function is likely to be in the growth of axons rather than in the synaptic alterations that occur in sensory-dependent plasticity after the axons have arrived at their targets and formed their initial connections.

Another criterion is that the concentration or activity of the substance should be altered in the same way that the critical period is affected by treatments, such as rearing in the dark, that affect the critical period. Rearing in the dark postpones both the start (Mower 1991) and the end (Cynader & Mitchell 1980) of the critical period. Thus, light-reared kittens are more plastic than dark-reared kittens at five to six weeks of age; at eight to nine weeks they are equally plastic; and at 12–20 weeks dark-reared kittens are more plastic than light-reared ones. This criterion has almost never been applied thoroughly, with comparisons made at all three ages, but it is a most important test.

Using the criteria listed above, we will evaluate the evidence that the NMDA and metabotropic glutamate receptors are crucial or critical factors in plasticity in the visual cortex.

NMDA receptors in plasticity in the visual cortex

Antagonists to the NMDA receptor reduce the ocular dominance shifts that usually occur after monocular deprivation (Kleinschmidt et al 1987). Control experiments show that this is not due to abolition of activity, as is the case when TTX is infused into the eye or the cortex. A reduction in the ocular dominance shift is found with an injection of the NMDA channel blocker, MK 801, such that the activity in the cortex is more than 50–75% of normal for at least 75% of the time (Daw 1994). However, the effect is not specific to ocular dominance shifts—the receptive field properties of the cells in the visual cortex are also degraded (Kleinschmidt et al 1987, Daw 1994).

The number of NMDA receptors in the visual cortex follows the critical period because there is a peak in the binding of [^{3}H]MK 801 at the same time as the peak of the critical period. However, there are still a substantial number of NMDA receptors in the adult (Gordon et al 1991, and Fig. 1), concentrated in layers II and III. Presumably, these receptors play a role in normal visual processing or in some form of plasticity other than ocular dominance plasticity.

One can make a functional assay of the contribution of NMDA receptors to the visual response of a cell by iontophoresis with APV (D-2-amino-5-phosphovaleric acid) and by measuring how much this reduces the visual response. The NMDA contribution to the visual response in layers II and III is high at all ages, on average between 60 and 80% (Fox et al 1989). On the other hand, the NMDA contribution to the visual response drops in layers IV, V and VI from 50–60% at three weeks of age to 10–15% at six weeks of age in the cat. This drop occurs at the same time that the afferents from the lateral geniculate nucleus to the cortex are segregating into separate left and right eye bands (LeVay et al 1978, see Fig. 2).

Rearing in the dark postpones this drop in the NMDA contribution to the visual response indefinitely (Czepita et al 1994). If cats are reared in the dark until six weeks of age, then brought into the light, the drop in the NMDA contribution to the visual response proceeds (Fox et al 1991). Moreover, what is affected is the NMDA contribution to the visual response, rather than the 2-(aminomethyl)phenylacetic acid (AMPA)/kainate contribution (Fox et al 1992).

In summary, NMDA receptors are on the synaptic pathway for plasticity because electrical activity carries the signals from retina to cortex and releases glutamate from the terminals into the cortex. This then activates all classes

Factors critical for plasticity

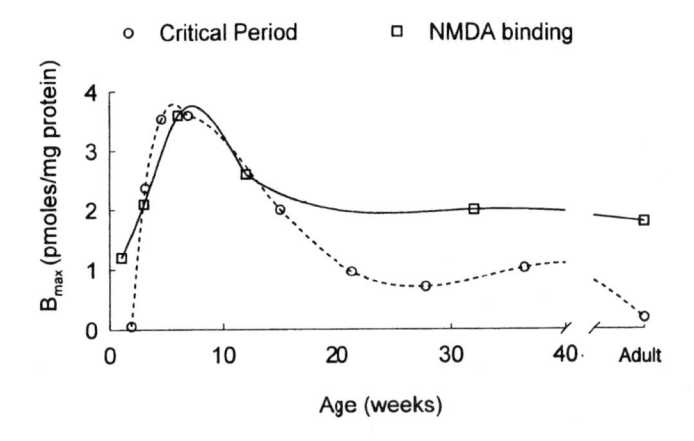

FIG. 1. Time-course of the critical period for ocular dominance plasticity in cat visual cortex, compared to the total number of binding sites (B_{max}) for binding of [^{3}H]MK 801 to NMDA (N-methyl-D-aspartate) receptors. Both rise at the same time, peak at the same time, and fall at the same time. However, ocular dominance plasticity is absent in the adult, although NMDA receptors remain. These are primarily in layers II and III (see Fig. 2). Curve for ocular dominance plasticity scaled to peak at the same level as NMDA receptor binding. From Daw (1994).

of glutamate receptor. Antagonists to the NMDA receptor reduce plasticity. The total number of NMDA receptors reaches a peak at the same time as the peak susceptibility for the effects of plasticity. There is a functional change in the NMDA contribution to the visual response in layers IV, V and VI, and this correlates with the segregation of left and right eye afferents in layer IV. Finally, rearing in the dark postpones both this functional change in the NMDA contribution to the visual response and the critical period.

Metabotropic glutamate receptors

Metabotropic glutamate receptors have also been implicated in the plasticity of the visual cortex. The first evidence came from measurements of ibotenate-stimulated production of phosphoinositides. This varies with age, and the time-course of its variation corresponds closely to the time-course of the critical period (Dudek & Bear 1989).

This experiment was performed before molecular biological techniques were applied to the study of metabotropic glutamate receptors. There are at least seven types of metabotropic glutamate receptor (Nakanishi 1992). Two of them (mGluR1 and mGluR5) increase the production of inositol trisphosphate

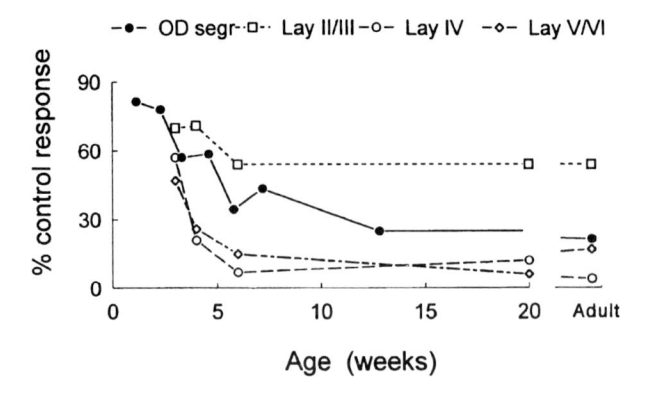

FIG. 2. NMDA (*N*-methyl-D-aspartate) contribution to the visual response in various layers of visual cortex as a function of age compared to ocular dominance segregation in layer (Lay) IV. NMDA contribution to the visual response was measured by iontophoresis with the NMDA antagonist APV (D-2-amino-5-phosphovaleric acid) and calculating how much the visual response was decreased (vertical axis represents percentage of control response for these points). Measurements from a number of single cells in a specific layer at a specific age were averaged to give the points shown. Ocular dominance segregation index calculated from grain counts from [^{3}H]proline transported from one eye to layer IV in the visual cortex (LeVay et al 1978). Grain counts were smoothed, and the area between grain counts and mean level was calculated then expressed as a fraction of a square wave going from minimum to maximum levels. This gives a number 0 for no segregation and 1 for complete segregation. Results are plotted in this figure as 100(1 − segregation index) to use the same scale as NMDA contribution to the visual response. From Daw (1994).

(InsP$_3$), and five decrease the production of cAMP (mGluR 2, 3, 4, 6 and 7). mGluR1 also increases the production of cAMP. The receptors can also be classified according to their agonist specificity: mGluR1 and mGluR5 are preferentially activated by quisqualate; mGluR2 and mGluR3 by trans-1-aminocyclopentane-1,3-dicarboxylate (ACPD); and mGluR4, mGluR6 and mGluR7 by L-2-amino-4-phosphonobutyrate (AP4).

We, therefore, studied the effect of metabotropic receptors on cAMP levels in the visual cortex, and how this effect varies with age in the rat (Flavin et al 1994). Cortical slices from young rats at age 24–25 days were compared with older rats at age 55–56 days. There was a substantial increase in cAMP production following treatment with 100 μM quisqualate, 8 μM 6-cyano-7-nitroquinoxaline-2,3-dione (CNQX) and 200 μM APV (Table 1). APV and CNQX were included to ensure that the observed effects of quisqualate were due to metabotropic glutamate receptor stimulation and not to any indirect actions of quisqualate via ionotropic glutamate receptors. Quisqualate increased cAMP production by a factor of 3.6 and 2.4 in young and older

TABLE 1 **Increase in cAMP[a] by an mGluR1 agonist**

Drug	Young rats (24–25 days old)	Older rats (55–56 days old)
None	18.3 ± 1.0	21.9 ± 2.0
Quisqualate + CNQX + APV	65.4 ± 11.9	51.9 ± 6.0
Quisqualate + MCPG	60.6 ± 7.7	47.2 ± 1.2

[a]cAMP expressed as pmol cAMP/mg protein.
APV, D-2-amino-5-phosphovaleric acid; CNQX, 6-cyano-7-nitroquinoxaline-2,3-dione; MCPG, (+)-α-methyl-4-carboxyphenylglycine.

rats, respectively. Addition of 0.5 mM (+)-α-methyl-4-carboxyphenylglycine (MCPG), which is an antagonist of a variety of metabotropic glutamate receptors, did not counteract this increase (Table 1).

There was also a decrease in cAMP production at some ages with ACPD and AP4 if high concentrations of these agonists were used (Table 2). In these experiments, 10 μM forskolin was used to increase production of cAMP. Adenosine deaminase (ADA) (1.3 units/ml) was also added to prevent any effects of forskolin on adenosine. 1 mM ACPD decreased cAMP concentration in older rats ($P < 0.01$, two-tailed test) but not significantly in young rats. 1 mM AP4 decreased cAMP production significantly in young rats ($P < 0.01$, two-tailed test) and slightly but not significantly in older rats. However, this decrease was not as substantial as the increase in cAMP observed with quisqualate. Thus, the main effect was observed with increases rather than decreases of cAMP levels. These increases were substantial, were found in young rats rather than old ones and were produced by specific concentrations of agonists.

mGluR1 and mGluR5 are, therefore, likely to be critical factors in plasticity because both of them increase InsP$_3$ production, and mGluR1 also increases cAMP levels. mGluR2 and mGluR3 decrease cAMP levels, but the agonist that preferentially activates them (ACPD) has a larger effect in older rats than in young ones. mGluRs 4, 6 and 7 also reduce cAMP, but the agonist that preferentially activates them (AP4) is not specific and only does this at high concentrations.

It is, therefore, interesting to see how the location and amount of mGluR1 and mGluR5 change with age in the cat visual cortex (Reid et al 1995). mGluR1 is concentrated in layers V, VI and I and does not change appreciably with age (Fig. 3). However, it is more abundant at the peak of the critical period than later on. The laminar pattern of mGluR5, on the other hand, does change with age (Fig. 4). It is distributed over layers I and III–VI early on, and later becomes restricted to layer IV. This pattern corresponds to the distribution of terminals of the afferents from the lateral geniculate nucleus to the cortex. In addition, the overall amount decreases with age.

TABLE 2 Decrease in cAMP[a] by mGluR2 and mGluR4 agonists

Drug	Young rats (24–25 days old)	Older rats (55–56 days old)
None	17.6 ± 1.8	22.2 ± 1.9
Forskolin + ADA	210 ± 16	197 ± 25
Forskolin + ADA + AP4	112 ± 7	138 ± 10
Forskolin + ADA + ACPD	164 ± 30	110 ± 11

[a]cAMP expressed as pmol cAMP/mg protein.
ACPD, trans-1-aminocyclopentane-1,3-dicarboxylate; ADA, adenosine deaminase; AP4, L-2-amino-4-phosphonobutyrate.

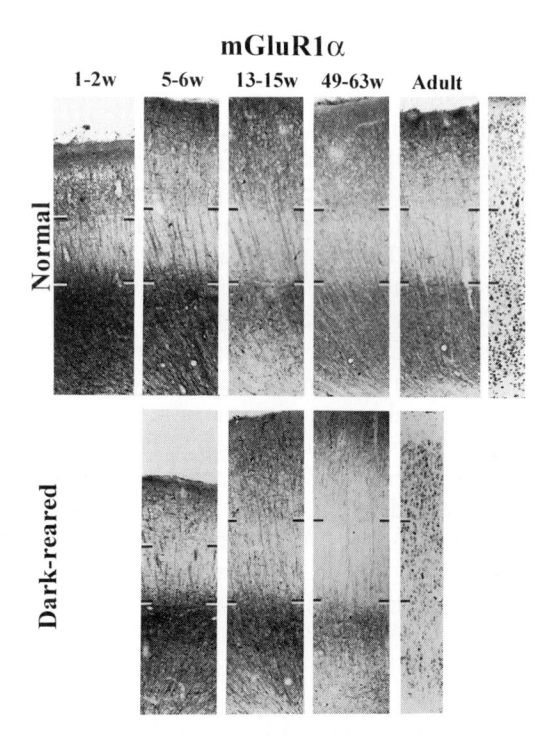

FIG. 3. Immunocytochemistry for mGluR1α in the visual cortex of cats of various ages. Ages given at the top; w, weeks. Nissl sections shown at the extreme right. Sections from dark-reared cats of the same age are shown below. Top, middle and bottom layers for normal cats are the same as those for dark-reared cats. Dark lines show the III/IV and IV/V boundaries. The layering pattern does not change with age, and dark rearing has no effect on the pattern.

Factors critical for plasticity

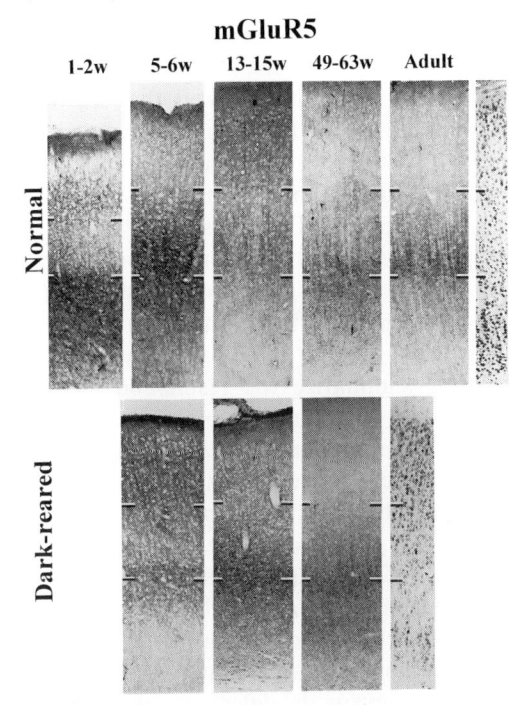

mGluR5

FIG. 4. Immunocytochemistry for mGluR5 in the same format as Fig. 3. Staining is concentrated in layers I and III–VI at early ages, and moves to layer IV (the adult pattern) between one to two weeks (w) and five to six weeks of age. Rearing in the dark postpones this shift in the layering pattern.

Rearing in the dark postpones this change in the laminar pattern of mGluR5 (Fig. 4). When kittens are reared in the dark until 49–63 weeks of age, mGluR5 remains at substantial levels in layers V and VI, but appears later in layer IV.

The specific role of mGluR1 as compared to mGluR5 is unclear. However, it is interesting to note that the laminar patterns of mGluR1 and mGluR5 are complementary to each other in the adult, with mGluR5 found primarily in layer IV and mGluR1 found in other layers. Moreover, substantial quantities of mGluR1 and mGluR5 are found in adult cortex after the critical period for ocular dominance plasticity is over. Consequently, like NMDA receptors, they must play a role in processing normal visual information, or in some other form of plasticity. The important point in relation to plasticity is that they have a greater effect on second messengers at the peak of the critical period than they do following the peak.

Experiments on whether metabotropic glutamate antagonists decrease or abolish ocular dominance shifts are controversial. Earlier experiments with antagonists that are not specific suggested that ocular dominance shifts could be decreased by these drugs (Bear & Dudek 1991). A more recent experiment with MCPG, which is believed to be the most specific antagonist for a variety of metabotropic glutamate receptors, showed that this antagonist does not decrease ocular dominance shifts (Hensch & Stryker 1994). However, the antagonist specificity of MCPG was tested in cells transfected with mGluR1 for InsP$_3$ increases, and in CHO (Chinese hamster ovary) cells transfected with mGluR2 or with mGluR4 for cAMP decreases (Hayashi et al 1994). It was not tested for cAMP increases produced by metabotropic glutamate receptors, and our evidence suggests that it is not a good antagonist for those effects (Table 1). Thus, the question concerning the effect of metabotropic glutamate antagonists on ocular dominance shifts remains to be investigated satisfactorily.

The following points summarize the results on metabotropic glutamate receptors: (1) metabotropic glutamate receptors are present in the cortex and are, like NMDA receptors, along the pathway for plasticity because glutamate is the excitatory transmitter in the cortex; (2) InsP$_3$ increases and cAMP increases produced by metabotropic glutamate receptors are larger at the peak of the critical period than they are in the adult; (3) immunocytochemistry shows more labelling for mGluR1 at the peak of the critical period (than after the peak) without a change in laminar distribution, and it also shows a change in the laminar distribution of mGluR5; and (4) as far as it has been tested, rearing in the dark postpones these changes. Whether mGluR antagonists reduce or abolish plasticity awaits the development of more potent and specific antagonists.

Interaction of metabotropic glutamate and NMDA receptors

NMDA receptors are hypothesized to play a role in plasticity either by increasing the possibility that the postsynaptic cell will fire, due to the voltage dependency of the NMDA channel; by letting Ca^{2+} into the cell to influence various enzymes within the cell; or both.

There are numerous steps at which metabotropic glutamate receptors may play a role in plasticity because they can activate G proteins, which activate phospholipase C to produce diacylglycerol and InsP$_3$. Diacylglycerol then increases cAMP, and InsP$_3$ releases Ca^{2+} from intracellular stores. One of the simplest possibilities is that metabotropic receptors may potentiate the response of the cell to the NMDA receptor. This has been found to occur in the cerebellum (Kinney & Slater 1993), hippocampus (Harvey & Collingridge 1993) and other areas. It also occurs in the visual cortex (Wang & Daw 1994).

Iontophoresis of NMDA near a neuron in the visual cortex depolarizes it. The metabotropic glutamate agonist ACPD has little effect on the polarization of the cell. However, when ACPD and NMDA are applied together, the response is significantly larger than when NMDA is applied alone (Fig. 5). This effect occurs within a few seconds, and decays within three minutes. It is,

Factors critical for plasticity

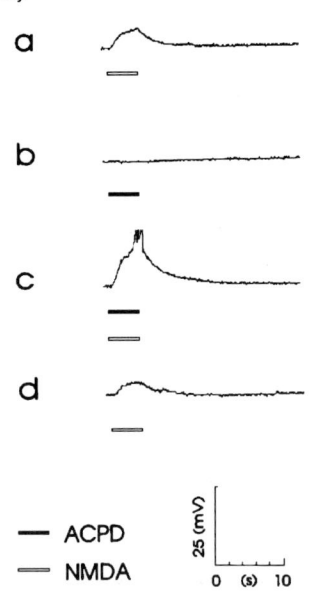

```
    a    ___/_____
         ═

    b    _____
         —

    c    ___/_____
         —
         ═

    d    ___/_____
         ═
```

— ACPD
— NMDA

25 (mV)

0 (s) 10

FIG. 5. Potentiation of the response to NMDA (*N*-methyl-D-aspartate) by ACPD (trans-1-aminocyclopentane-1,3-dicarboxylate) in rat visual cortex. Recording from a cell in layer V (resting potential −69 mV, input resistance 110 MΩ). (a) Iontophoresis of NMDA near the cell gives a depolarization; (b) iontophoresis of ACPD gives no response; (c) iontophoresis of ACPD together with NMDA increases the response, to 2.8 times the control (calculated as the area under the depolarization); (d) response to NMDA recovers to less than control within two minutes. Scale, as shown.

therefore, a short-term effect on the time-scale of plasticity in the visual cortex. If this is the route by which metabotropic receptors affect plasticity, then the primary effect has to be one of the two effects of NMDA receptors. However, there may be more than one route by which both NMDA and metabotropic glutamate receptors have an effect.

Steps further down the path

The final outcomes in the visual cortex after closing one eye of a young cat are that axonal terminals from the deprived eye retract, terminals from the good eye expand and the properties of cells in the cortex change. This clearly involves activation of enzymes by second messengers within the cell and changes in the proteins synthesized by the cell. More investigations will be needed before we know all the steps in the pathway and which steps are more active in young cats compared with older ones. Both NMDA and metabotropic glutamate receptors affect the level of Ca^{2+} within the cell. At the present time, one can only say that more Ca^{2+} enters the cell at the peak of the critical period because there are more NMDA receptors. There is little evidence that Ca^{2+} entry or release per receptor is increased (Feldman et al 1990), or that Ca^{2+} channels are more active. Indeed, one of the few measurements that has been made on the Ca^{2+} system shows that dihydropyridine binding drops between 14 and 28 days of age in the cat, as does the concentration of GAP-43, before the critical period starts (Bode-Greuel & Singer 1988). One of the few steps that is likely to be along the pathway that has been measured quantitatively involves protein kinase C (Sheu et al 1990), because the cytosolic fraction of protein kinase C follows the critical period in the cat quite closely. The membrane fraction follows the critical period less closely.

One of the final goals of research in this field is to bring back plasticity in older animals. It may not be easy to do this until all the steps in the path have been identified because more than one step may be missing or less active in the adult. Consequently, attempts to bring back plasticity have not been tried with many of the putative agonists of steps in the path. Where it has been tried, e.g. with NGF, the results are often controversial (Carmignoto et al 1993, Gu et al 1994). However, rapid progress is being made. One step in the pathway, that involving the NMDA receptor, fulfils all the criteria listed for a critical step in the path, and another, involving the metabotropic receptor, fulfils two of the criteria. A third step, involving protein kinase C, has so far been shown to fulfil one of the criteria. Hopefully, the complete pathway will be established soon, with evidence for all the steps in relation to all the criteria.

Acknowledgements

We would like to thank Carl Romano for supplying the mGluR5 antibody, and Craig Blackstone and Richard Huganir for the mGluR1α antibody. Doug Gregory provided facilities and advice for the cAMP experiments. This work was supported by a grant from the National Eye Institute, EY 00053; a grant from the National Institute of Neurological Diseases and Stroke, PO1 29343; and a grant from the Human Frontier Science Program. S. N. M. R. was supported by a fellowship EY 06443, and H. J. F. by EY 06591.

References

Bear MF, Dudek SM 1991 Stimulation of phosphoinositide turnover by excitatory amino acids: pharmacology, development, and role in visual cortical plasticity. Ann N Y Acad Sci 627:42–56

Bear MF, Singer W 1986 Modulation of visual cortical plasticity by acetylcholine and noradrenaline. Nature 320:172–176

Bode-Greuel KM, Singer W 1988 Developmental changes of the distribution of binding sites for organic Ca^{++} channel blockers in cat visual cortex. Brain Res 70:266–275

Carmignoto G, Canella R, Candeo P, Comelli MC, Maffei L 1993 Effects of nerve growth factor on neuronal plasticity of the kitten visual cortex. J Physiol 464:343–360

Cynader MS, Mitchell DE 1980 Prolonged sensitivity to monocular deprivation in dark-reared cats. J Neurophysiol 43:1026–1040

Czepita D, Reid SNM, Daw NW 1994 Effect of longer periods of dark-rearing on NMDA receptors in cat visual cortex. J Neurophysiol 72:1220–1226

Daw NW 1994 Mechanisms of plasticity in the visual cortex. Invest Ophthalmol 35:4168–4179

Daw NW, Sato H, Fox K, Carmichael T, Gingerich R 1991 Cortisol reduces plasticity in the kitten visual cortex. J Neurobiol 22:158–168

Dudek SM, Bear MF 1989 A biochemical correlate of the critical period for synaptic modification in the visual cortex. Science 246:673–677

Feldman D, Sherin JE, Press WA, Bear MF 1990 N-methyl-D-aspartate-evoked calcium uptake by kitten visual cortex maintained *in vitro*. Exp Brain Res 80:252–259

Flavin HJ, Daw NW, Gregory D, Reid SNM 1994 Metabotropic glutamate receptor stimulated cAMP is implicated in visual cortical plasticity. Soc Neurosci Abstr 20:1171

Fox K, Sato H, Daw NW 1989 The location and function of NMDA receptors in cat and kitten visual cortex. J Neurosci 9:2443–2454

Fox K, Daw NW, Sato H, Czepita D 1991 Dark-rearing delays the loss of NMDA-receptor function in kitten visual cortex. Nature 350:342–344

Fox K, Daw NW, Sato H, Czepita D 1992 The effect of visual experience on development of NMDA receptor synaptic transmission in kitten visual cortex. J Neurosci 12:2672–2684

Freeman RD, Bonds AB 1979 Cortical plasticity in monocularly deprived immobilized kittens depends on eye movement. Science 206:1093–1095

Gordon B, Mitchell B, Mohtadi K, Roth E, Tseng Y, Turk F 1990 Lesions of nonvisual inputs affect plasticity, norepinephrine content and acetylcholine content of visual cortex. J Neurophysiol 64:1851–1860

Gordon B, Daw NW, Parkinson D 1991 The effect of age on binding of MK-801 in the cat visual cortex. Dev Brain Res 62:61–67

Gu Q, Liu Y, Cynader MS 1994 Nerve growth factor-induced ocular dominance plasticity in adult cat visual cortex. Proc Natl Acad Sci USA 91:8408–8412

Harvey J, Collingridge GL 1993 Signal transduction pathways involved in the acute potentiation of NMDA responses by 1S, 3R-ACPD in rat hippocampal slices. Br J Pharmacol 109:1085–1090

Hayashi Y, Sekiyama N, Nakanishi S et al 1994 Analysis of agonist and antagonist activities of phenylglycine derivatives for different cloned metabotropic glutamate receptor subtypes. J Neurosci 14:3370–3377

Hensch TK, Stryker MP 1994 Postsynaptic metabotropic glutamate receptors do not mediate ocular dominance plasticity. Soc Neurosci Abstr 20:216

Kinney GA, Slater NT 1993 Potentiation of NMDA receptor-mediated transmission in turtle cerebellar granule cells by activation of metabotropic glutamate receptors. J Neurophysiol 69:585–594

Kleinschmidt A, Bear MF, Singer W 1987 Blockade of 'NMDA' receptors disrupts experience-dependant plasticity of kitten striate cortex. Science 238:355–358

LeVay S, Stryker MP, Shatz CJ 1978 Ocular dominance columns and their development in layer 4 of cat's visual cortex: quantitative study. J Comp Neurol 179:223–244

LeVay S, Wiesel TN, Hubel DH 1980 The development of ocular dominance columns in normal and visually deprived monkeys. J Comp Neurol 191:1–51

McIntosh H, Daw NW, Parkinson D 1990 GAP-43 in the cat visual cortex during postnatal development. Visual Neurosci 4:585–594
Mower GD 1991 The effect of dark rearing on the time course of the critical period in cat visual cortex. Dev Brain Res 58:151–158
Nakanishi S 1992 Molecular diversity of glutamate receptors and implications for brain function. Nature 258:597–603
Reid SNM, Romano C, Hughes T, Daw NW 1995 Immunohistochemical study of two phosphoinositide-linked metabotropic glutamate receptors (mGluR1α and mGluR5) in the cat visual cortex before, during and after the peak of the critical period for eye-specific connections. J Comp Neurol 355:470–477
Shaw C, Cynader MS 1984 Disruption of cortical activity prevents ocular dominance changes in monocularly deprived kittens. Nature 308:731–734
Sheu FS, Kasamatsu T, Routtenberg A 1990 Protein kinase C activity and substrate (F1/GAP-43) phosphorylation in developing cat visual cortex. Brain Res 524:144–148
Singer W 1982 Central core control of developmental plasticity in the kitten visual cortex. I. Diencephalic lesions. Exp Brain Res 47:209–222
Stryker MP, Harris WA 1986 Binocular impulse blockade prevents the formation of ocular dominance columns in cat visual cortex. J Neurosci 6:2117–2133
Wang XF, Daw NW 1994 Potentation of the response to NMDA by ACPD in rat visual cortex. Soc Neurosci Abstr 20:489
Wiesel TN, Hubel DH 1963 Single cell responses in striate cortex of kittens deprived of vision in one eye. J Neurophysiol 26:1003–1017

DISCUSSION

Rubenstein: Does rearing in the dark change the time of onset of puberty?
Daw: Possibly, although puberty is at the very tail-end of the critical period.
Rubenstein: The critical period looked biphasic, so puberty might be associated with the second phase.
Daw: We tried to find a relationship between the critical period and the onset of puberty using cortisol. We found that when cortisol was given before puberty, the cortex became less plastic. However, when we looked at endogenous rises in cortisol levels, we found that cats don't have adrenarche (Daw et al 1991). Therefore, although cortisol does reduce ocular dominance plasticity, the natural rise in cortisol that occurs just before puberty is not adequate to explain the reduction in ocular dominance plasticity that occurs around that time.
Rakic: Results from cataract operations in humans (Mitchell & Timney 1984) suggest that the critical period is extended after puberty because recovery can only occur if eye surgery is performed before puberty.
Bolz: You showed that the contribution of the *N*-methyl-D-aspartate (NMDA) receptor decreases in parallel with the segregation of ocular dominance columns, and that rearing in the dark prolongs the decrease. Is ocular dominance segregation also delayed in cats reared in the dark?

Daw: There's some controversy over the effects of rearing in the dark and ocular dominance segregation. Mike Stryker (Stryker & Harris 1986) points out that rearing in the dark does not have as large an effect as tetrodotoxin (TTX), whereas George Mower and Nick Swindale say that rearing in the dark does reduce ocular dominance segregation (Mower et al 1985, Swindale 1988). The general consensus is that rearing in the dark reduces ocular dominance segregation, but not as much as TTX. No one has compared this with NMDA receptors.

Blakemore: I have a general comment about the assay that is used to assess cortical plasticity. I suggest that it is inappropriate to consider as equivalent both changes in the segregation of thalamic axons in layer IV and physiological shifts in ocular dominance outside layer IV. Some of your evidence came from anatomical changes in layer IV, for which the sensitive period is quite brief. In monkeys segregation is complete and the sensitive period is over by about six weeks of age (LeVay et al 1980). In contrast, physiological changes in ocular dominance outside layer IV can be seen after monocular deprivation starting as late as a year of age in cats and two years of age in monkeys.

The algorithms of synaptic plasticity must surely be quite different inside and outside layer IV, despite the fact that they receive the same input from the two eyes. With normal binocular vision, layer IV segregates: in other words, there must be selective removal of synaptic input from one eye or the other. But outside layer IV there's normally a maintenance, and even an enhancement, of binocularity, implying that neurons are strengthening synaptic input from both eyes.

Daw: I agree. However, most people think about ocular dominance segregation in terms of thalamocortical synapses. The fundamental rationale is that conjunction of input between the two eyes retains binocularity, and lack of conjunction of input increases monocularity. As you just pointed out, layer IV starts off as binocular then becomes monocular, and the question is how this could occur. If one turns the argument around and says that thalamocortical synapses are driving everything, one comes to the conclusion that there is more binocularity and conjunction of input early on in layer IV than later. Taking that a step further, one can ask the question of how you could get more conjunction of input in layer IV early on than later.

Innocenti: What is the relationship between synaptogenesis and the critical period in the cat, and also in the monkey, where synaptogenesis has been studied in detail?

Daw: Three groups have looked at that: Cragg (1975) and Winfield (1983) in the cat; and Bourgeois & Rakic (1993) in the monkey. They all find that there's a higher density of synapses at the peak of the critical period, and that this density decreases later on. It would also be useful to have an assay of how fast the synapses are turning over, but no one has developed one as far as I know.

Rakic: The number of synapses in the monkey decreases during adolescence, around the time of puberty (Bourgeois & Rakic 1993). This time coincides well with the end of the critical period. Jean-Pierre Bourgeois from Institut Pasteur, Paris, has spent several years in my laboratory working on this problem and has examined literally thousands of electron micrographs. However, he did not see any unequivocal ultrastructural evidence of synaptic turnover.

Innocenti: We have looked at the formation of boutons in callosal axons terminating near the area 17/18 border. We have electron microscopic evidence that they are probably synapses (Aggoun-Zouaoui & Innocenti 1994). We observe an increase in the number of boutons that peaks at approximately postnatal day 70. During this period, the distribution is topographically specific and little selectivity, if any, is gained by the selective elimination of synapses. Consequently, the critical period may be related to synaptogenesis at least as much as to synaptic exuberance and elimination.

Parnavelas: One should be careful in interpreting the total number of synapses. The total number includes synapses from a variety of cell types, and many of these may have nothing to do with plasticity or learning. The increase in the total number of synapses may, therefore, not correlate with anything else other than the maturation of the system. In the rat, when we classified synapses into presumptive excitatory (type 1) and inhibitory (type 2), we found that the number of type 2 synapses first increased then decreased significantly after the third postnatal week (Blue & Parnavelas 1983). We also found that the presence of these type 2 synapses coincided with the presence of hypertrophic non-pyramidal neurons in the cortex. These observations, together with the observation of transient dendritic spines during this period in the Golgi-stained material, are compatible with the suggestion that the early postnatal period is a time of great activity and plasticity in the non-pyramidal cell population.

Daw: Winfield (1983) didn't find a difference in the decline of synaptogenesis between the layers. We need an assay of turnover because the decline of the critical period may not be related to the number of synapses.

Blakemore: There is a problem in simply relating synapse density with conventional measurements of ocular dominance. The ocular dominance classification describes only the ratios of responsiveness through two eyes, and not the absolute levels of response. Consequently, large changes in the absolute excitability of cells can go unnoticed. For instance, the ocular dominance distribution of striate cortical cells in young kittens is similar to that in adult cats (Blakemore & Van Sluyters 1975). Presumably, the general increase in responsiveness of cells with age is correlated with the increase in synaptic density, and the plastic changes in ocular dominance may be due to small modulations in this overall pattern of increasing innervation.

Rakic: Looking at the total numbers of synapses is not misleading if you ask the right question. It can tell us that synaptogenesis occurs simultaneously in all layers and in other areas of the cortex. We're interested in the visual system, but we should recognize that vision is not dependent on events in area 17 alone.

O'Leary: Edward Jones has published work in collaboration with Stewart Hendry which shows that activity blockades or peripheral deprivations can dramatically decrease γ-aminobutyric acid (GABA) content in GABAergic cells and various other cells in the adult visual cortex. Does this also occur during the critical period with these deprivations? And if it does, how would it impact on plasticity?

Daw: Up-regulation and down-regulation of transmitters occurs at all ages, but I'm not sure that it has been looked at as a function of age in the visual cortex.

Jones: We looked at it in monkeys (Hendry & Jones 1988, Benson et al 1991, 1994, Huntsman et al 1994). When we attempted to do it in adult cats, we found that the effects were far less robust (Benson et al 1989). I assumed that the effects would be dramatic.

Schmutz: I would like to comment on NMDA receptor activity. You showed that binding of NMDA to its receptor occurs in parallel with the critical period. This correlates well with results showing an increase in NMDA receptor density during the critical period. You also showed that 6-cyano-7-nitroquinoxaline-2,3-dione (CNQX) had no effect. Are similar density and binding results available for 2-(aminomethyl)phenylacetic acid (AMPA)/kainate receptors with respect to the critical period?

Daw: Yes, Barbara Gordon (personal communication) has studied this extensively. The overall quantity of all receptors increases, but I couldn't summarize what all the layering changes are.

Schmutz: Do the numbers of receptors increase constantly?

Daw: Fairly constantly.

Ghosh: I have a question on the role of NMDA receptors in plasticity. Reiter & Stryker (1988) showed that inhibition of the postsynaptic elements in a monocular deprivation experiment allows the deprived eye to become dominant. If the postsynaptic cell is suppressed in those experiments, is the NMDA receptor still involved?

Daw: The NMDA receptors are probably no longer functional in Reiter & Stryker's muscimol (5-aminomethyl-3-hydroxyisoxazole) experiments. The actual result of this experiment was that the deprived eye starts to drive more cells than the good eye, so the result is the reverse of the usual result from monocular deprivation. NMDA receptors are active up to about -70 mV. We don't know precisely in Reiter & Stryker's experiments how much the membrane potential of the cells was changed. They did not measure the membrane potential of the cells. In the visual cortex *in vivo*, normal cells are probably resting at about -55 mV (Sato et al 1991).

Blakemore: Is it not possible that the effect of MK 801 is simply due to an overall reduction in postsynaptic excitation in young cats because a substantial fraction of the excitatory postsynaptic potential is mediated via

NMDA receptors? In that case, MK 801 might just be reducing all activity in the cortex rather than specifically preventing synaptic plasticity. Did you measure the effect of MK 801 on the activity of cortical cells in adult or young cats?

Daw: For young cats, the experiment started when they were 46 days of age, results were recorded at 56 days of age and there were 10 days of monocular deprivation. The reduction in cortical cell firing produced by MK 801 was measured at 56 days of age.

Blakemore: It may be a trivial point, but wouldn't a better control be to look at the effect of MK 801 on the level of firing at 46 days rather than 56 days?

Daw: Observations at both ages would be best.

Blakemore: F. Sengpeil, P. Kind and myself have some preliminary results on the effect of the NMDA antagonist 2-amino-5-phosphovaleric acid (APV) on synaptic plasticity in the cat cortex. Application of an Elvax sheet (DuPont UK Ltd, Hemel Hempstead) impregnated with APV to the surface of the cortex in a monocularly deprived kitten blocks ocular dominance shifts in the upper layers, but not the lower layers. This is compatible with the concentration gradient set up by APV release from the Elvax sheet. However, we are concerned that this result may simply be due to the reduction in overall activity in the upper layers because more of the post-synaptic current is mediated by NMDA receptors at an early age.

Daw: What age is that?

Blakemore: We applied APV from about four weeks of age to about six weeks of age or more.

Daw: So there could be substantial changes in that period of time?

Blakemore: Yes.

D. Price: Does APV block segregation or interfere with plasticity? Dennis O'Leary suggested that APV does not affect segregation in his system (O'Leary et al 1995, this volume). How can this be reconciled?

Bonhoeffer: This is a totally different story. Nigel Daw is looking at changes arising from closing one eye, whereas Dennis O'Leary is looking at normal development of the whiskers.

D. Price: Didn't one of the assays just look at segregation without deprivation?

Bonhoeffer: No, I don't think anybody has done that.

O'Leary: We did an equivalent experiment to ocular deprivation where we removed whiskers during the critical period and applied APV continuously to the cortex. We found that the plasticity in response to the peripheral whisker lesion was diminished by about 75% (Schlaggar et al 1993). The segregation phenomena during normal development were different in terms of sensitivity to activity blockade, but the anatomy of the system is distinct—at all times during segregation, the geniculocortical afferents are in a continuous distribution. In contrast, in the barrel system, at a certain point during their segregation ventroposterior thalamocortical afferents are no longer continuous, and other afferent systems become intercalated between them. Agmon and colleagues (1993) have shown that there is more directed ingrowth into layer IV in the barrel system than what people have assumed is occurring in the geniculocortical system. This could also contribute to the differences.

References

Aggoun-Zouaoui D, Innocenti GM 1994 Postnatal development of callosal arbors near the 17/18 border in kittens. Eur J Neurosci Suppl 7:51

Agmon A, Yang LT, O'Dowd DK Jones EG 1993 Organized growth of thalamocortical axons from the deep tier of terminations into layer IV of developing mouse barrel cortex. J Neurosci 13:5365–5382

Benson DL, Isackson PJ, Hendry SHC, Jones EG 1989 Expression of glutamic acid decarboxylase mRNA in normal and monocularly deprived cat visual cortex. Mol Brain Res 5:279–287

Benson DL, Isackson PJ, Gall CM, Jones EG 1991 Differential effects of monocular deprivation on glutamic acid decarboxylase and type II calcium-calmodulin-dependent protein kinase gene expression in the adult monkey visual cortex. J Neurosci 11:31–47

Benson DL, Huntsman MM, Jones EG 1994 Activity-dependent changes in GAD and preprotachykinin mRNAs in visual cortex of adult monkeys. Cereb Cortex 4:40–51

Blakemore C, Van Sluyters RC 1975 Innate and environmental factors in development of kitten's visual cortex. J Physiol 248:663–716

Blue ME, Parnavelas JG 1983 The formation and maturation of synapses in the visual cortex of the rat. II. Quantitative analysis. J Neurocytol 12:697–712

Bourgeois JP, Rakic P 1993 Changes in synaptic density in the primary visual cortex of the macaque monkey from fetal to adult stage. J Neurosci 13:2801–2820

Cragg BG 1975 The development of synapses in the visual system of the cat. J Comp Neurol 160:147–166

Daw NW, Sato H, Fox K, Carmichael T, Gingerich R 1991 Cortisol reduces plasticity in the kitten visual cortex. J Neurobiol 22:158–168

Hendry SHC, Jones EG 1988 Activity-dependent regulation of GABA expression in the visual cortex of adult monkeys. Neuron 1:701–712

Huntsman MM, Isackson PJ, Jones EG 1994 Lamina-specific expression and activity-dependent regulation of seven $GABA_A$ receptor subunit mRNAs in monkey visual cortex. J Neurosci 14:2236–2259

LeVay S, Wiesel TN, Hubel DH 1980 The development of ocular dominance columns in normal and visually deprived monkeys. J Comp Neurol 191:1–51

Mitchell DE, Timney B 1984 Postnatal development of function in the mammalian visual system. In: Darian-Smith L (ed) Handbook of physiology, section I: The nervous system, part I: Sensory processes. American Physiological Society, Bethesda, MD, p 507–555

Mower GD, Caplan CJ, Christen WG, Duffy FH 1985 Dark rearing prolongs physiological but not anatomical plasticity of the cat visual cortex. J Comp Neurol 235:448–466

O'Leary DDM, Borngasser DJ, Fox K, Schlaggar BL 1995 Plasticity in the development of neocortical areas. In: Development of the cerebral cortex. Wiley, Chichester (Ciba Found Symp 193) p 214–230

Reiter HO, Stryker MP 1988 Neural plasticity without postsynaptic action potentials: less active inputs become dominant when kitten visual cortical cells are pharmalogically inhibited. Proc Natl Acad Sci USA 85:3623–3627

Sato H, Daw NW, Fox K 1991 An intracellular recording study of stimulus-specific response properties in cat area 17. Brain Res 544:156–161

Schlaggar BL, Fox K, O'Leary DDM 1993 Postsynaptic control of plasticity in developing somatosensory cortex. Nature 364:623–626

Stryker MP, Harris WA 1986 Binocular impulse blockade prevents the formation of ocular dominance columns in cat visual cortex. J Neurosci 6:2117–2133

Swindale NV 1988 Role of visual experience in promoting segregation of eye dominance patches in the visual cortex of the cat. J Comp Neurol 267:472–488

Winfield DA 1983 The postnatal development of synapses in the different laminae of the visual cortex in the normal kitten and in kittens with eyelid suture. Dev Brain Res 9:155–169

Cortical development and neuropathology in schizophrenia

E. G. Jones

Department of Anatomy and Neurobiology, University of California, Irvine, CA 92717, USA

Abstract. Epidemiological studies suggest that perturbations occurring during pregnancy can increase the incidence of schizophrenia among offspring. Examination of the neuropathology of the brains of some schizophrenics suggests that a defect in the later phases of cerebral cortical development, notably the last phases of neuronal migration and the establishment and refinement of patterns of cortical connections, may be involved. Most of these studies are conjectural, and the relationship between primary lesions and potential secondary retrograde and anterograde effects in the circuitry linking the prefrontal cortex, basal forebrain, mediodorsal thalamus and medial temporal cortex is unknown. Our hypothesis, based on neuromorphological and gene expression studies, is that a disturbance of migration or in the pattern of preprogrammed cell death in the subplate zone of the developing cerebral cortex causes a failure to establish normal patterns of connections in the overlying cortex. This compromised circuitry subsequently decompensates, leading to schizophrenic symptoms and activity-dependent manifestations of altered gene expression for neurotransmitter- and receptor-related molecules.

1995 Development of the cerebral cortex. Wiley, Chichester (Ciba Foundation Symposium 193) p 277–295

The belief that some or all forms of schizophrenia have a basis in disordered brain development has enjoyed some popularity for a number of years (Bender 1947). Recent epidemiological studies have supported this hypothesis. A number of potential risk factors have also been identified that may lead to an increase in the incidence of schizophrenia (Mednick & Cannon 1991). One of the most compelling of these is the correlation between increased incidence and maternal influenza infection in the second trimester (Mednick et al 1988, Kendell & Kemp 1989, Barr et al 1990, Crow et al 1991, O'Callaghan et al 1991, Sham et al 1992). A perturbation during the second trimester would interfere with some of the fundamental processes of forebrain (particularly cerebral cortex) maturation. Defective function of the cerebral cortex is one of the hallmarks of schizophrenia. Certain forms of neuropathology in the brains of adult patients with schizophrenia can be interpreted by a developmental hypothesis, but many observations are not consistent and their correlations with developmental insults remain tenuous.

Potential targets of second trimester disturbances

Early phases of cortical development include the generation of large numbers of neurons by mitosis of precursor cells in the neuroepithelium lining the walls of the lateral ventricles and the formation of a cortical plate by migration of these cells along neuroglial guides to the surfaces of the hemispheres (Rakic 1972, 1974). These proliferative and early migration stages are completed in the second trimester. Interference with the proliferative process in experimental rats results in severe disruptions of cortical cytoarchitecture and loss of cortex (Yurkewicz et al 1984). Severe disruptions are not apparent in schizophrenia, suggesting that the disease does not stem from altered cell proliferation or a disruption of early migration. Early in the second trimester, however, the final phases of migration, involving cells destined for the superficial layers of the cortex, are being completed and the connectivity of the cortex is being established. Disruption of either of these events may potentially lead to severe disruption of cortical function in later life.

Cerebral cortical pathology in schizophrenia potentially results from disturbed development

A number of cellular and cytoarchitectonic anomalies in the medial areas of the temporal lobe in the brains of some schizophrenics have been attributed to altered cortical development. They include abnormal positioning and orientation of cells and their dendrites in the area of the cortex between the presubiculum and the hippocampus (Kovelman & Scheibel 1986, Altschuler et al 1987, Conrad et al 1991, Christison et al 1989). In addition, altered lamination patterns in the anterior areas of the entorhinal cortex have been described (Jakob & Beckmann 1986, Arnold et al 1991a). This malorientation of cells is subtle, depending largely upon techniques that do not reveal the full morphology of a cell. It is difficult to specify what kind of developmental perturbation could induce them, other than perhaps a reduction in the number of their afferent fibres. This would imply the presence of a lesion elsewhere, leading to degeneration of the parent cells of such fibres; however, there is no current evidence for this. Anomalies of entorhinal cortex architecture are found in approximately 50% of the brains examined, especially in those from non-paranoid schizophrenics. They are more prominent in the left hemisphere and they have attracted attention because they may suggest that the migration of young neurons to this region of the cerebral cortex during development has been disturbed. Many of the cells characteristic of the islands of layer II are displaced into layer III, resulting in a poorly developed lamination and loss of cells in all layers (Jakob & Beckmann 1986). These observations, although largely qualitative, suggest that a migratory disturbance has occurred. The sampling interval between sections examined, however, was not quoted in this paper. This is important because the anterior entorhinal cortex is divided into at least five areas that are oriented obliquely across the uncus and hippocampal gyrus (Beall & Lewis 1992). All of these areas possess cell islands in layer II but their number, size and discreteness varies in each area. Normally, these cells express the α type II Ca^{2+}/calmodulin-dependent protein kinase, the glutamate receptor subunit GluR2 and the γ_2-subunit of the $GABA_A$ (γ-aminobutyric acid$_A$) receptor (Fig. 1). An adequate assessment of pathological variation requires examination of serial sections and comparison of homologous areas between control and affected brains.

Other forms of neuropathology in schizophrenia

Two other pathological changes in the brains of schizophrenics are less readily explicable in terms of the developmental hypothesis. These are: a moderate dilatation of the ventricles (Hyde & Weinberger 1990, Johnstone et al 1976, 1989, Roberts & Bruton 1990), particularly the inferior horn of the lateral ventricle; and reduction in volume of the thalamus (Andreasen et al 1994), particularly of the mediodorsal nucleus in which there is a 40% loss of cells (Pakkenberg 1990). The latter is probably not a consequence of neuroleptic treatment (Pakkenberg 1992). There is no obvious explanation for the dilatation of the ventricles. It is not accompanied by gliosis in the surrounding white matter (Benes et al 1986, Roberts & Bruton 1990, Purohit et al 1993) and is thus unlikely to reflect recent death of substantial numbers of axons. However, it is present from the earliest phases of the disease and does not progress (Hyde & Weinberger 1990, Johnstone et al 1976, 1989, Roberts & Bruton 1990), suggesting that axons which are produced excessively and pruned in the course of normal development up to the perinatal period (LaMantia & Rakic 1990, Meissirel et al 1991, Webster et al 1991) could have been eliminated in larger numbers than usual. This process would not be associated with gliosis and would potentially reduce functional connectivity between the temporal and frontal lobes. Disruption of these connections is thought to be a major contributing factor in the fragmentation of thought that occurs in schizophrenia. Excessive axon loss should, however, be accompanied by a substantial reduction in the numbers of neurons at the sources of the reduced projections. Unfortunately, there is little reproducible evidence for a major reduction in cell numbers, or volume, of the prefrontal, entorhinal or hippocampal cortical areas (Falkai et al 1988, Heckers et al 1991, Benes et al

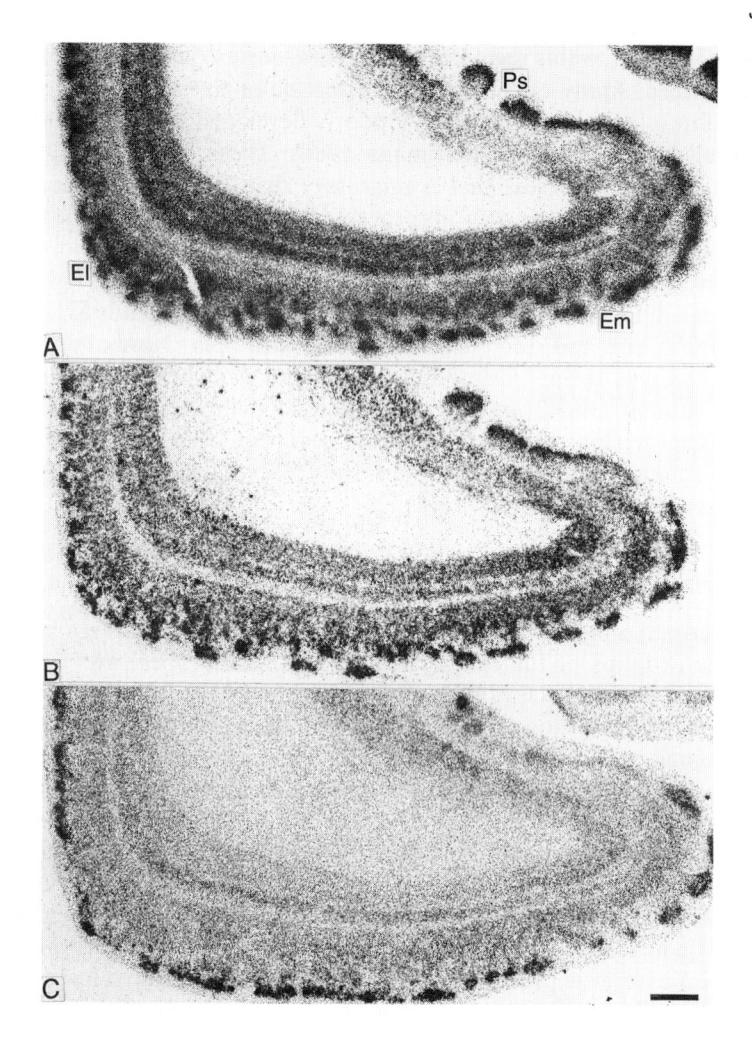

FIG. 1. Photomicrographs of autoradiograms of sections through human entorhinal cortex showing the localization of mRNAs encoding: (A) α type II Ca^{2+}/calmodulin-dependent protein kinase; (B) the glutamate receptor subunit, GluR2; (C) the γ_2-subunit of the γ-aminobutyric acid$_A$ receptor. Note labelling of cell islands of presubiculum (Ps) and of medial (Em) and lateral (El) entorhinal areas. Bar = 1 mm.

1991, Daviss & Lewis 1993, Selemon et al 1993, Akbarian et al 1995a, Bunney et al 1993, Pakkenberg 1993).

The loss of cells in the mediodorsal nucleus is associated with hypoactivity of the prefrontal cortex and a reduction in the numbers of neurons in the nucleus accumbens (Buchsbaum et al 1992, Weinberger et al 1992). These observations are interesting because the prefrontal cortex represents the major thalamic input to the mediodorsal nucleus, and the nucleus accumbens provides a source of afferent input to the mediodorsal nucleus via GABAergic cells of the ventral pallidum. Therefore, the mediodorsal nucleus is at a key junction in the proposed 'psychosis circuitry' joining the prefrontal cortex, limbic striatum and medial temporal cortex, which is thought to be functionally compromised in schizophrenia (Goldman-Rakic 1987). It is not yet known if the mediodorsal nucleus is the only thalamic nucleus affected in schizophrenia or whether the reported reduction in volume in the thalamus stems from loss of cells in all thalamic nuclei. The loss of cells without gliosis may represent a primary condition or a retrograde or anterograde transneuronal degeneration resulting from pathology elsewhere. The absence of gliosis does not support the existence of a toxic insult or acute retrograde degeneration, but it may suggest a retrograde effect or a slow transneuronal effect operating over a protracted period. Gliosis is not prominent in either case. If the loss of cells in the mediodorsal nucleus is due solely to a retrograde effect, one would anticipate the presence of primary pathology in the prefrontal cortex. Loss of mediodorsal cells would, in turn, result in a reduced afferent drive to the prefrontal cortex, possibly accounting for the hypofrontality of schizophrenics. The lack of integration of thought processes in schizophrenics may also be explained by a lack of thalamocortical integration in which normally large populations of cortical and thalamic neurons are thrown into coherent oscillations that underlie changes in conscious state (Llinás & Ribary 1993). A further feature of altered patterns of activity in the psychosis circuitry would be activity-dependent changes in the expression of genes encoding neurotransmitters, receptors and other neuronal molecules that could have major consequences for integrative brain function (Jones 1990, 1993). These changes may be independent of detectable changes in cytoarchitecture.

Interstitial neurons of the white matter and transmitter-related gene expression in schizophrenia

Interstitial neurons are abundant in the telencephalic white matter of larger mammalian brains, and they are survivors of a more substantial population of neurons that play an important role in cortical development. They persist from the cortical subplate, which includes the first neurons that migrate to the surface of the cerebral hemisphere and form the primordial cortex. Neurons that arrive later migrate through the subplate, accumulate beneath the molecular layer in the classical inside out sequence and form the definitive cortex (Marin-Padilla 1988, Rakic 1974).

All afferent axons arriving at the cortex and efferents destined for extrinsic targets must grow through the subplate. Interactions are established during this migration, some apparently synaptic, that help to set up the complex patterns of cortical circuitry. These include defining the zones of dominance of afferent fibres in the overlying cortex, and guidance of afferent axons to subcortical targets (Shatz et al 1988, Friauf et al 1990, Ghosh et al 1990, McConnell et al 1989, O'Leary & Koester 1993). After performing these tasks, many subplate neurons undergo a programme of cell death (Chun & Shatz 1989a,b). Neurons that escape this programme persist as interstitial neurons of the white matter. They can be identified by immunostaining with antisera raised against specific enzymes, peptides, receptors and structural proteins, or by *in situ* hybridization of the corresponding mRNAs (Hendry et al 1984, Jones et al 1994, Akbarian et al 1993a,b, 1995a).

A developmental disturbance of the subplate would give rise to an abnormal distribution of interstitial neurons and potentially to alterations in the structure and connectivity of the overlying cortex. Such a disturbance could result from a migratory defect in the subplate neurons or an alteration in their pattern of preprogrammed cell death. The latter would probably begin in the second trimester in humans.

We have found that interstitial neurons of the white matter are displaced systematically in the frontal and temporal lobes of the brains of schizophrenics (Akbarian et al 1993a,b, 1995b). Interstitial neurons stained for NADPH diaphorase are normally concentrated in layers II and VI of the cortex and in the underlying white matter. In schizophrenics these cells are displaced deeper into the white matter (Fig. 2). There are statistically significant reductions in the densities of NADPH diaphorase-positive cells in the cortex and superficial compartments of the white matter (less than 2 mm deep to the cortex) and corresponding increases in white matter compartments (up to 8 mm deep to the cortex). Larger populations of interstitial neurons, such as those immunoreactive for microtubule-associated protein 2 (MAP-2) and for non-phosphorylated neurofilament proteins (Fig. 2) show the same distorted distribution in schizophrenics (Akbarian et al 1995a). The maldistribution of NADPH diaphorase-positive former subplate neurons is unlikely to reflect a major disturbance of neuronal migration to the cortex because the number and density of neurons, and the cytoarchitecture of the overlying prefrontal cortex, are the same as those of normal brains (Akbarian et al 1995b). The maldistribution may instead stem from an alteration in the programme of cell death in the subplate or from other unknown factors.

An alteration in the subplate that affects apoptotic cell death is likely to be a relatively late event in cortical development. In this case, the connectivity of the cortex should be compromised. Defective cortical circuitry ensuing from a subplate defect may be relatively modest in schizophrenics and only manifested

Schizophrenia and brain development

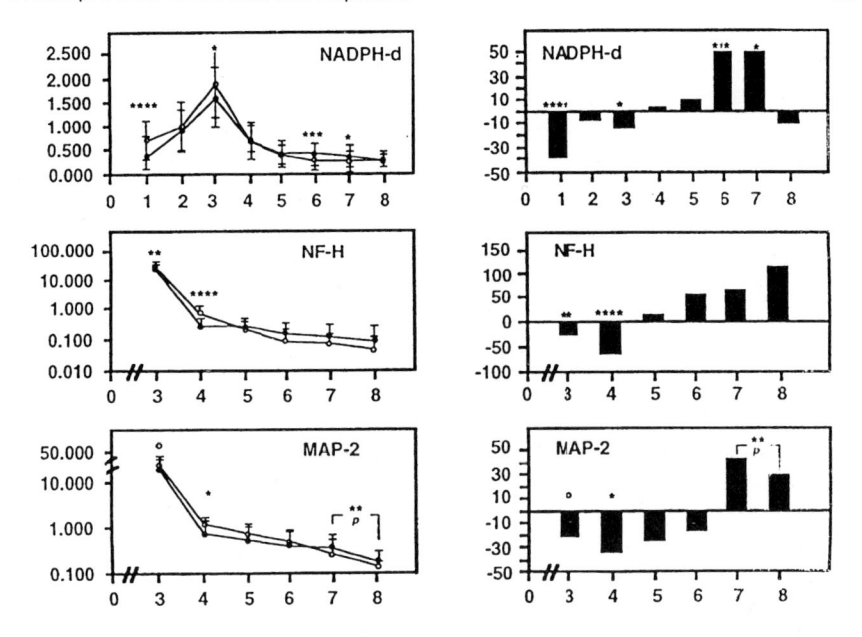

FIG. 2. In schizophrenics the numbers of cells stained for nicotinamide-adenosine dinucleotide phosphate diaphorase (NAPDH-d), for non-phosphorylated neurofilament epitopes (NF-H) and for microtubule-associated protein 2 (MAP-2) decrease in superficial layers, but increase in deep compartments of the white matter beneath the prefrontal cortex. Results are compared with matched control brains. In both columns, x-axes indicate compartments of cortex (1,2) and compartments (3–8) of increasing depth in white matter. Left column, y-axes: density of cells per compartment. Filled circles, schizophrenics; open circles, matched controls. Right column, y-axes: % deviation of cell density per compartment in schizophrenics relative to matched controls. ****$P < 0.0001$, ***$P < 0.001$, **$P < 0.01$, *$P < 0.1$, °$P < 0.05$.

when subjected to stress during life events in adolescence and young adulthood, when schizophrenic symptoms become evident.

As yet, there is no way of confirming that the circuitry of the cerebral cortex is anatomically defective in schizophrenia. An anatomical defect occurring at the synaptic level, such as the elimination of excessive synapses or defective stabilization of established synapses (Rakic et al 1986, Missler et al 1993), may be difficult to pinpoint. A number of observations, however, suggest that alterations in gene expression for certain neurotransmitter-related molecules may be involved. Expression of these molecules may depend on the activity of afferent connections. Indeed, activity-dependent events during development are known to play an important role in forming and stabilizing patterns of

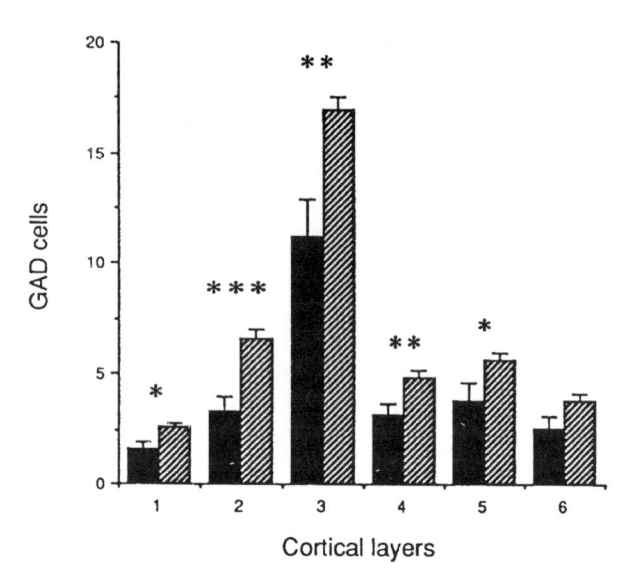

FIG. 3. Reductions in cells showing expression of mRNA encoding the 67 kDa glutamic acid decarboxylase (GAD cells) which occurs without loss of cells in prefrontal cortex of schizophrenics. Filled bars, schizophrenics; hatched bars, controls. ***$P<0.001$, **$P<0.01$, *$P<0.1$. (From Akbarian et al 1995c.)

connections in the nervous system and in maintaining functional connectivity throughout life (Shatz 1990, Jones 1990, 1993).

Glutamic acid decarboxylase (GAD) is necessary for GABA synthesis. A selective reduction in the levels of GAD mRNA occurs in parallel with changes in the distribution of interstitial neurons of the white matter (Akbarian et al 1995c) (Fig. 3). This reduction occurs without a loss of neurons in the prefrontal cortex and without statistically significant changes in the levels of mRNAs encoding the more abundant $GABA_A$ receptor subunits (Akbarian et al 1995a) or the α type II Ca^{2+}/calmodulin-dependent protein kinase (Akbarian et al 1995c) associated with excitatory synapses (Jones et al 1994). Schizophrenics have also been reported to exhibit a reduction in $GABA_A$ receptor binding in the cingulate cortex (Benes et al 1992), an increase in dopamine D4 receptor binding in frontal cortex (Seeman et al 1993) and a decrease in the level of mRNA encoding the dopamine D3 receptor in the parietal and motor cortex (Schmauss et al 1993). These effects are not, therefore, non-specific effects of the overall reduced frontal cortex activity that is demonstrable in schizophrenics (Buchsbaum et al 1992, Weinberger et al 1992). The effects may depend on reductions in afferent input caused by altered cortical circuitry; for example, similar to those which occur in the cortex of experimental animals under conditions of reduced or enhanced activity (Hendry & Jones 1986, 1988, Huntsman et al 1994, Benson et al 1991, 1994,

Gall & Isackson 1989, Bronstein et al 1992). This would result in an alteration of the balance of excitation and inhibition in the frontal cortex, which could in turn have a similar activity-dependent effect in regions projecting from the frontal cortex, e.g. the limbic striatum, mediodorsal thalamus and medial temporal regions. Activity-dependent alterations in cortical circuitry may include alterations in dopamine receptor function in the nucleus accumbens (Grace 1991), down-regulation of microtubule-associated proteins in the hippocampal formation (Arnold et al 1991b) and cell loss in the mediodorsal nucleus of schizophrenics. These may all occur following a relatively minor perturbation of cortical development.

Acknowledgements

This work was supported by Grants numbers MH 44188 and MH 53674 from the National Institutes of Health, United States Public Health Service, by a Stanley Award from the National Alliance for the Mentally Ill and by the Frontier Research Program in Brain Mechanisms of Mind and Behavior, at the Institute for Physical and Chemical Research (Riken) Japan. I thank Dr S. Akbarian for valuable contributions and Drs W. E. Bunney Jr and S. G. Potkin for assistance in obtaining pathological material.

References

Akbarian S, Bunney WE Jr, Potkin SG et al 1993a Altered distribution of nicotinamide-adenine-dinucleotide-phosphate diaphorase cells in frontal lobe of schizophrenics implies disturbances of cortical development. Arch Gen Psychiatry 50:169–177

Akbarian S, Vinuela A, Kim JJ, Potkin SG, Bunney WE Jr, Jones EG 1993b Distorted distribution of nicotinamide-adenine dinucleotide phosphate-diaphorase neurons in temporal lobe of schizophrenics implies anomalous cortical development. Arch Gen Psychiatry 50:178–187

Akbarian S, Kim JJ, Tafazolli A et al 1993c Altered neurocellular composition of prefrontal cortex in schizophrenia. Neurosci Abstr 19:200

Akbarian S, Huntsman MM, Kim JJ et al 1995a $GABA_A$ receptor subunit gene expression in human prefrontal cortex: comparison of schizophrenics and controls. Cereb Cortex, in press

Akbarian S, Kim JJ, Potkin SG, Bunney WE Jr, Jones EG 1995b Maldistribution of interstitial neurons in prefrontal white matter of schizophrenics. Arch Gen Psychiatry, in press

Akbarian S, Kim JJ, Potkin SG et al 1995c Gene expression for glutamic acid decarboxylase is reduced without loss of neurons in prefrontal cortex of schizophrenics. Arch Gen Psychiatry 52:258–266

Altschuler LL, Conrad A, Kovelman JA, Scheibel A 1987 Hippocampal pyramidal cell orientation in schizophrenia. Arch Gen Psychiatry 44:1904–1908

Andreasen NC, Arndt S, Swayze V et al 1994 Thalamic abnormalities in schizophrenia visualized through magnetic resonance image averaging. Science 266:294–298

Arnold SE, Hyman BT, Van Hoesen GW, Damasio AR 1991a Some cytoarchitectural abnormalities of the entorhinal cortex in schizophrenia. Arch Gen Psychiatry 48:625–632

Arnold SE, Lee VM-Y, Gur RE, Trojanowski JQ 1991b Abnormal expression of two microtubule-associated proteins (MAP2 and MAP5) in specific subfields of the hippocampal formation in schizophrenia. Proc Natl Acad Sci USA 88:10850–10854

Barr CE, Mednick SA, Munk-Jorgensen P 1990 Exposure to influenza during gestation and adult schizophrenia. Arch Gen Psychiatry 47:869–874

Beall MJ, Lewis DA 1992 Heterogeneity of layer II neurons in human entorhinal cortex. J Comp Neurol 321:241–266

Bender L 1947 Childhood schizophrenia: clinical study of 100 schizophrenic children. Am J Orthopsychiatry 17:40–56

Benes FM, Davidson J, Bird ED 1986 Quantitative cytoarchitectural analyses of the cerebral cortex of schizophrenics. Arch Gen Psychiatry 43:31–35

Benes FM, McSparren J, Bird ED, SanGiovanni JP, Vincent SL 1991 Deficits in small interneurons in prefrontal and cingulate cortices of schizophrenic and schizoaffective patients. Arch Gen Psychiatry 48:996–1001

Benes FM, Vincent SL, Alsterberg G, Bird ED, SanGiovanni JP 1992 Increased GABA A receptor binding in superficial layers of cingulate cortex in schizophrenics. J Neurosci 12:924–929

Benson DL, Huntsman MM, Jones EG 1994 Activity-dependent changes in GAD and preprotachykinin mRNAs in visual cortex of adult monkeys. Cereb Cortex 4:40–51

Benson DL, Isackson PJ, Gall CM, Jones EG 1991 Differential effects of monocular deprivation on glutamic acid decarboxylase and type II calcium–calmodulin-dependent protein kinase gene expression in the adult monkey visual cortex. J Neurosci 11:31–47

Bronstein JM, Micevych P, Popper P, Huez G, Farber DB, Wasterlain CG 1992 Long lasting decreases of type II calmodulin kinase expression in kindled rat brains. Brain Res 584:257–260

Buchsbaum MS, Haier RJ, Potkin SG et al 1992 Frontostriatal disorder of cerebral metabolism in never-medicated schizophrenics. Arch Gen Psychiatry 49:935–942

Bunney WE Jr, Akbarian S, Kim JJ, Hagman JO, Potkin SG, Jones EG 1993 Gene expression for glutamic acid decarboxylase is reduced in prefrontal cortex of schizophrenics. Neurosci Abstr 19:199

Christison GW, Casanova MF, Weinberger DR, Rawlings R, Kleinman JE 1989 A quantitative investigation of hippocampal pyramidal cell size, shape, and variability of orientation in schizophrenia. Arch Gen Psychiatry 46:1027–1032

Chun JJM, Shatz CJ 1989a The earliest-generated neurons of the cat cerebral cortex: characterization by MAP2 and neurotransmitter immunohistochemistry during fetal life. J Neurosci 9:1648–1667

Chun JJM, Shatz CJ 1989b Interstitial cells of the adult neocortical white matter are the remnant of the early generated subplate neuron population. J Comp Neurol 282:555–569

Conrad AJ, Abebe T, Austin R, Forsythe S, Scheibel AB 1991 Hippocampal pyramidal cell disarray in schizophrenia as a bilateral phenomenon. Arch Gen Psychiatry 48:413–417

Crow TJ, Done DJ, Johnstone EC 1991 Influenza and schizophrenia after prenatal exposure to 1957 A2 influenza epidemic. Lancet 337:1248–1250

Daviss SR, Lewis DA 1993 Calbindin- and calretinin-immunoreactive local circuit neurons are increased in density in the prefrontal cortex of schizophrenic subjects. Neurosci Abstr 19:201

Falkai P, Bogerts B, Rozumek M 1988 Limbic pathology in schizophrenia: the entorhinal region—a morphometric study. Biol Psychiatry 24:515–521

Friauf E, McConnell SK, Shatz CJ 1990 Functional synaptic circuits in the subplate during fetal and early postnatal development of cat visual cortex. J Neurosci 10:2601–2613

Gall CM, Isackson PJ 1989 Limbic seizures increase neuronal production of mRNA for nerve growth factor. Science 245:758–761

Ghosh A, Antonini A, McConnell SK, Shatz CJ 1990 Requirement for subplate neurons in the formation of thalamocortical connections. Nature 347:179–181

Goldman-Rakic PS 1987 Circuitry of primate prefrontal cortex and regulation of behavior by representational memory. In: Brookhart JM, Mountcastle VB (eds) Handbook of physiology: the nervous system, vol V. American Physiological Society, Washington DC, p 373–417

Grace AA 1991 Phasic versus tonic dopamine release and the modulation of dopamine system responsivity: a hypothesis for the etiology of schizophrenia. Neuroscience 41:1–24

Heckers S, Heinsen H, Geiger B, Beckmann H 1991 Hippocampal neuron number in schizophrenia. Arch Gen Psychiatry 48:1002–1008

Hendry SHC, Jones EG 1986 Reduction in number of immunostained GABAergic neurones in deprived-eye dominance columns of monkey area 17. Nature 320:750–753

Hendry SHC, Jones EG 1988 Activity-dependent regulation of GABA expression in the visual cortex of adult monkeys. Neuron 1:701–712

Hendry SHC, Jones EG, Emson PC 1984 Morphology distribution and synaptic relations of somatostatin and neuropeptide Y-immunoreactive neurons in rat and monkey neocortex. J Neurosci 4:2497–2517

Huntsman MM, Isackson PJ, Jones EG 1994 Lamina-specific expression and activity-dependent regulation of seven $GABA_A$ receptor subunit mRNAs in monkey visual cortex. J Neurosci 14:2236–2259

Hyde TM, Weinberger DR 1990 The brain in schizophrenia. Semin Neurol 10:276–286

Jakob H, Beckmann H 1986 Prenatal development disturbances in the limbic allocortex in schizophrenics. J Neural Transm 65:303–326

Johnstone EC, Frith CD, Crow TJ, Husband J, Kreel L 1976 Cerebral ventricular size and cognitive impairment in chronic schizophrenia. Lancet 2:924–926

Jones EG 1990 The role of afferent activity in the maintenance of primate neocortical function. J Exp Biol 153:155–176

Jones EG 1993 GABAergic neurons and their role in cortical plasticity in primates. Cereb Cortex 3:361–372

Jones EG, Huntley GW, Benson DL 1994 Alpha calcium/calmodulin-dependent protein kinase II selectively expressed in a subpopulation of excitatory neurons in monkey sensory-motor cortex: comparison with GAD-67 expression. J Neurosci 14:611–629

Kendell RE, Kemp IW 1989 Maternal influence in the etiology of schizophrenia. Arch Gen Psychiatry 46:878–882

Kovelman JA, Scheibel AB 1986 A neurohistologic correlate of schizophrenia. Biol Psychiatry 19:1601–1621

Lamantia A-S, Rakic P 1990 Axon overproduction and elimination in the corpus callosum of the developing rhesus monkey. J Neurosci 10:2156–2175

Llinás R, Ribary U 1993 Coherent 40-Hz oscillation characterizes dream state in humans. Proc Natl Acad Sci USA 90:2078–2081

Marin-Padilla M 1988 Early ontogenesis of the human cerebral cortex. In: Peters A, Jones EG, (eds) Cerebral cortex, vol 7: Development and maturation of cerebral cortex. Plenum Press, New York, p 1–30

McConnell SK, Ghosh A, Shatz CJ 1989 Subplate neurons pioneer the first axon pathway from the cerebral cortex. Science 245:978–982

Mednick SA, Cannon TD 1991 Fetal development, birth and the syndromes of adult schizophrenia. In: Mednick SA, Cannon TD (eds) Fetal neural development and adult schizophrenia. Cambridge University Press, New York, p 3–13

Mednick SA, Machon RA, Huttunen M, Bone D 1988 Adult schizophrenia following prenatal exposure to an influenza epidemic. Arch Gen Psychiatry 45:189–192

Meissirel C, Dehay C, Berland M, Kennedy H 1991 Segregation of callosal and association pathways during development in the visual cortex of the primate. J Neurosci 11:3297–3316

Missler M, Wolff A, Merker HJ, Wolff JR 1993 Pre- and postnatal development of the primary visual cortex of the common marmoset. II. Formation, remodelling, and elimination of synapses as overlapping processes. J Comp Neurol 333:53–67

O'Callaghan E, Sham P, Takei N, Glover G, Murray RM 1991 Schizophrenia after prenatal exposure to 1957 A2 influenza epidemic. Lancet 337:1248–1250

O'Leary DDM, Koester SE 1993 Development of projection neuron types, axon pathways, and patterned connections of the mammalian cortex. Neuron 10:991–1006

Pakkenberg B 1990 Pronounced reduction of total neuron number in mediodorsal thalamic nucleus and nucleus accumbens in schizophrenics. Arch Gen Psychiatry 47:1023–1028

Pakkenberg B 1992 The volume of the mediodorsal thalamic nucleus in treated and untreated schizophrenics. Brain Res 7:95–100

Pakkenberg B 1993 Total nerve cell number in neocortex in chronic schizophrenics and controls estimated using optical dissectors. Biol Psychiatry 34:768–772

Purohit DP, Davidson M, Perl DP et al 1993 Severe cognitive impairment in elderly schizophrenic patients: a clinicopathological study. Biol Psychiatry 33:255–260

Rakic P 1972 Mode of cell migration to the superficial layers of the fetal monkey neocortex. J Comp Neurol 45:61–84

Rakic P 1974 Neurons in rhesus monkey visual cortex: systematic relation between time of origin and eventual disposition. Science 183:425–427

Rakic P, Bourgeois J-P, Eckenhoff MF, Zecevic N, Goldman-Rakic PS 1986 Concurrent overproduction of synapses in diverse regions of the primate cerebral cortex. Science 232:232–235

Roberts GW, Bruton CJ 1990 Notes from the graveyard: neuropathology and schizophrenia. Neuropathol Appl Neurobiol 16:3–16

Schmauss C, Haroutunian V, Davis KL, Davidson M 1993 Selective loss of dopamine D3-type receptor mRNA expression in parietal and motor cortices of patients with chronic schizophrenia. Proc Natl Acad Sci USA 90:8942–8946

Seeman P, Guan H-C, Van Tol HHM 1993 Dopamine D4 receptors elevated in schizophrenia. Nature 365:441–444

Selemon LD, Rajkowska G, Goldman-Rakic PS 1993 Cytologic abnormalities in area 9 of the schizophrenic cortex. Neurosci Abstr 19:200

Sham PC, O'Callaghan E, Takei N, Murray GK, Hare H, Murray RM 1992 Schizophrenia following prenatal exposure to influenza epidemics between 1939 and 1960. Br J Psychiatry 160:461–466

Shatz CJ 1990 Impulse activity and the patterning of connections during CNS development. Neuron 5:745–756

Shatz CJ, Chun JJM, Luskin MB 1988 The role of the subplate in the development of the mammalian telencephalon. In: Peters A, Jones EG (eds) Cerebral cortex, vol 7. Development and maturation of cerebral cortex. Plenum Press, New York, p 35–58

Webster MJ, Ungerleider LG, Bechevalier J 1991 Connections of inferior temporal areas TE and TEO with medial temporal-lobe structures in infant and adult monkeys. J Neurosci 11:1095–2116

Weinberger DR, Berman KF, Suddath R, Torrey EF 1992 Evidence of dysfunction of a prefrontal–limbic network in schizophrenia: a magnetic resonance imaging and regional cerebral blood flow study of discordant monozygotic twins. Am J Psychiatr 149:890–897

Yurkewicz L, Valentino KL, Floeter MK, Fleshman JW Jr, Jones EG 1984 Effects of cytotoxic deletions of somatic sensory cortex in fetal rats. Somatosens Res 1:303–327

DISCUSSION

Molnár: Your hypothesis suggests that there's an abnormal pattern of cell death in the cortical subplate that could lead to a disturbance in the adult cortical organization. However, if the subplate and the cortical plate form a temporary functional unit during cortical circuit formation and this transient circuitry is essential for normal development, then perhaps it is equally conceivable that the cortex itself is abnormal and the subplate cannot function. Therefore, your hypothesis could be turned around to propose that there is an abnormal development within the cortical plate resulting in an abnormal interplay between the cortex and subplate. This may be the reason why the subplate cells survive in unexpectedly high numbers.

Jones: I do not disagree with that. I'm suggesting that there is a disturbance, but this does not necessarily imply that there is not an accompanying disturbance at an early stage in development. Preprogrammed cell death has been suggested because this is one obvious aspect that we can examine in postmortem human material with the available markers. I don't want to imply that it is only preprogrammed cell death that is involved.

Parnevalas: What percentage of interstitial cells are γ-aminobutyric acid (GABA)ergic?

Jones: We've never quantified that. It's a relatively small number.

Parnavelas: Is it smaller than the population of cells expressing NADPH diaphorase?

Jones: No, it's larger.

Parnavelas: Why haven't you quantified the GABAergic cell populations? You would then have had a larger population to examine.

Jones: We have quantified cells expressing NADPH diaphorase and microtubule-associated protein 2, and stained by SM132. This quantification has taken S. Akbarian, myself and a large number of undergraduate students about four years to do, so give us another year and we'll have quantified the GABAergic cells.

Parnavelas: A large percentage of schizophrenics have a family history of schizophrenia. Have you noticed any difference in the structure of the subplate or interstitial cells in the brains of patients who had a family history compared to those who didn't?

Jones: It's my impression that if there is a family history followed by a maternal perturbation in the second trimester, the incidence of schizophrenia is increased considerably. We don't have any schizophrenic brains that come from families in which there was another known schizophrenic, so I have no way of answering your question at the moment.

Rakic: You proposed two scenarios: one involving the subplate and the other involving defective neuronal migration. If only one perturbation is involved, widespread and dramatic changes may not be observed in the entire cortex and only the cells that move to a given layer may be perturbed. The absence of changes in the total numbers of cells is not incompatible with this idea because there may be differential increases in cell number or in the rate of cell death in different regions. I believe that your results are compatible with an effect on migration rather than on the subplate. The surviving cells in the white matter may, therefore, be cortical cells that did not move to the cortex.

Jones: We first thought that we had demonstrated an alteration in the migratory process, but I feel less comfortable with that now. The situation is more multidimensional than the way I first presented it, so I'm trying to work with the smallest number of parameters that can be quantified at the moment. Your proposal implies that the cells that normally die in the cortex are preserved to compensate for the cells which didn't make it there. It also implies that migration is disturbed only at one particular phase in development. It's difficult to conceive that if you affect migration at one particular stage, the subsequent migration is not going to be affected.

Rakic: Migration can be blocked by drugs, viruses or X-irradiation. For example, we irradiated the monkey at early and late stages of corticogenesis, and we found that deeper and more superficial layers were depleted of neurons and that many ectopic cells were present in the subjacent white matter (O. Algan & P. Rakic, unpublished results).

Jones: My studies cannot be compared with your studies on X-irradiated monkeys or our studies on methylazoxymethanol-treated rats (Yurkewicz et al 1984) because changes resulting from these treatments are more radical than those in the brains of schizophrenics.

Bolz: Do X-irradiated monkeys have the same number of cells in the superficial layers?

Rakic: The density and numbers of cells change, although it is difficult to assess the changes in cell number. The same situation occurs in schizophrenia where the density of cells changes, although Edward Jones proposes that there's no evidence of cell loss.

Jones: This varies depending on whose results you believe. There are studies in which a reduction in cells has been claimed (Benes et al 1991) and others

(Selemon et al 1993) where the opposite has been claimed. Our results (Akbarian et al 1995) are in between because we state that there's no change.

Walsh: It is possible that this may be extended to other common types of pathology that have been reported in epileptic brains (Meencke & Veith 1992, Meencke 1989). Meencke refers to a common pathology as microdysgenesis. Microdysgenesis often represents persistent neurons in layer I and layer VI. It is possible that these are preplate and subplate cells that did not die when they were supposed to. This could be a primary effect or a secondary marker.

Jones: Yes, that's a good point. It may also be extended to developmental dyslexia, where there have been some reports of disturbances in the angular gyrus region (Galaburda et al 1985).

Krubitzer: Did you look in the sensory regions of the cortex for similar differences?

Jones: We have looked at the visual cortex, although not quantitatively but purely for control purposes. It looks as though the visual cortex may not be as radically affected as the prefrontal regions of the cortex. The problem with the visual cortex (and the hippocampus) is that the underlying white matter is thin, so it is difficult to stretch it out into compartments for comparing the superficial and deep compartments. We have not looked at the somatosensory or auditory cortex.

Krubitzer: Is it possible that different areas of the cortex may be affected at different times during development?

Jones: Yes, it's possible.

Levitt: Hazel Murphy's laboratory has looked at prenatal cocaine exposure in a rabbit model, and their results support the concept that an alteration at a restricted period in development can lead to long-term changes (Wang et al 1995). They showed that *in utero* exposure to cocaine results in long-term, postnatal changes in the number of GABA-positive and glutamic acid decarboxylase (GAD)-positive cells without any changes in total neuron number. These changes are also specific to certain cortical areas, so there are selected areas in which prenatal cocaine is more likely to induce permanent changes in GABA neurons.

Blakemore: Were the effects on GAD expression primary or secondary to the hypoactivity of the frontal cortex in schizophrenia?

Jones: It is likely that the disruption of circuit formation leads to an alteration in the balance of excitation and inhibition, and that this then affects GAD expression. I have no way of assessing whether changes in GAD expression promote the hypoactivity of the prefrontal cortex or whether they stem from the hypoactivity.

Rubenstein: The cerebral cortex seems to have been the focus for studying most of these human diseases, but my instinct is that the subcortical areas, particularly the amygdala, may be involved in some types of schizophrenia. One of the primary symptoms of schizophrenia is fear, and it is possible that this fear drives the secondary behavioural/cognitive symptoms. Consequently, it may be beneficial to look at the amygdala and other subcortical areas.

Jones: You're absolutely right. Changes in the numbers of NADPH diaphorase-positive cells in the pons have been reported in schizophrenics. Some of the best documented quantitative studies of cell loss have also been reported in the mediodorsal thalamus and the nucleus accumbens (Pakkenberg 1990), although changes in the amygdala are much less consistent. My feeling is that in schizophrenia, there's an inability to switch between circuits involving particular areas of the cortex and particular subcortical structures, including particular thalamic nuclei. Consequently, the switch between accessing an input and producing an output fails to occur. The cortex alone is not solely responsible, loops and oscillations between different subcortical components are also involved.

Parnavelas: There was also a recent article in *Science* that emphasized the role of the thalamus (Andreason et al 1994).

Jones: This was an imaging study, based upon normalizing images of schizophrenic brains to a standard template of the thalamus. They found that the thalamus is reduced in size in schizophrenics. However, my interpretation of their results is that there is a loss of the white matter surrounding the thalamus. This may represent a loss of cells in the thalamus with a subsequent loss of thalamocortical axons, and it would agree with Pakkenberg's results which suggest that a 40% loss in the number of cells occurs in the mediodorsal nucleus (although this has never been assessed in terms of other thalamic nuclei, so it is not known if it is a general phenomenon). Alternatively, it may mean that a cortical defect arises from an alteration in the number of corticofugal fibres.

Roberts: But some of Pakkenberg's patients had leucotomies, and people have shown that leucotomies themselves can cause a loss of neurons in the thalamus.

Jones: I'm not sure that those patients had leucotomies. Many of the patients that were looked at by Arnold et al (1991) in the Yakovlev collection, from the point of view of medial temporal cortex anomalies, had leucotomies. Pakkenberg examined two series of patients, one of which had been medicated and the other that was taken during the years in which the patients didn't receive neuroleptics.

Schmutz: Your hypothesis is based on the observation that a perturbation in the second trimester plays a role in schizophrenia. Therefore, your study design should reflect this hypothesis. Was the inclusion criterion that the mothers had a stress in the second trimester? And was the exclusion criterion the occurrence of stresses in the second trimester?

Jones: Those are both important considerations that we have obviously thought about. However, we have only a limited number of schizophrenic brains, so we have just evaluated matched pairs regardless of the history, which

Schizophrenia and brain development

is often not available anyway. As time proceeds and as we accumulate more brains, we will re-evaluate the results in terms of the crucial factors that you have suggested. The other division that we have not yet made is into different types of schizophrenia, e.g. into paranoid and non-paranoid patients. The severity of the defect and the localization of the defect may also play a role in the symptomatology, but at the moment, influenza is attractive to us, even though we have no way of knowing that the mothers of these people whose brains we have examined were exposed to influenza.

Kennedy: Developmental theories of schizophrenia need to take into account the late onset of the disease. For instance, it has been suggested that failure of late synapse elimination may be involved (Feinberg 1982). We have looked at NADPH diaphorase-positive neurons in monkeys and we have found that they mature very late in postnatal life. Your results indicated that there were no NADPH diaphorase-positive cells in the upper layers of schizophrenic brains. Is this because of a failure to migrate or excess elimination?

Jones: You have raised a number of different questions here. The age of onset is important because overt symptoms manifest themselves mainly in late adolescence and early adulthood. This probably indicates that the circuitry has been compromised, but not to such an extent that it can't handle the basic day to day events up to puberty and beyond. However, following the stresses of life events that occur at puberty and in young adulthood, the circuitry decompensates in some way. This may be associated with an increase in synaptic loss above the normal pattern that occurs over time. Alternatively, it may be associated with insufficient axon terminal ramifications on certain cells that are unable to maintain fully functional connectivity. There is also increasing evidence, mainly anecdotal, which suggests that schizophrenics are not completely normal during childhood. Consequently, the overt symptoms observed later may occur on top of what was already a compromised cortex. However, this needs to be assessed further.

You also mentioned the absence of NADPH diaphorase-positive cells. In the section that I showed, there were no positive cells in the upper layers, but this was just one section, and their distribution is quite patchy. Consequently, it is possible that there are no positive cells in any one section of the upper layer. However, I do not want to imply that there is a total loss from the upper layers, rather that the numbers are reduced in schizophrenics.

Molnár: We looked at the developmental expression of NADPH diaphorase in the rodent (Molnár et al 1993). Expression of NADPH diaphorase started around embryonic day 20 to postnatal day (P) 0, but the stained polymorphous neurons were scattered sparsely in the lowest part of the cortical plate and the adjacent subplate. They were clearly not restricted to the subplate. During the first postnatal week, the NADPH diaphorase-expressing multiform and bipolar or bistellate cells appeared progressively throughout upper cortical layers in a similar manner to the peptidergic and GABAergic neurons (see Parnavelas et al

1988). The expression of NADPH diaphorase peaked at around P7 then declined slightly at around P13 in all layers. In the rat the stained cells were distributed equally over the different cortical layers with the exception of layer I. Did you observe NADPH diaphorase-positive cells in the marginal zone?

Jones: I have found that in monkeys NADPH diaphorase-positive afferent cells don't actually express NADPH diaphorase until early in the postnatal period (unpublished observations). Expression continues throughout life, and is also found in the molecular layer where the cells have the appearance of Cajal–Retzius cells. You also intimated that cells may migrate into the cortex while expressing NADPH diaphorase. Our studies in monkeys indicate that these cells do not switch on NADPH diaphorase expression until after this migration is complete.

Molnár: That's what I implied. The cells are already in their final position when the NADPH diaphorase expression moves up progressively throughout the cortical layers.

Jones: This does not exclude that they're not a valid marker for former subplate cells, but I would like to think that they represent remnants of the early preplate.

Kennedy: The subplate is not restricted to the white matter. The subplate is defined as a population of cells that are generated early. Cells that are generated early are mostly found in the white matter, but they are also found in layers V and VI. The occasional cell is also found in the upper layers. They are similar in numbers and perhaps in distribution to the cells that express NADPH diaphorase.

References

Akbarian S, Kim JJ, Potkin SG et al 1995 Gene expression for glutamic acid decarboxylase is reduced without loss of neurons in prefrontal cortex of schizophrenics. Arch Gen Psychiatry 52:258–266

Andreason NC, Arndt S, Swayze V et al 1994 Thalamic abnormalities in schizophrenia visualised through magnetic resonance image averaging. Science 266:294–298

Arnold SE, Hyman BT, Van Hoessen GW, Damasio AR 1991 Some cytoarchitectural abnormalities of the entorhinal cortex in schizophrenia. Arch Gen Psychiatry 48:625–632

Benes FM, McSparren J, Bird ED, San Giovanni JP, Vincent SL 1991 Deficits in small interneurons in prefrontal and cingulate cortices of schizophrenic and schizoaffective patients. Arch Gen Psychiatry 48:996–1001

Feinberg I 1982 Schizophrenia: caused by a fault in programmed synaptic elimination during adolescence. J Psychiatr Res 17:319–334

Galaburda AM, Sherman GF, Rosen GD, Aboitiz F, Geschwind N 1985 Developmental dyslexia: four consecutive patients with cortical anomalies. Ann Neurol 18:222–233

Meencke H-J 1989 Pathology of childhood epilepsies. Clevel Clin J Med 56:111–120

Meencke H-J, Veith G 1992 Migration disturbances in epilepsy. Epilepsy Res Suppl 9:31–40

Molnár Z, Mitrofanis J, Blakemore C 1993 Neurons producing nitric oxide in the developing rat cortex. J Physiol 59:124P

Pakkenberg B 1990 Pronounced reduction of total neuron number in mediodorsal thalamic nucleus and nucleus accumbens in schizophrenics. Arch Gen Psychiatry 47:1023–1028

Parnavelas JG, Papadopoulos GC, Cavanagh ME 1988 Changes in neurotransmitters in development. In: Peters A, Jones EG (eds) Cerebral cortex, vol 7: Development and maturation of cerebral cortex, Plenum Press, New York, p 177–209

Selemon LD, Rajkowska G, Goldman-Rakic PS 1993 Cytologic abnormalities in area 9 of the schizophrenic cortex. Soc Neurosci Abstr 19:200

Wang X-H, Levitt P, Grayson DR, Murphy EH 1995 Intrauterine cocaine exposure of rabbits: persistent elevation of GABA immunoreactive neurons in anterior cingulate cortex but not visual cortex. Brain Res, in press

Yurkewicz L, Valentino KL, Floeter MK, Fleshman HW Jr, Jones EG 1984 Effects of cytotoxic deletions of somatic sensory cortex in fetal rats. Somatosens Res 1:303–327

Pathology of cortical development and neuropsychiatric disorders

G. W. Roberts, M. C. Royston* and M. Götz

Department of Molecular Neuropathology, SmithKline Beecham Pharmaceuticals, New Frontiers Science Park, Harlow, Essex CM19 5AW, and *Departments of Psychiatry & Anatomy, Charing Cross & Westminster Medical School, St Dunstan's Road, London W6 8RP, UK

> *Abstract.* Epilepsy is a well-documented consequence of about 150 rare genetic syndromes and malformations of the central nervous system. These syndromes are generally associated with fairly gross defects within the central nervous system and they were thought to be responsible for a small minority of cases. However, improved methods of neuropathological investigations and extensive magnetic resonance imaging studies have revealed a range of disturbances in cortical cytoarchitecture in patients with epileptic seizures previously considered as idiopathic (up to 70% of epilepsy). Structural abnormalities have also been demonstrated in the brain in schizophrenia. These consist of disturbed cortical cytoarchitecture (best described in the temporal lobe) and a diffuse loss of grey matter. The absence of the pathological stigma characteristic of degenerative processes indicates that these structural changes are the result of an abnormal pattern of brain development. The relationship between the type and location of developmental abnormality and the subsequent clinical syndrome (e.g. generalized or localized epilepsy) and the effects of aberrant cortical development on the functional integrity of the adult brain require definition.
>
> *1995 Development of the cerebral cortex. Wiley, Chichester (Ciba Foundation Symposium 193) p 296–321*

Neuropsychiatric disorders are hugely expensive to society both in terms of the cost of treatment and care and in the extent of suffering and social stigma they cause. Epilepsy and schizophrenia are two of the most common neuropsychiatric disorders and between them affect over 10 000 000 people in the western hemisphere. Therapeutic approaches that offer a degree of symptom control are available for both conditions. However, there are significant problems with treatment resistance and unwanted side effects, leaving considerable scope for improvement. The drive for improved therapies has been handicapped by a limited (in the case of epilepsy) or non-existent (in the case of schizophrenia) understanding of the aetiology and pathophysiology of these disorders (see Meldrum & Bruton 1992, Roberts 1991 for reviews).

Recent studies on the aetiology and pathology of epilepsy and schizophrenia have produced results that describe an unexpected commonality between these two disorders. Despite profound differences in epidemiology and pathophysiology, abnormal cortical development appears to be a significant pathological process in both diseases. This paper summarizes the evidence supporting the involvement of abnormal cortical development, and it indicates how the spectacular advances in neurodevelopmental biology could help identify the genetic features and developmental processes that underlie both conditions.

Epilepsy

Epilepsy is an episodic disorder of the nervous system that arises from the excessive synchronous and sustained discharge of a group of neurons. An epileptic seizure is a brief and usually unprovoked stereotyped disturbance of behaviour, emotion, motor function or sensation, resulting from such cortical neuronal discharge. An unequivocal diagnosis of epilepsy is appropriate only when seizures have recurred in an apparently spontaneous fashion.

Epidemiology

Epilepsy is the most common serious neurological disorder, with a lifetime prevalence of up to 5% (World Health Organization estimate) of the population. Therefore, one in 20 of the population will have an epileptic seizure (excluding a febrile convulsion) during their life. Following this initial seizure, about 0.5% of the population will develop epilepsy. However, the exact figures for the extent of epilepsy within the population are difficult to determine because methods of collecting cases and criteria for diagnosis vary widely between studies. A recent survey of the literature (Shorvon 1990) reported incidence rates of 20–70 cases/100 000 people per year (range 11–134 cases/100 000 people per year) and prevalence rates of 4–100 cases/1000 people in the general population (range 1.5–30 cases/1000 people) are reported (Fig. 1). The incidence and prevalence of epilepsy varies with age. Rates are greatest in early childhood, decline between the ages of five and 60, and rise again in elderly people. The incidence and prevalence of epileptic disorders increases steeply after the age of 60. Typical rates of incidence and prevalence of those aged 40–59 are 12 cases/100 000 people and 7.3 cases/1000 people, respectively; whereas for those over 60, these figures rise to 80 cases/100 000 people and 10.2 cases/1000 people, respectively. About 25% of the total number of patients with epilepsy in the community are over 60. In addition, epilepsy rates differ according to sex, socioeconomic class, race and geographical location. Males are affected more often than females and rates of epilepsy are higher in lower socioeconomic groups.

Aetiology

The complexity of the epidemiology, although daunting at first sight, can, in part, be explained by the known causes and pathophysiology of the disease (Roberts et al 1993). Almost all epileptic disorders can be related to genetic disorders, birth trauma, febrile convulsions, head injury, toxic insults or organic disease (in some cases several factors might contribute). Males are more likely than females to have neurodevelopmental abnormalities and to experience neurological difficulties as a result of birth trauma. Low socioeconomic class is related to increased rates of birth difficulties. In addition, although the rates decline in early adulthood, increased social activity results in an increased risk of accident (especially head injury in males), and age increases the risk of organic degenerative disease. Together, these features explain the U-shaped nature of the curve in Fig. 1.

Virtually any process, either genetic or acquired (e.g. trauma or disease), that damages or alters brain structure or disturbs the electrical functions of the brain can give rise to epilepsy. This is illustrated by the observation that there are almost 150 rare genetic disorders which have epilepsy as a common feature. However, these known genetic disorders together account for less than 1% of all epilepsy. They can be summarized as follows:

(1) approximately 25 autosomal dominant conditions (eg. tuberous sclerosis, acute intermittent porphyria, neurofibromatosis);
(2) about 100 types of autosomal recessive phenotypes (eg. Krabbe's disease, Batten's disease);
(3) chromosome X-linked disorders (eg. Aircardi syndrome).

Neuropathology

Extensive neuropathological studies have been carried out to identify and/or define the neuropathological basis of epilepsy. A simple common neuropathology has not been found, and many surveys have reported that a specific neuropathology is not apparent in up to 70% of cases (Bruton 1988). However, more recent studies utilizing improved imaging and pathological methodologies indicate a higher proportion of cases with a discrete pathology. For example, congenital malformations, such as porencephaly, microgyria, cortical dysplasia and arteriovenous malformations, are significantly more common than previously documented (Kuzniecky 1994).

The neuropathology associated with epileptic seizures is described according to its general morphology rather than its association with any particular type

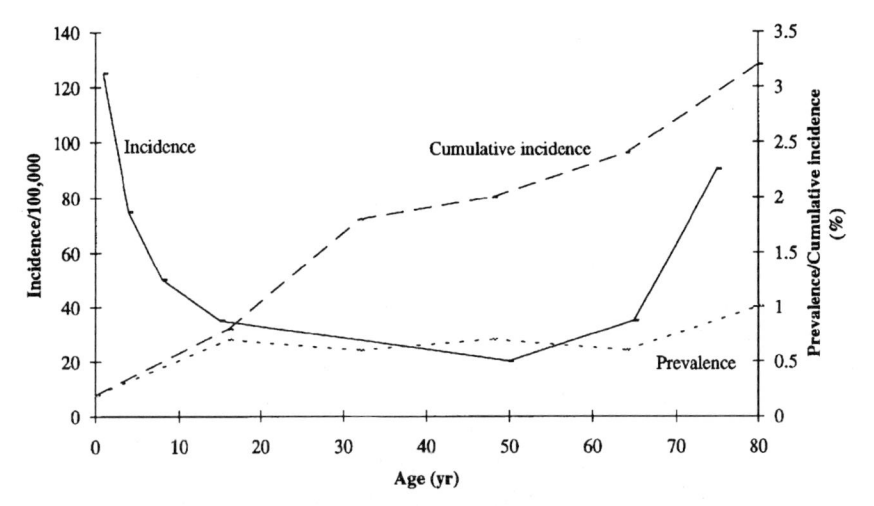

FIG. 1. Age-specific cumulative incidence and prevalence rates of epilepsy from a community study (adapted from Roberts et al 1993).

of seizure. In pathological terms, epilepsy can be divided into three categories (from Meldrum & Bruton 1992):

(1) symptomatic, in which the identified pathology (eg. tumour) is considered to give rise to seizure activity (this is also called secondary epilepsy);
(2) idiopathic, in which no clearly identified causal pathology is found (this is also called cryptogenic or primary epilepsy);
(3) special forms of epilepsy.

Brain structure is frequently abnormal in cases of idiopathic epilepsy where no clearly identified primary causal pathology is present. This apparent paradox is related to the fact that the physiological processes involved in seizure activity can themselves result in pathology. Several types of pathology are associated with long-standing seizure activity, e.g. Ammon's horn sclerosis (due to febrile convulsions and/or repetitive seizure activity). Ischaemic damage due to repeated episodes of seizure activity or particularly prolonged seizure (e.g. in status epilepticus) is also seen frequently. Although pathology is present in these cases, it can be identified as a likely consequence of the epilepsy and not its primary cause.

Disordered cortical cytoarchitecture

Recent studies using sophisticated imaging and neuropathological techniques have started to document the importance of disordered cortical cytoarchitecture as a primary causal factor in epilepsy (Bruton 1988, Palmini

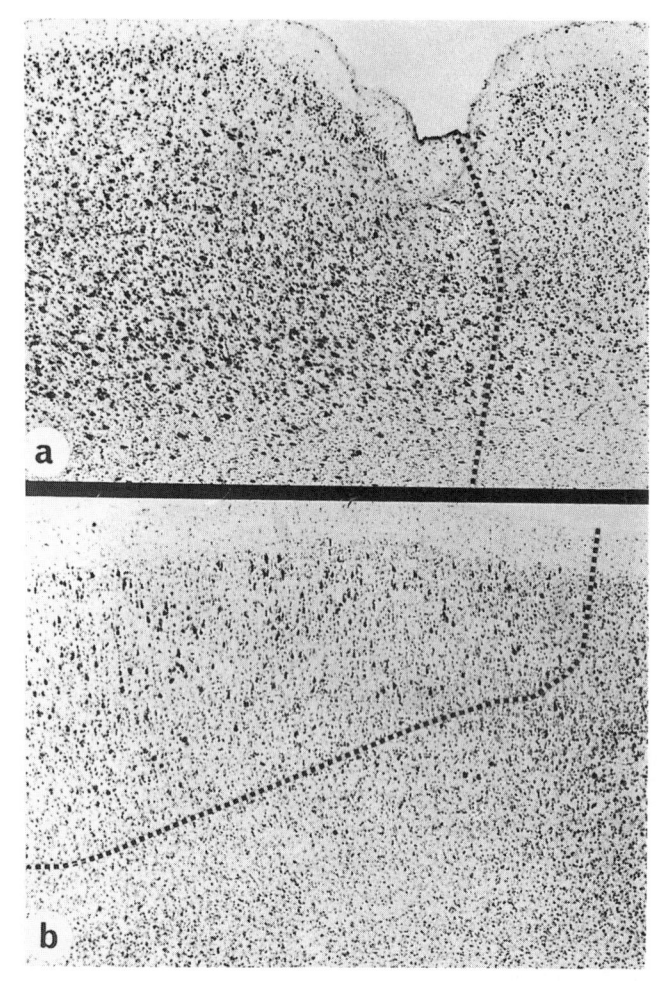

FIG. 2. Resected cortex from two cases (a and b) of drug-resistant epilepsy. The junction between the normal and the abnormal cortex is abrupt. The presence of large, bizarre nerve cells can be seen clearly even at relatively low magnification, although the area of abnormal cortex is often not visible to the naked eye. Cresyl violet stain. Magnification ×4230. (Courtesy of Dr C. J. Bruton, Corsellis collection, Runwell Hospital, Essex, UK.)

et al 1994, Kuzniecky 1994, Barkovitch & Kjos 1992). Systematic pathological surveys have reported the presence of disordered cortical cytoarchitecture in regions of cortex that have been resected during surgery to alleviate chronic epilepsy. Such morphological abnormalities (graphically described as architectural follies) have been labelled as hamartomas (small benign tumours of developmental origin) or cortical dysplasia (Bruton 1988). These types of lesions appear as clusters of abnormally large neurons in the cortex, as disordered regions of cortical lamination or as dystopic groups of neurons in the subcortical white matter (Fig. 2). The site and extent of these lesions is variable, and a close match between clinical symptomatology and the type of pathology has not been identified (Wylie et al 1994). These reports have been supported by imaging studies, such as magnetic resonance imaging (MRI), that have documented the presence of abnormal density patterns on images thought to represent regions of disturbed cortical cytoarchitecture (Kuzniecky 1994, Barkovitch & Kjos 1992). These types of cortical abnormalities are thought to occur during the neurogenesis of the fetal cortex in the second half of pregnancy (Palmini et al 1994). Clinicopathological studies indicate that these kinds of lesions are common in epilepsy that arises early in childhood and that they are resistant to treatment (Bruton 1988). This type of pathology is common in paediatric localization-related epileptic syndromes (eg. temporal lobe epilepsy) and is often found after resection of the cortex, even when imaging studies did not originally detect an abnormality.

Disordered cortical cytoarchitecture is the pathology thought to be a factor in a variety of epileptic syndromes including primary generalized epilepsy, West's syndrome and temporal lobe epilepsy. The precise relationship between the extent and types of cortical abnormality and epilepsy is debated. This is, in part, due to the growing realization that variants of cortical architecture are comparatively common in normal brains. Therefore, some pathologists consider ectopic neurons as minor variations of the cytoarchitectural theme, whereas others interpret them as the morphological substrate of increased susceptibility to epilepsy. The threshold whereby the former becomes the latter has not been defined.

Epilepsy and schizophrenia

The relationship between epilepsy and schizophrenia has been examined from a number of different standpoints. These include the occurrence of epilepsy in patients with schizophrenia, the development of schizophrenia in patients with established epilepsy, the similarity between the clinical phenomena seen in temporal lobe epilepsy and in schizophrenia, and the antagonism hypothesis, i.e. that epilepsy and schizophrenia are mutually incompatible (this led to the introduction of convulsive therapy as a treatment for schizophrenia, which is now considered ineffective). More recent studies have examined the neuropathology of patients with temporal lobe epilepsy and psychotic symptoms. Roberts et al (1991) demonstrated that the psychoses of temporal lobe epilepsy are associated with neuronal or glial developmental abnormalities which are located in medial temporal lobe structures (entorhinal cortex, hippocampus, amygdala), suggesting that the psychotic symptoms arise as a consequence of dysfunction in this region. This observation has considerable importance in terms of which areas of the brain may be abnormal in schizophrenia.

Schizophrenia

Schizophrenia is characterized by fundamental distortions of thinking. These distortions include delusions, altered perceptions, blunted or inappropriate emotional responses and personality changes (Roberts et al 1993). These symptoms are experienced in a setting of clear consciousness and (generally) undiminished intellectual capacity. Typically, a diagnosis of schizophrenia is made when the patient is in their early twenties. Several studies have indicated that patients who receive a firm diagnosis of schizophrenia will follow one of the following three patterns of illness:

(1) a small number of psychotic episodes followed by a return to their pre-morbid level of function (good prognosis): occurs in 20% of cases;
(2) an episodic course characterized by psychotic episodes that may be precipitated by negative life events or adverse social circumstances. Deterioration in personality and inadequate social function are present: occurs in 60% of cases;
(3) personality deterioration and severely impaired social function/integration (poor prognosis: occurs in 20% of cases.

This gloomy prognosis can only be partially ameliorated by the therapies currently available.

Epidemiology

Adults have a 0.5–1% risk of developing schizophrenia in their lifetime, and the incidence is about 15–20 new cases/100 000 of the population (Roberts et al 1993). A large World Health Organization study has demonstrated that the risk of schizophrenia globally is relatively constant and that the symptoms of the syndrome are similar in each country. There are few exceptions to this pattern of homogeneity, although a high prevalence of schizophrenia has been reported in Croatia and the west of Ireland. The disease can occur at any age between early teens and late seventies, although the vast majority of cases have an age of onset of 15–45. In Europe, the average peak incidence in men is at 27, three years earlier than that in women, although both sexes are affected equally. Studies in other countries using more rigid diagnostic criteria have reported earlier ages of onset.

Aetiology

The cause of schizophrenia is not known. However, several factors have been the subject of speculation in the development of aetiological hypotheses. These include:

(1) genetics: schizophrenia is often familial, and monozygotic twins develop the disease more frequently than dizygotic twins;
(2) season of birth: winter months show an excess of births;
(3) viral pandemics: reported excess of patients with schizophrenia born three to six months after influenza pandemics;
(4) fetal, birth or perinatal infection or trauma;
(5) neurological or neuropsychiatric conditions: can give rise to schizophrenia-like symptoms;
(6) abnormal development: psychological surveys, and imaging and neuropathological studies suggest a failure in brain development.

The details underpinning these proposals have been discussed extensively in the literature (Lewis & Murray 1987, Roberts 1991, Done et al 1991, 1994, Jones et al 1995).

At a simplistic level, the risk of developing schizophrenia increases with genetic proximity. However, the interpretation of these epidemiological results and their conversion into 'genetic models' of inheritance is fraught with difficulties: parents are often unaffected and thus have a lower risk than siblings; if one of the parents has schizophrenia, it is more likely to be the mother; the risk in offspring is greatly increased if both parents have schizophrenia; and the exact symptoms (penetrance) can vary greatly within families. The genetics of schizophrenia, therefore, do not conform to a single simple genetic model. However, the relative importance of the various aetiological factors in relation to the percentage of cases of schizophrenia they explain (Roberts 1991) are roughly summarized as follows (taking the current literature as a whole):

(1) genetic, 80%;
(2) neurological disorder, less than 5%;
(3) season of birth and viral pandemics, less than 5%;
(4) drug induced/others, 5%;
(5) fetal/birth/perinatal events, less than 3%.

Treatment of psychotic symptoms with drugs was discovered serendipitously in the 1950s. The first such drug was reserpine and this was quickly followed by the class of drugs known as the typical antipsychotics or major tranquillizers. This class of drugs includes the phenothiazines, e.g. chlorpromazine; the butyrophenones, e.g. haloperidol; and the thioxanthenes, e.g. chlorprothixene.

These drugs all have an affinity for dopamine receptors, which is related to their clinical efficacy. This important observation has led to the development of the 'dopamine hypothesis', which suggests that overactivity of dopamine receptors plays a fundamental role in the aetiology of schizophrenia. However, despite considerable research, there are unanswered questions concerning the specific role of dopamine (see Royston & Simpson 1993 for review). Furthermore, the recent introduction of the atypical antipsychotics, such as clozapine and risperidone, adds to the difficulties because they do not conform to the dopamine affinity/clinical efficacy rule, yet have profound effects on the symptomatology of schizophrenia. These results suggest that multiple neurotransmitter systems are involved in schizophrenia.

Pathology

Recent *in vivo* imaging techniques have dramatically altered our concept of schizophrenia and have led to the rejection of the idea that schizophrenia is a functional or non-organic disorder. Computed tomography (CT) studies have demonstrated that ventricular enlargement occurs in the brains of patients with chronic schizophrenia. MRI studies have replicated this finding and they have also documented the presence of a diffuse reduction in cortical grey matter (Suddath et al 1990, Breier et al 1992). This reduction, although affecting many cortical areas, is concentrated on the area of the temporal lobes. Controlled studies on monozygotic twins discordant for schizophrenia have demonstrated that a 15% loss of temporal lobe grey matter and an increase in the size of the lateral ventricle occurs in schizophrenics. These results provide the first firm evidence of the biological (as opposed to psychological or functional) nature of schizophrenia. Meta-analysis of the pooled results from many CT studies indicates that the phenomenon of larger ventricles is unimodal. This finding is significant because it suggests that the phenomenon of brain abnormality in schizophrenia is a unitary one and, by extrapolation, that the various syndrome manifestations are simply reflections of a common pathological process (Roberts 1991).

Despite the increased sensitivity of MRI for detecting minor pathological damage (e.g. periventricular hyperintensities and white-matter damage), there is little evidence from either CT or MRI studies of increased amounts of other pathological damage in the brain in schizophrenia. However, positron emission tomography (PET) studies show the presence of relative hypometabolism and reduced blood flow in the dorsolateral prefrontal cortex of patients who are given psychological tasks (e.g. the Wisconsin Card Sort) that activate this brain region (Buchsbaum 1990). In contrast, patients required to perform reasoning tasks that do not activate the prefrontal cortex do not show abnormalities in cerebral blood flow. A similar result is seen when comparing discordant monozygotic twins. Hypofrontality appears to be independent of neuroleptic treatment and it is present in untreated patients on their first admission. Preliminary reports indicate that one effect of neuroleptic treatment is to diminish the degree of hypofrontality.

Neuropathology

Ventricular enlargement in the brain of patients with chronic schizophrenia is often visible to the naked eye. Brain weight reductions of about 5% and a reduced (fixed) brain length of 0.8 cm have been observed bilaterally in men and women in some, but not all, studies. Brains otherwise appear normal and do not display marked structural aberrations. Sulcogyral patterns are also within normal limits (Bruton et al 1990). Schizophrenia has not been associated with any other gross pathological phenomena (Roberts 1991). However, a diagnosis of schizophrenia does not offer immunity from a supervening disease process and patients examined in post-mortem studies are often in their sixth, seventh or eighth decade. This may account for the sporadic case reports that link schizophrenia to various types of pathology (Bruton et al 1990, reviewed in Roberts 1991).

Quantitative morphological studies of schizophrenia replicate the CT/MRI findings of increased ventricular size, and they extend the documentation of structural changes. These changes include a pattern of tissue loss particularly affecting entorhinal cortex, hippocampus, temporal lobe grey matter and central grey matter (Bogerts et al 1985, Brown et al 1986, Jakob & Beckmann 1986, Falkai et al 1988, Jeste & Lohr 1989, Arnold et al 1991).

More detailed neuropathological studies have revealed that neuronal loss occurs in the insula, entorhinal cortex and prefrontal cortex. Quantitative studies indicate that about a 20% reduction in neuronal number occurs in the anterior hippocampus and entorhinal cortex, and smaller reductions occur in certain cortical layers of the cingulate (layer V) and prefrontal (layer VI) cortex (Benes & Bird 1987). In addition, changes in the cytoarchitectural arrangements of neurons in the entorhinal cortex, hippocampus and insula have been observed.

One interesting finding that has been reported by several groups is a disturbance in the organization of and loss of layer II (pre-α) neurons (Jakob & Beckmann 1986, Arnold et al 1991). These modified pyramidal neurons are interesting for several reasons (Roberts 1991, Roberts et al 1993, Fig. 3):

(1) they are modified pyramidal cells organized into unique clusters;
(2) they contain large amounts of the enzyme nitric oxide synthase;
(3) their projections form part of the perforant pathway;
(4) they receive inputs from almost all cortical areas.

This finding is important because cytoarchitectural disturbances in this region could lead to a disturbance in the functioning of distant brain regions that are synaptically connected (e.g. dorsolateral prefrontal cortex, cingulate cortex, hippocampus). This could result in considerable compromise and distortion of the input and integration of information from many sensory modalities.

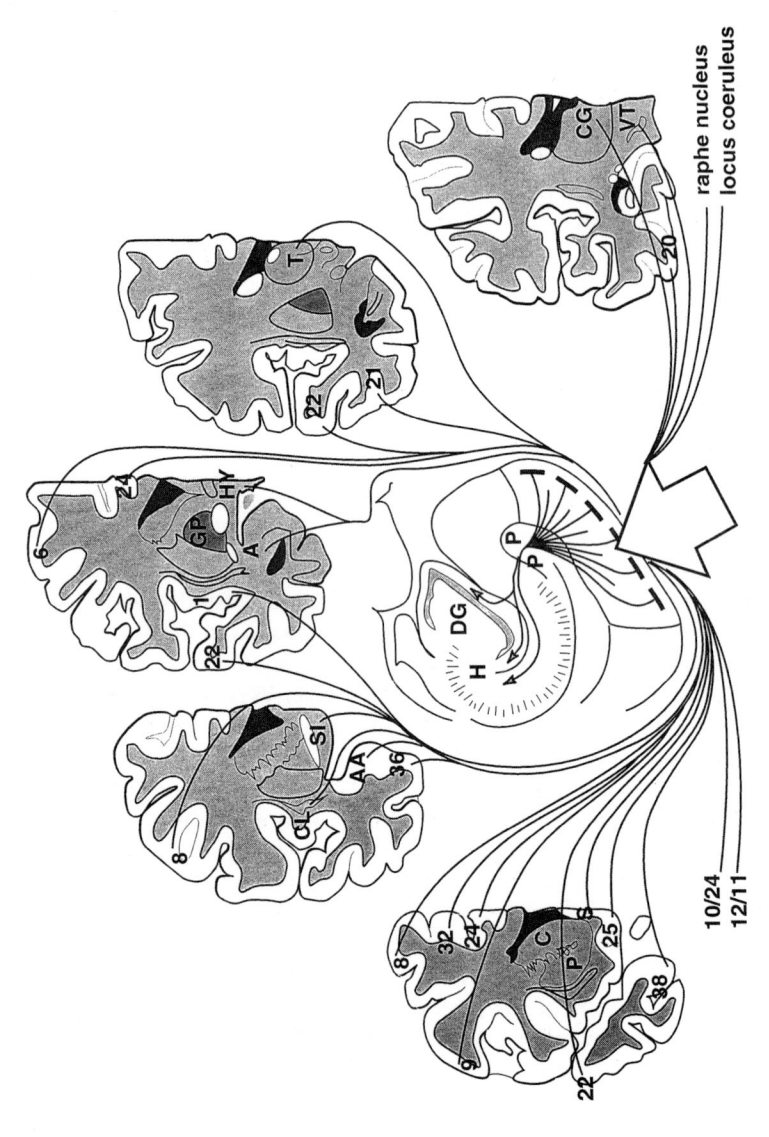

FIG. 3. Links between various cortical and subcortical areas and the parahippocampal gyrus are shown diagrammatically. The projections shown are those from the same hemisphere. Note that similar (but often smaller) projections from the opposite hemisphere also exist. Large arrow, parahippocampal gyrus; ■, white matter; □, ventricle; A, amygdala; AA, anterior amygdala; C, caudate; CG, central grey; CL, claustrum; DG, dentate gyrus; GP, globus pallidus; H, hippocampus; HY, hypothalamus; I, insula; P, putamen; PP, perforant path, S, septum; SI, substantia inominata; T, thalamus; VT, ventral tegmentum; 6–46, Brodman areas. (Adapted from Roberts et al 1991.)

Jones (1995, this volume) has reported a disturbance in the distribution of a subpopulation of NADPH diaphorase-positive neurons in specific frontal cortical areas. These studies suggest that, in addition to overt structural changes, a diffuse and subtle alteration in the cytoarchitecture of the cortex is present, which is fundamentally important in terms of the development of schizophrenic symptoms.

Developmental versus degenerative pathology

These cytoarchitectural changes occur as a consequence of aberrant development rather than as a result of a degenerative process (Roberts 1991). Evidence to support this statement comes from neuropathological studies which have examined the brain for degenerative changes, i.e. the presence of excess gliosis. Large systematic studies, using an antibody marker against glial fibrillary acidic protein, did not find any evidence of increased gliosis in any region of the temporal lobe. Subsequent studies, attempting to demonstrate a relationship between structural brain changes and quantitative increases in gliosis, have also been negative (reviewed in Roberts 1991).

Summary

The report above reviews the neuropathological evidence supporting the hypothesis that schizophrenia and many types of epilepsy arise fundamentally as a consequence of abnormal cortical development. This can be summarized as follows:

(1) sophisticated imaging and pathological techniques have shown that a significant proportion of cases of epilepsy have a clearly identifiable abnormality of cortical cytoarchitecture;
(2) the abnormalities commonly seen in epilepsy include clusters of abnormally large neurons, regions of disordered cortical lamination or dystrophic groups of neurons in the subcortical white matter;
(3) a distinction can be made, in pathological terms, between causal pathology due to aberrant cortical development, and acquired pathology consequent upon repeated/prolonged epileptic activity;
(4) there are subtle alterations in the cytoarchitecture of key cortical regions in schizophrenics;
(5) neuropathological studies of schizophrenia do not reveal any evidence for an ongoing degenerative process.

Developmental neurobiology and neuropsychiatric disease

To understand how an underlying process of aberrant cortical development may identify the aetiologies of epilepsy and schizophrenia, we need to consider the current understanding of the process of normal brain development. The complex pattern of cortical arrangement and interconnection arises from the orchestration of a number of different processes: proliferation, migration and elimination/refinement. Abnormalities of mechanisms that underlie both the migration of neurons into their correct position and functional specialization might produce schizophrenic and epileptic abnormalities, and therefore merit closer inspection.

Mechanisms of cortical development

The adult mammalian cortex is composed of different regions that originated sequentially during evolution. The most recently evolved cortex is the six-layered neocortex, and the three-layered allocortex and mesocortex are the oldest. Functional subdivisions, composed of specific neurons, are apparent within each of these structures. These neurons are involved with particular aspects of sensory or motor innervation. Moreover, a further degree of specialization is the arrangement of cortical neurons in horizontal layers. Neurons in each layer share certain morphological, neurochemical and physiological traits. The question arises as to how this positional specification is acquired during cortical development. The current hypotheses polarize into two suggestions:

(1) regional differences in the cortex are predetermined at early developmental stages: this assumes that a proto-map exists in the proliferative zone, which precedes the functional map present in adult cortex (Rakic 1988);
(2) cortical precursor cells are equipotential, and neurons acquire their positional specification later in development (O'Leary 1989).

Neuronal migration drives both of these concepts of cortical development. The critical features are the mechanisms by which cells choose their final position and how cell fate decisions are related to this choice, i.e. whether neurons are specified before migration and then select their correct position, or whether neurons acquire their phenotype after they arrive randomly at their respective positions. These alternatives predict that displaced cells would have different cell fate decisions. If specification occurred before migration, displaced cells would express cellular properties different from their normal environment; conversely, specification by the environment after migration would lead to cells expressing phenotypes corresponding to their new environment.

Transplantation studies reveal that laminar position is specified before the onset of migration (McConnell & Kaznowski 1991). This is more controversial for the tangential position. The radial unit hypothesis was proposed on the basis of the anatomy of radial glial cells, along which cortical neurons migrate, and the radial arrangement of cells in the developing cortex (Rakic 1988). This hypothesis predicts that migration along the radial glial fibres maintains a radial alignment between the ventricular zone, where neurons are generated, and their final position in the adult cortex. Radial migration could also explain the existence of the proto-map. Alternatively, cells could be specified at much later developmental stages, generating a protocortex (O'Leary 1989). There now is increasing evidence for early positional specification during cortical development, although some degree of cellular dispersion also occurs (at least in rodent cortex).

Migration in the developing cortex

Migration of cortical cells occurs radially in that cells migrate from the ventricular zone to the pial side. However, a variety of techniques suggest that tangential dispersion also occurs (for review see Götz 1995). Therefore, the prediction that cells migrate in radial units is not met as precisely in rodent cortex as it is in other parts of the central nervous system, e.g. the retina, where clonally related cohorts of migrating cells have been found. A higher degree of radial organization seems to occur during the development of primate cortex, but the pattern of migration, which may be interesting, has not yet been worked out. Kennedy & Dehay (1993) have pointed out that there are important and potentially instructive differences in the mechanisms used to construct circuits during very prolonged (primates) or very short (rodents) periods of time.

Studies of migration in rodents have revealed that cells cross different telencephalic domains (for review see Götz 1995). This observation is striking because of the different organization and layering described above. The cells also disperse freely across functional boundaries within each of these regions (Walsh & Cepko 1992). For example, a quantitative analysis of clonal dispersion in the hippocampus revealed a free, stochastic dispersion of neuronal clones across area boundaries (Grove et al 1993). However, the majority of cortical neurons do not disperse more than a few hundred micrometres (in rodents), so that most cells remain within a given area. Radial location is better visualized when the distribution of large numbers of cells, rather than an individual cell, is examined. When marker genes, such as *lacZ*, are inserted in one X chromosome, they are stochastically inactivated in female embryos, generating a mosaic of precursor cells. Studies of such chimeras have provided striking visual evidence for radial columnar organization in the developing cortex. Tan & Breen (1993) found that the ratio of blue (chromosome with *lacZ* activated) to white (chromosome with *lacZ* inactivated) cells in the radial columns of the cortex was 2:1. Thus, the majority of cortical cells remain within a radial column, but the column also incorporates some dispersed cells.

Failure of the genetic control of proliferation or radial migration would be expected to have profound effects on cortical structure. These effects may include smaller surface areas, altered cortical thickness or lamination, and focal or areal regions of neuronal ectopias (such as is seen in lissencephaly and pachygyria). These aberrant developmental processes almost certainly contribute to the aberrant cortical dysmorphologies seen in familial forms of epilepsy.

Positional specification in the developing cortex

A critical question in schizophrenia is why cortical abnormalities are found in specific and consistent cortical regions. The process of positional specification in the developing cortex may assist our understanding of this.

There is increasing evidence for heterogeneity of the ventricular zone, both at cellular and positional levels, despite its homogenous histological appearance. Precursor cells in the ventricular zone differ in their specification and generate distinct cell types: astrocytes, oligodendrocytes or neurons (Price et al 1995, this volume). Molecular correlates to this functional specification have not been discovered, although the positional areal heterogeneity of the ventricular zone is highlighted by some molecular markers. The expression pattern of these markers corresponds to specific functional regions in the adult cortex. (Levitt 1984, Arimatsu et al 1992, Dehay et al 1993, Cohen-Tannoudji et al 1994). This suggests that positional specification has occurred in the ventricular zone, as suggested by the proto-map hypothesis, even though most of these molecules are not expressed in the ventricular zone itself.

The limbic system-associated membrane protein (LAMP) is a cell surface molecule (Levitt 1984) that is particularly relevant to areas most affected in schizophrenic brains. LAMP is detected in limbic areas comprising allocortical, mesocortical and neocortical regions of the cortex. It is a glycoprotein and a member of the immunoglobulin superfamily. It is expressed from early developmental stages on postmitotic neurons, and not on glial or precursor cells (Levitt 1984, Zacco et al 1990). Therefore, only neurons located at particular (limbic) positions express LAMP. *In vitro* studies revealed that precursor cells are specified as to whether they express LAMP or not. This depends on their position in the developing cortex (Ferri & Levitt 1993). Cells from limbic cortex that are dissociated before the onset of LAMP immunoreactivity become LAMP positive after a few days *in vitro*. In contrast, when cells are taken from more medial, neocortical regions, few acquire LAMP immunoreactivity *in vitro*. Therefore, precursor cells isolated *in vitro* regulate LAMP expression in the same manner as they would have done *in vivo*. These results indicate that LAMP expression is specified in precursor cells and is not dependent on further environmental influences.

LAMP is involved in axonal growth and mediates appropriate connectivity. LAMP is located on the cell body and dendrites of adult neurons, whereas during development it is also transiently localized on axonal processes (Zacco et al 1990). Therefore, during formation of neuronal connectivity, LAMP is present on afferent fibres from specific thalamic nuclei and their appropriate target cells in the limbic cortex. A pivotal role for LAMP in mediating the match between limbic afferents and their target cells has been supported by various experimental approaches (Levitt et al 1995, this volume). LAMP could also be involved in the segregation of LAMP-positive cells from LAMP-negative cells, and may therefore set up limbic and non-limbic telencephalic domains because it is a homophilic adhesion molecule. This is consistent with recent reaggregation experiments that have revealed differences in the adhesive properties between cells derived from different telencephalic compartments. Limbic (hippocampal) and non-limbic (neocortical) cells segregate *in vitro*, suggesting that they do so *in vivo* (Götz et al 1994). Therefore, LAMP could be a molecular correlate for the combined alterations of cell migration and projection patterns found in limbic areas of schizophrenic brains.

Early regional specification in the cortex also occurs in the transgenic mouse line H-2Z1. Cohen-Tannoudij et al (1994) showed that the expression of *lacZ*, driven by specific regulatory elements of a major histocompatibility complex class I gene, occurred exclusively in primary somatosensory cortical areas of the telencephalon. When prospective somatosensory cortex was transplanted to a different cortical region or to the cerebellum prior to the generation of layer IV cells (in which the transgene is predominantly expressed later), the expression pattern corresponded to the original location. Conversely, transgene expression could not be initiated in visual cortex after transplantation into somatosensory cortical areas. This suggests that precursor cells are irreversibly specified as to whether they express the transgene or not, and that this is dependent on their position. These results also suggest that a single defect in a promoter region can affect cells selectively at specific positions in the cortex.

The positional expression pattern of some marker molecules is already specified in ventricular zone cells; however, they do not yet express these molecules. Studies of cell cycle dynamics in developing primate cortex also demonstrate a prominent difference between cortical precursor cells located at specific positions: ventricular zone cells in the presumptive primary visual area 17 proliferate faster than those which generate the secondary visual area 18 (Dehay et al 1993). Area 17 contains a higher number of neurons than area 18, so the cell cycle difference detected in the ventricular zone is likely to contribute to the difference in cell number of these areas in the adult. The observation that differences in the mitotic rate during development are graded, whereas the transition in cell number between cortical areas in the adult is abrupt, is particularly relevant (Dehay et al 1993). Comparably, LAMP is expressed in a developmental gradient—an increase in numbers of LAMP-positive cells from medial to more lateral areas (Ferri & Levitt 1993).

How are smooth transitions during development transformed into abrupt transitions in the adult? Cellular minorities are found at the ends of gradients and are caused by cellular dispersion. The elimination of these cells transforms a smooth boundary to an abrupt boundary. This may be achieved either by respecification of cells or by their death. Both of these mechanisms are involved in the formation of areal differences during cortical development.

Plasticity of positional specification

The specification of cellular properties can be revealed by isolation *in vitro*. The specified phenotypes, however, could still be plastic and amenable to environmental influences. Alternatively, they could be irreversibly specified, i.e. determined. Plasticity or determination of cell fate is best assessed by manipulating environmental influences, either *in vitro* by varying the culture conditions or *in vivo* by transplantation. Plasticity observed after transplantation does not reveal the lack of specification, but it shows the instructive role of environmental influences and the lack of determination. In the case of LAMP expression, cortical cells are specified whether or not to express LAMP according to their position, as shown by *in vitro* isolation (Ferri & Levitt 1993). However, when they are transplanted, fate decision can be altered in accordance to their new location (Barbe & Levitt 1991). When cortex is taken at embryonic day 12 from limbic areas and transplanted to neocortical areas, the transplanted cells do not express LAMP even though they would have done so at their normal location or when isolated *in vitro*. Conversely, when neocortical tissue is transplanted to limbic areas, it becomes LAMP immunoreactive according to its new position (Barbe & Levitt 1991). These results show that LAMP expression can be respecified at a different cortical position at a time when it is specified and would occur autonomously *in vitro*. The plasticity in LAMP expression is restricted to early developmental stages, and its specification cannot be altered in transplants taken two days later (Barbe & Levitt 1991).

This critical period raises the issue of whether other positional fate decisions are determined simultaneously or sequentially. For example, transplantation of the determined transgenic somatosensory cortex was performed at a time when LAMP expression was already determined. Is transgene expression plastic at an earlier time, and is this timing similar for other aspects of areal specification? Many phenotypic aspects of cortical cells are determined at their final mitotic division, e.g. the cell type, the laminar position, the transmitter specification and the axonal projection (for reviews see Götz & Price 1994, Götz 1995). The timing observed for the area-specific expression patterns to date is consistent with determination during the final mitosis.

The observed plasticity of cell fate decisions suggests a model of areal specification that combines the evidence for early positional information in the cortex (as suggested by the proto-map hypothesis) with the evidence for cellular dispersion and respecification (as suggested by the protocortex hypothesis). The majority of cells that maintain their position could preserve positional information, and subsequently respecify minorities of displaced cells. Alternatively, displaced cells could be eliminated by cell death.

Taken together, developmental studies establish that a close relationship exists between migration, positional information and cell fate decisions. Aberration of these processes may account for the patterns of disarray of particular cell types identified in both epilepsy and schizophrenia. Adhesion molecules that are expressed only in specific cortical regions and are involved in the formation of specific projection patterns may also affect cell migration. The cell type that links all these developmental aspects is the radial glial cell population. Radial glial cells guide neuronal migration and are themselves, or are at least closely related to, the first neuroepithelial stem cells (Gray & Sanes 1992). There is an inverse relationship between migrational patterns and cell fate decisions. In parts of the central nervous system, where descendants of the same precursor cell migrate radially, clonal descendants differentiate into a wide variety of different cell types; whereas in telencephalic areas, where migration is less radially restricted, precursor cells generate predominantly a single cell type (Grove et al 1993, Price et al 1995, this volume). Therefore, it is tempting to speculate that changes in radial glial cells affect migration, and regional and cell type specification.

Conclusions

Epilepsy and schizophrenia share elements of a common pathogenetic process in that disturbances of neurodevelopmental mechanisms might give rise to circuit/neuronal/synaptic or transmitter disturbances which are sufficient to alter normal brain function. However, there are some limitations and differences in the magnitude of the developmental disturbance documented in each condition. Neurodevelopmental anomalies may be a common cause of epilepsy, but they are not the exclusive cause. In addition, neurodevelopmental disturbances may be widespread or focal and appear in almost any brain region. Our present understanding suggests that developmental disturbances sufficient to provoke epileptic seizures generally involve significant dislocations of the cortical cytoarchitecture. This profile of the developmental disorders which can cause epilepsy indicates that abnormalities at any stage of the proliferative or migrational phases could give rise to an epileptogenic lesion.

In contrast, the parameters defining the profile of the developmental disorder in schizophrenia are more restrictive. The absence of a catastrophic or severe structural change in the schizophrenic brain argues strongly against a disorder of the earliest stages of proliferation or migration. In addition, the pattern of structural abnormalities appears more pathognomonic and focal. Temporal lobe structures seem to be consistently affected, although other widespread but subtle changes have also been recorded. These changes may be a consequence of a restricted abnormality in development having effects on the extent and pattern of regional connectivity (Roberts 1991). This profile suggests that the mechanisms responsible for positional specification are a productive area of investigation in the search for the causes of schizophrenia.

Acknowledgements

M. C. R. is supported by the Mental Health Foundation, the Stanley Foundation and the Wellcome Trust.

References

Arimatsu Y, Miyamoto M, Nihonmatsu I et al 1992 Early regional specification for a molecular neuronal phenotype in the rat neocortex. Proc Natl Acad Sci USA 89:8879–8883

Arnold SE, Hyman BT, Van Hoesen GW, Damasio AR 1991 Some cytoarchitectural abnormalities of the entorhinal cortex in schizophrenia. Arch Gen Psychiatry 48:625–632

Barbe MF, Levitt P 1991 The early commitment of fetal neurons to the limbic cortex. J Neurosci 11:519–533

Barkovich AJ, Kjos BO 1992 Gray matter heterotopias: MR characteristics and correlation with developmental & neurologic manifestations. Radiology 182:493–499

Benes FM, Bird ED 1987 An analysis of the arrangement of neurons in the cingulate cortex of schizophrenic patients. Arch Gen Psychiatry 44:608–616

Bogerts B, Meertz E, Schonfeldt-Bausch R 1985 Basal ganglia and limbic system pathology in schizophrenia. Arch Gen Psychiatry 42:784–791

Breier A, Buchanan RW, Elkashef A, Munson RC, Kirkpatrick B, Gellad F 1992 Brain morphology and schizophrenia: a magnetic resonance imaging study of limbic, prefrontal cortex, and caudate structures. Arch Gen Psychiatry 49:921–926

Brown R, Colter N, Corsellis JAN et al 1986 Postmortem evidence of structural brain changes in schizophrenia. Differences in brain weight, temporal horn area and parahippocampal gyrus compared with affective disorder. Arch Gen Psychiatry 43:36–42

Bruton CJ 1988 The neuropathology of temporal lobe epilepsy (Maudsley monograph No. 31). Oxford University Press, Oxford

Bruton CJ, Crow TJ, Frith CD, Johnstone EC, Owens DGC, Roberts GW 1990 Schizophrenia and the brain: a prospective clinico-neuropathological study. Psychol Med 20:285–304

Buchsbaum MS 1990 The frontal lobes, basal ganglia and temporal lobes as sites for schizophrenia. Schizophrenia Bull 16:379–389

Cohen-Tannoudji M, Babinet C, Wassef M 1994 Early determination of a mouse somatosensory cortex marker. Nature 368:460–463

Dehay C, Giroud P, Berland M, Smart I, Kennedy H 1993 Modulation of the cell cycle contributes to the parcellation of the primate visual cortex. Nature 366:464–466

Done DJ, Johnston EC, Frith CD, Golding J, Shephard PM, Crow TJ 1991 Complications of pregnancy and delivery in relation to psychosis in adult life: data from the British perinatal mortality survey. BMJ 302:1576–1580

Done DJ, Crow TJ, Johnston EC, Sacker A 1994 Childhood antecedents of schizophrenia and affective illness: social adjustment at ages 7 and 11. BMJ 309:699–703

Falkai P, Bogerts B, Rozumek M 1988 Limbic pathology in schizophrenia: the entorhinal region—a morphometric study. Biol Psychiatry 24:515–521

Ferri R, Levitt P 1993 Cerebral cortical progenitors are fated to produce region-specific neuronal populations. Cereb Cortex 3:187–198
Götz M 1995 Getting there and being there in the cerebral cortex. Experientia 51: 301–316
Götz M, Price J 1994 Cell fate and axonal projections from the cerebral cortex. Semin Dev Biol 5, in press
Götz M, Wizenmann A, Lumsden A, Price J 1994 Specific adhesion of telencephalic precursor cells. Soc Neurosci Abstr 20:877
Gray GE, Sanes JR 1992 Lineage of radial glia in the chicken optic tectum. Development 114:271–283
Grove EA, Williams BP, Li DQ, Hajihosseini M, Friedrich A, Price J 1993 Multiple restricted lineages in the embryonic cerebral cortex. Development 117:553–561
Jakob H, Beckmann H 1986 Prenatal developmental disturbances in the limbic allocortex in schizophrenics. J Neural Transm 65:303–326
Jeste DV, Lohr JB 1989 Hippocampal pathologic findings in schizophrenia: a morphometric study. Arch Gen Psychiatry 46:1019–1024
Jones EG 1995 Cortical development and neuropathology in schizophrenia. In: Development of the cerebral cortex. Wiley, Chichester (Ciba Found Symp 193) p 277–295
Jones P, Rodgers B, Murray R, Marmot M 1994 Child developmental risk factors for adult schizophrenia in the British 1946 birth cohort. Lancet 344:1398–1402
Kennedy H, Dehay C 1993 Cortical specification of mice and men. Cereb Cortex 3:171–186
Kuzniecky RI 1994 Magnetic resonance imaging in developmental disorders of the cerebral cortex. Epilepsia 35:S44–S56
Levitt P 1984 A monoclonal antibody to limbic system neurons. Science 223:299–301
Levitt P, Ferri R, Eagleson K 1995 Molecular contributions to cerebral cortical specification. In: Development of the cerebral cortex. Wiley, Chichester (Ciba Found Symp 193) p 200–213
Lewis SW, Murray RM 1987 Obstetric complications, neurodevelopmental deviance and risk of schizophrenics. J Psychiatr Res 21:413–421
Meldrum BS, Bruton CJ 1992 Epilepsy. In: Adams JH, Duchen LW (eds) Greenfields neuropathology, 5th edn. Edward Arnold, London, p 1246–1283
McConnell SK, Kaznowski CE 1991 Cell cycle dependence of laminar determination in developing neocortex. Science 254:282–285
O'Leary DDM 1989 Do cortical areas emerge from a protocortex? Trends Neurosci 12:400–406
Palmini A, Andermann E, Andermann F 1994 Prenatal events and genetic factors in epileptic patients with neuronal migration disorders. Epilepsia 35:965–976
Price JP, Williams BP, Götz M 1995 The generation of cellular diversity in the cerebral cortex. In: Development of the cerebral cortex. Wiley, Chichester (Ciba Found Symp 193) p 71–84
Rakic P 1988 Specification of cerebral cortical areas. Science 241:170–176
Roberts GW 1991 Schizophrenia: a neuropathological perspective. Br J Psychiatry 158:8–17
Roberts GW, Done DJ, Bruton CJ, Crow TJ 1990 A 'mock up' of schizophrenia: temporal lobe epilepsy and schizophrenia-like psychosis. Biol Psychiatry 28:127–143
Roberts GW, Leigh PN, Weinberger DR 1993 Neuropsychiatric disorders. Wolfe Medical, London
Royston MC, Simpson MDC 1993 Post-mortem neurochemistry of schizophrenia. In: Kerwin RD (ed) Cambridge medical reviews, vol 1: Neurobiology and psychiatry. Cambridge University Press, Cambridge, p 1–15

Shorvon SD 1990 Epidemiology, classification, natural history and genetics of epilepsy. Lancet 336:93–96

Suddath RL, Christison GW, Torrey EF, Casanove MF, Weinberger DR 1990 Anatomical abnormalities in the brains of monzygotic twins discordant for schizophrenia. N Engl J Med 322:789–794

Tan SS, Breen S 1993 Radial mosaicism and tangential cell dispersion both contribute to mouse neocortical development. Nature 362:638–640

Walsh C, Cepko CL 1992 Widespread dispersion of neuronal clones across functional regions of the cerebral cortex. Science 255:434–440

Wylie E, Baumgartner C, Prayson R et al 1994 The clinical spectrum of focal cortical dysplasia and epilepsy. J Epilepsy 7:303–312

Zacco A, Cooper V, Chantler PD, Fisher-Hyland S, Horton HL, Levitt P 1990 Isolation, biochemical characterization and ultrastructural analysis of the limbic system-associated membrane protein (LAMP), a protein expressed by neurons comprising functional neural circuits. J Neurosci 10:73–90

DISCUSSION

Innocenti: That a disturbance of cortical conductivity exists in the brains of schizophrenics is an interesting hypothesis. Has electroencephalography coherence in schizophrenia been studied? If so, how consistent are these studies in schizophrenics and what do they show?

Roberts: The electroencephalography results have been particularly difficult to interpret. Brain electrical activity mapping studies demonstrate that the most robust changes occur within the temporal lobe and frontal cortex. However, these results are also difficult to interpret because simply performing the mapping in these patients causes them to be anxious and frightened and they are, therefore, probably in a different state of mind.

Jones: Do you exclude completely all the Scandinavian studies (Mednick & Cannon 1991) which suggest that influenza has an effect in schizophrenia?

Roberts: Even within the group that supports the influenza idea, there are people who can't replicate their own findings. For example, Torrey et al (1991) analysed the effects of influenza on over 10 000 patients from the USA. They didn't see a significant effect.

Kennedy: Why does there have to be a uniquely neurological basis for schizophrenia? Control adults with disturbed thinking processes probably don't exist. There seems to be a lot of variability in the schizophrenia results. What is the predictive power of these studies? If it is based on the psychological profiling of children, this could correlate perfectly with a psychological cause of schizophrenia. The only thing which would be convincing to a non-believer would be to find a good correlation. If schizophrenia could be predicted, on the basis of a brain scan, before the patient developed any symptoms, then this would be consistent with a neurological defect. If you could do this only after the symptoms had developed, it would not be inconsistent with a psychological cause.

Rakic: A clear-cut correlation would not be expected because schizophrenia is not a single disease. We are just looking at symptoms, and the faulty connections between areas may arise because certain receptors are not working, the second-messenger pathways are not operating or certain cells are missing. These may all have similar symptoms but different causes.

Roberts: In relation to Henry Kennedy's second point, there are a couple of anecdotal reports of patients who have been scanned for other reasons and who have subsequently developed schizophrenia. When they looked back at the previous nuclear magnetic resonance imaging scans, they observed an increased ventricular size (see Roberts 1991).

Blakemore: I have problems with Henry Kennedy's suggestion that individuals may have grossly disturbed psychological behaviour without any underlying physical difference between their brains and the brains of people without such disorders.

Kennedy: Different people have different levels of intelligence, but there are no reported morphological characteristics of the brain that determines intelligence.

Blakemore: True, none have been reported. But that's not to say that more subtle differences do not exist. We just don't know what to look for.

Kennedy: But differences in intelligence may exist that are not related to structural differences. Twenty years ago, it was a popular idea that schizophrenia didn't have a physiological basis and that it was a psychological phenomenon.

Blakemore: But that was before the evidence for a genetic factor in schizophrenia. The same applies to unipolar depression, which also appears to have a genetic component.

Kennedy: Depression runs in families, as does incest. There are families that have high rates of incest, but it doesn't mean that it has a biological basis.

Roberts: The strongest point is the genetics. The concordance rate for monozygotic twins is 60–70%.

Rubenstein: What about non-identical twins?

Roberts: It's lower—about 50%. The closer an affected biological relative is to you, the higher your chances are of developing the disease (Roberts et al 1993).

Kennedy: Yes, but they may have the same family environment.

Blakemore: In the Weinberger study (Roberts et al 1993), identical twins shared the same family environment, shared the same uterus and experienced similar birth events, yet demonstrable cerebral hypoactivity and ventricular enlargement were present only in the twin that had schizophrenia. How can you relate this to the view that schizophrenia is due to a genetically determined developmental problem?

Roberts: One of the problems with schizophrenia is that it's not a clear-cut autosomal dominant disorder. The Weinberger study is an example of incomplete penetrance (Roberts et al 1993). There are also some provisional results which indicate that an unaffected twin is not the same as an age-matched control, but it is not quite different enough to be sick.

Daw: So you can argue from the twin studies that influenza cannot be the sole cause.

Rakic: Schizophrenia may be caused by an interaction between genetics and the environment.

Walsh: Or different genetic events may be responsible. There are some autosomal dominant diseases that require a second somatic mutation, either during development or at a later date, to express the phenotype. Tuberous sclerosis is an example in the brain, where a germline mutation is inherited, and where specific lesions result from mutations that occur later in the remaining normal cells. The second mutation appears to be the immediate cause of clonal lesions in the brain and cysts in other organs. If there is a dispersion of sibling cells in the development of the cortical plate or other parts of the brain, then the clonal pathology following the second hit in the brain progenitor cells could lead to a distributed pathology. This could make the pathology more difficult to find because the cells that inherit the loss of heterozygosity are scattered over large distances.

Blakemore: Gareth Roberts, what is the role of the dopamine system in schizophrenia?

Roberts: This is a difficult story to deal with. Neuroleptics work, but nobody really knows why. They seem to block dopamine D2 receptors, although clozapine doesn't interact primarily with dopamine D2 receptors and is better at dealing with symptoms.

Ghosh: I have a question on the large study of 60 schizophrenics described by Jones et al (1994) and Done et al (1994). I imagine that they have looked extensively to see whether that group has a shared history or whether they can be correlated with exposure to influenza, because they have all the records during pregnancy and early childhood. Is there an event in pregnancy or in early childhood that defines them as a separate group?

Roberts: The study group comprised people scattered over the UK who were all born at the same time. Other than that, they do not have anything else in common.

Innocenti: Is there an animal model for schizophrenia?

Roberts: If a neonatal lesion is made in the rat medial temporal lobe, and the adult is subsequently examined, changes are observed when they're stressed. However, if the same lesion is made in the adult rat, the same effect is not observed. This suggests that both the lesion and the process of development are required to produce an effect.

J. Price: What is the significance of enlarged ventricles in schizophrenia?

Rakic: Enlarged ventricles may occur because there is less white matter present, and there is less white matter because there are less cortical–cortical and cortical–subcortical connections.

Jones: In addition, the likelihood of that loss being developmental is high because it is not accompanied by gliosis. If it was a secondary anterograde or retrograde effect, it would have to be accompanied by gliosis, unless it occurred over a very protracted timescale. But if that were the case, one would expect the ventricular enlargement to be increased over time, whereas all the evidence suggests that the ventricular dilatation is present from the beginning and never advances.

Ghosh: In subplate lesion experiments the thickness of the cortical plate is normal, but the ventricles may be enlarged. Immediately after the ablation there is a short period of gliosis, but there is no long-term gliosis.

Blakemore: Edward Jones, is it possible that the displacement of the NADPH diaphorase-positive cells that you see may not be due to an error in their distribution, but to a difference in the way in which axons are subsequently laid down in the white matter? If some parts of the white matter are reduced or increased in thickness, the position of the cells may vary depending on where the missing axons should have been.

Jones: Yes, it's possible. Normally, presumptive subplate cells are carried deep into the white matter, which reflects the enlargement of the white matter. It is difficult to correlate the increased density deep in the white matter in schizophrenics with a loss of axons, but it doesn't rule out that this is a rearrangement which is related to the rearrangement of axons.

Innocenti: A smaller white matter does not necessarily mean fewer axons. Studies in the size of the corpus callosum have shown that it is difficult to predict the number of axons from the size of the corpus callosum (LaMantia & Rakic 1990). Enlarged ventricles do not necessarily imply loss of axons either. The diameter of the corpus callosum increases at least four times from the onset of myelination onwards both in cats (Berbel & Innocenti 1988) and in the monkey (LaMantia & Rakic 1990), whereas the total number of axons decreases slightly (or remains stable). Thus, myelination and/or the size of axons may be the most important parameter affecting the size of white matter in the central nervous system.

Jones: We have stressed that a loss of axons may be involved, but equally the reduction in volume of the white matter could be due to reduced myelination. There's no direct evidence that I know of, but some people have claimed it.

Blakemore: Does anyone know of any data on hemispheric differences?

Roberts: If the two hemispheres are compared, the biggest changes are usually found in the left hemisphere. Therefore, one problem is explaining a mechanism for the generation of asymmetrical pathology (Roberts 1991).

Walsh: It is possible that either hemisphere may be affected, but that one cannot detect any abnormalities in people where the pathology affects the right hemisphere.

Blakemore: Tim Crow and others (see Crow 1995) have reported that schizophrenia is associated with reduced cerebral asymmetry.

Levitt: There have been studies of different populations that have led to false conclusions about the linkage of sibling order, sex, handedness and even stuttering to schizophrenia. Ken Kidd's group at Yale looked at a number of the stuttering cases and eliminated many of the linkage tales (Kidd et al 1981, Cox et al 1984).

Roberts: We put forward a simple idea to explain the asymmetry. The human brain develops asymmetrically—the left hemisphere lagging behind the right. Any factor that affected both hemispheres at the same time may give an asymmetrical effect, i.e. only on the left hemisphere (Roberts 1991).

Bolz: What is the evidence for asymmetrical development?

Roberts: People have looked at fetal development in humans and plotted it out (see Roberts 1991 for review). A good question is why it occurs.

Walsh: Asymmetry of the pathological findings could also result from ascertainment bias. Focal cortical heterotopia is more common in the left hemisphere than the right hemisphere, and more common in the frontal lobe than in the parietal lobe. The presumed reason for that is that the frontal lobe is more liable to seizures for other reasons. Consequently, the frequency of pathology may be equal everywhere, but the same pathology may cause more severe symptoms if it's on the left hemisphere rather than on the right hemisphere.

Schmutz: One has to be cautious when judging whether a relationship exists between lesions and cases of schizophrenia or epilepsy that have been detected in people over the age of 40. There is a tendency for oversimplification. For example, you stated that the commonest cause of epilepsy was a cortical dysmorphology; however, epilepsy is not a single disease but a conglomerate of diseases with a number of different underlying causes. Indeed, some cases of epilepsy cannot be correlated with any morphological lesion. The most prominent morphological correlations are observed in patients with complex partial seizures. One should, therefore, try to combine identified types of seizures with the respective morphological lesions. It is dangerous to state that the lesion is the cause—it might be the other way around. Also, from an epileptogenic point of view, physiological studies may reveal more relevant findings. During & Spencer (1993) have published some work on brain microdialysis in patients with epilepsy. They show that during an epileptic seizure, the concentration of glutamate is higher in the epileptogenic hippocampus, and that during an epileptic seizure, glutamate release is increased. These physiological effects are probably more significantly related to the origin of the disease than any morphological lesion one can detect.

Walsh: It is possible that genetic causes of epilepsy involve neurotransmitter pathways that are active in both the developing brain and the adult brain. It is possible that there is a disorder in the development of the brain or, on the other hand, there may be some disorders that cause abnormal neurotransmitter function in the adult, and the developmental disorder that you see is just a convenient marker for identifying the disease.

Rubenstein: We observed the loss of a specific class of interneurons in the forebrain of a mouse lacking a homeobox gene (M. Qui, A. Bulfone, M. Martinez, S. Anderson & J. Rubenstein, unpublished results). Thus, it is possible that mutations may cause subtle histological abnormalities, and they may also predispose the mouse to pathological disorders. For instance, the loss of interneurons in critical areas such as the hippocampus may lead to the spread of excitation. This could result in epilepsy.

References

Berbel P, Innocenti GM 1988 The development of the corpus callosum in cats: a light- and electron-microscopic study. J Comp Neurol 276:132–156

Cox NJ, Kramer PL, Kidd KK 1984 Segregation analyses of stuttering. Genet Epidemiol 1:245–253

Crow TJ 1995 The relation between morphologic and genetic findings in schizophrenia: an evolutionary perspective. In: Fog R, Gerlach J, Hemmington R (eds) Schizophrenia. Munksgaard, Copenhagen p 15–30

Done DJ, Crow TJ, Johnston EC, Sacker A 1994 Childhood antecedents of schizophrenia and affective illness: social adjustment at ages 7 and 11. BMJ 309:699–703

During MJ, Spencer DD 1993 Extracellular hippocampal glutamate and spontaneous seizure in the conscious human brain. Lancet 341:1607–1610

Jones P, Rodgers B, Murray R, Marmot M 1994 Child developmental risk factors for adult schizophrenia in the British 1946 birth cohort. Lancet 344:1398–1402

Kidd KK, Heimbuch RC, Records MA 1981 Vertical transmission of susceptibility to stuttering with sex-modified expression. Proc Natl Acad Sci USA 78:606–610

LaMantia A-S, Rakic P 1990 Cytological and quantitative characteristics of four cerebral commissures in the rhesus monkey. J Comp Neurol 291:520–537

Mednick SA, Cannon TD 1991 Fetal development, birth and the syndromes of adult schizophrenia. In: Fetal neural development and adult schizophrenia. Cambridge University Press, New York

Roberts GW 1991 Schizophrenia: a neuropathological perspective. Br J Psychiatry 158:8–17

Roberts GW, Leigh PN, Weinberger DR 1993 Neuropsychiatric disorders. Wolfe Medical, London

Torrey EF, Bowler AE, Rawlings R 1991 An influenza epidemic and the seasonality of schizophrenic births. In: Kurstak E (ed) Psychiatry and biological factors. Plenum, New York, p 109–116

Final discussion

Determination and specification

Innocenti: There seems to be some confusion over the terms specificity, determination and plasticity. Could Jack Price review the precise meanings of these terms?

J. Price: These terms, as far as developmental biology is concerned, have formal meanings. Specificiation means that a cell has an identity, but that this identity can be changed. The classical definition is that if you put a cell or a piece of tissue in a neutral environment and it carries on doing what it was originally doing, then it is specified. Nonetheless, the behaviour of a specified cell can be altered by putting it into a different environment. If the behaviour of a cell cannot be altered by changing the environment, it is considered to be determined. The use of the term cell lineage agitates me the most. Haematologists like to talk about 'cells changing their lineages', which is impossible. In my opinion, the meaning of this term should be restricted to a description of the presumptive fate of a cell, i.e. what it does under normal circumstances. It cannot, therefore, be applied to situations in which cells are grown in culture. In my opinion, there is no such thing as a cell lineage experiment in a culture dish, but different people disagree about that.

Bartlett: We've inherited a model of precursor differentiation that relies heavily on a sequential lineage map introduced by haematologists and others many years ago. This model portrays differentiation as the stepwise progression of a cell down a certain lineage. When one tries to identify the steps in this progression, as many people have done in the neural crest over the last 20 years, one obtains many possible phenotypes through which a cell may progress. In general, we should not waste our time looking for the intermediate stages. We should focus more on the signalling that gives rise to a particular cell phenotype, rather than trying to decide exactly where that cell is in the pathway at that time. This will not be possible until we know more about the state of that cell in terms of a repertoire of transcriptional factors.

Innocenti: There are two types of determination—conditional and unconditional. What is the difference between these types?

J. Price: Jonathan Slack's undergraduate textbook defines these terms precisely, and I recommend that you read it (Slack 1991). What you're calling 'conditional determination' is what he calls 'specification'.

Rakic: All the cells in an organism, with the exception of a few lymphocytes, have the same DNA composition, so when you said determination cannot be changed, you really mean that it cannot be changed under the circumstances that you tested.

J. Price: That is correct. You could never be sure that you tried all the possible circumstances.

Rakic: Therefore, you cannot prove determination.

J. Price: No. In that sense it's an operational definition.

Rubenstein: There are some instances where recombination may be involved. For example, recombination occurs in the manufacture of B cells, so that DNA from one B cell lineage is different from another. It is possible that this might also occur in the brain, although such hypotheses have been disproven so far. Nevertheless, it is formally possible that recombination occurs during the development of the brain to produce lineages that are irreversibly different.

Rakic: That would be a good definition for determination.

Boncinelli: There is no doubt that progressive determination occurs. A cell can be changed easily at the beginning of this progression, but not so easily near the end. We should, therefore, refer to the degree of determination.

J. Price: I agree with that. It's also important to specify the parameters precisely when one says something is determined. For example, confusion arises with the use of terms like multipotential. People often say a precursor cell is multipotential without specifying which parameters they are describing, i.e. they mean the cell can make different types of neurons, but not other types of cells.

Boncinelli: Positional information and histological information must go together, even if it is evident that the genetic control of positional information is significantly different from histological information. In this case, if one follows the classical terminology, specification should be confined to positional information, and determination should be confined to histological information. The study of the genetics of *Drosophila* has resulted in a wealth of data on positional information. For example, apical regions of different appendages come from different imaginal discs, but they possess the same positional information for the tip or base of the appendage (Cohen & Jürgens 1990). Positional information must also be important in the brain because of the complex connectivity.

Krubitzer: A number of people refer to development as a decision-making process instead of just a series of events that can unravel in a number of ways to produce variant individuals which comprise a species. Some of these events are selected for or against. For instance, a cell can move right or left, or a neuron can exercise a particular option of differentiation dependent upon genetic or epigenetic factors. However, one cannot globally describe development as a decision-making process because it's not.

Blakemore: Do you mean that cells themselves do not make decisions, rather that the environment determines what they do?

Krubitzer: A particular cell doesn't necessarily choose an option. It can develop according to various options that are either selected for or against. In this way, the developmental process of a particular species is built up but it's not a decision-making process moving towards a predetermined point.

J. Price: It's not clear to me why it's wrong to talk about a cell reaching a point where it has to do A or B, and it chooses to do A. Obviously, there's no mental process involved but there must be a mechanism whereby a certain outcome is selected over another.

Rakic: Decision making implies that a cell responds to other factors which make that change. I prefer to use the term differential gene expression, which involves both specification and determination.

J. Price: But specification and determination do not necessarily involve differential gene expression. Changes in the organization of the cytoskeleton or changes in motility can occur without invoking new gene transcription. Decision making implies that an interaction is involved between the environment and the cells that are contained within it.

Levitt: Decisions can be made within a continuum of choices that are restricted as a cell proceeds through development. This doesn't necessarily mean that if one choice is made over another, a particular cell will not reach the same end point in terms of basic phenotype.

J. Price: As we understand more about the molecular biology, it will become more discontinuous. One of the problems with the terms of specification and determination is that they are used to cover up our ignorance. What we would really like to say is that events in the cytoplasm or the nucleus of this cell caused certain genes to be turned on or off, or one pattern of cellular behaviour to be excluded and another to be activated.

Levitt: But a particular neuron has a complex phenotype and does not acquire all those phenotypes at the same time. We're not going to find a single genetic switch that changes a cell instantly so that it acquires, in some specified or determined way, all the characteristics that it will express later on.

J. Price: There may be a crucial point as a cell goes through its final mitosis when a number of decisions are made. These decisions may all be simultaneous. It is possible that a whole series of events are switched on or off, and this has multiple downstream consequences.

Kennedy: The notion of whether certain events are simultaneous or not is important. A sequence of genetically controlled steps results in stepwise inheritance and maximizes the potential for environmental effects.

Innocenti: Is it necessary to assume that genetic steps are involved in determination and specification? It is possible that they are caused by local events, such as regulation via interactions with the cytoskeleton, for example.

J. Price: I agree, which is why it's important to know when decisions are actually made and not just when the consequences of them are acted out. If a particular decision is made early, events that occur later would be downstream consequences of that early decision. One of the problems, in terms of the control of neurogenesis, is that a growth cone has to enact complicated decisions when it is a long distance away from its transcriptional machinery. It is likely, therefore, that any switches at the genetic level which influence growth cones must be set up at an earlier stage.

Innocenti: An analogous situation is when a neuron chooses which synapses to maintain and which to eliminate. One could think of a complicated targeting system where the genome sends information to the right synapse, but it would be much easier if one could think of a locally controlled mechanism.

Rubenstein: We seem to be talking about synapse formation, cell identity and regionalization in the same conversation, which is impossible. We should try to organize our thoughts along the lines that regionalization of the nervous system is genetically controlled, and then, once we specify a region, differentiation of specific cell types occurs. The latter step is also genetically controlled. However, although some processes (migration, axon projection and synaptogenesis, for example) will be genetically controlled, others will have epigenetic effects. We need to specify what process we're talking about so that we don't confuse the issues.

Boncinelli: I agree. We cannot say that everything occurs at the same time. There are operational definitions of reversibility and irreversibility. If one takes a nucleus out of a cell and tries to make a frog embryo, the success rate decreases as the nucleus is taken later and later in development (see Gilbert 1994). Regionalization of the brain stops around embryonic day 10 in the mouse, when there is a decrease in potentiality and an increase in restriction. Neurobiologists usually speak about perinatal and postnatal periods which are distinct from the establishment of genetic or epigenetic differential expression.

Rakic: Irreversibility may depend on the limitations of the technique. It is possible that in 10 or 100 years from now, an entire frog may be made from a single tectal cell.

References

Cohen SM, Jürgens G 1990 Mediation of *Drosophila* head development by gap-like segmentation genes. Nature 346:482–485
Gilbert SF 1994 Developmental biology, 4th edn. Sinauer, Sunderland, MA
Slack JMW 1991 From egg to embryo: regional specification in early development, 2nd edn. Cambridge University Press, Cambridge

Index of contributors

Non-participating co-authors are indicated by asterisks. Entries in bold type indicate papers; other entries refer to discussion contributions.

Indexes compiled by Liza Weinkove

*Bagnard, D., 173
*Bailey, K. A., 85
*Barrett, G., 85
Bartlett, P., 37, 38, 68, 69, 82, 83, **85**, 95, 96, 97, 98, 114, 123, 124, 125, 209, 322
Blakemore, C., **1**, 52, 55, 59, 60, 62, 118, 119, 120, 122, **127**, 142, 143, 144, 146, 147, 169, 188, 189, 190, 191, 192, 193, 196, 197, 211, 246, 247, 248, 254, 271, 272, 273, 274, 291, 317, 318, 319, 320, 324
Bolz, J., 54, 59, 60, 63, 81, 110, 119, 141, 168, 170, **173**, 186, 187, 188, 189, 190, 191, 192, 193, 194, 195, 196, 225, 270, 290, 320
Boncinelli, E., 56, **100**, 110, 111, 112, 114, 115, 122, 123, 323, 325
Bonhoeffer, T., 118, 120, 123, 168, 187, 209, 226, 274
*Borngasser, D. J., **214**
*Broccoli, V., **100**
*Brooker, G. J., 85

*Cheema, S. S., 85
*Clausen, J., **231**

Daw, N., 117, 121, 165, 166, 197, 223, 224, 247, 248, 253, **258**, 270, 271, 272, 273, 274, 318
*Dutton, R., 85

*Eagleson, K., **200**

*Ferri, R., **200**
*Flavin, H. J., **258**

*Ford, M. O., 85
*Fox, K., **214**

Ghosh, A., 56, 57, 58, 68, 81, 95, 113, 124, 125, 148, **150**, 165, 166, 167, 168, 169, 170, 187, 188, 193, 195, 196, 208, 210, 245, 273, 318, 319
*Götz, M., **71**, **296**
*Gulisano, M., **100**

Innocenti, G. M., 54, 57, 60, 96, 97, 147, 171, 187, 209, 229, 271, 272, 316, 318, 319, 322, 324, 325

Jones, E. G., 53, 54, 80, 81, 125, 144, 189, 190, 226, 227, 255, 273, **277**, 289, 290, 291, 292, 293, 294, 316, 319

Kennedy, H., 52, 62, 68, 112, 113, 119, 120, 195, 224, 229, 246–247, 251, 253, 254, 255, 293, 294, 316, 317, 324
*Kilpatrick, T. J., 85
*Koblar, S. A., 85
*Kossel, A., **173**
Krubitzer, L. A., 113, 119, 207, 254, 291, 323, 324

LaMantia, A. S., 35, 55, 59, 64, 79, 80, 81, 82, 96, 97, 110, 112, 113, 118, 122, 123, 145, 146, 147, 166, 167, 170, 189, 191, 195, 196, 210, 211, 227, 228, 248, 255, 256
*Lavdas, A., **41**

Index of contributors

Levitt, P., 37, 38, 39, 59, 79, 97, 113, 142, 143, 144, 146, 147, 169, **200**, 208, 209, 210, 211, 212, 224, 228, 229, 246, 247, 255, 256, 291, 324
*Likiardopoulos, V., **85**
*Lotto, R. B., **231**

Maffei, L., 83, 189, 190, 244, 247
*Magowan, G., **231**
*Mione, M. C., **41**
Molnár, Z., 55, 118, 119, 123, **127**, 140, 141, 142, 143, 144, 145, 146, 147, 148, 167, 169, 170, 187, 192, 194, 196, 197, 224, 225, 227, 246, 248, 249, 254, 255, 289, 293, 294
*Murphy, M., **85**

*Nurcombe, V., **85**

O'Leary, D. D. M., 34, 35, 39, 56, 61, 63, 118, 144, 166, 171, 188, 193, 194, 195, **214**, 223, 224, 225, 226, 227, 228, 229, 253, 255, 273, 274

Parnavelas, J. G., 37, **41**, 52, 53, 54, 55, 56, 57, 61, 62, 67, 80, 83, 95, 121, 170, 247, 272, 289, 292
Price, D. J., 79, 80, 110, 115, 140, **231**, 245, 246, 247, 248, 249, 274
Price, J., 52, 55, 57, 58, 62, 69, **71**, 79, 80, 81, 82, 95, 96, 118, 121, 123, 194, 195, 209, 210, 212, 245, 318, 322, 323, 324, 325

Rakic, P., 36, 55, 56, 62, 64, 67, 68, 83, 98, 112, 114, 119, 120, 121, 147, 170, 192, 194, 226, 227, 228, 248, 252, 253, 254, 255, 270, 272, 290, 317, 318, 319, 323, 324, 325
*Reid, C., **21**
*Reid, S. N. M., **258**
*Richards, L. R., **85**
Roberts, G. W., 292, **296**, 316, 317, 318, 319, 320
*Royston, M. C., **296**
Rubenstein, J. L., 61, 79, 96, 110, 111, 112, 114, 121, 122, 123, 188, 245, 270, 291, 317, 321, 323, 325

*Schlaggar, B. L., **214**
Schmutz, M., 273, 292, 320
*Spada, F., **100**

*Talman, P. T., **85**

Walsh, C., **21**, 34, 35, 36, 37, 38, 39, 52, 55, 61, 63, 64, 67, 68, 94, 110, 112, 117, 118, 122, 186, 196, 211, 212, 291, 318, 320, 321
*Wang, X.-F., **258**
*Warren, N., **231**
*Williams, B. P., **71**

Subject index

acetylcholine, 259
ACPD (trans-1-aminocyclopentane-1,3-dicarboxylate), 262, 263, 264, 266–267
activity, neural, 13, 166–168
  growth factor production and, 240, 247–248
  reduced, in schizophrenia, 284–285
  spiny stellate dendritic morphology and, 180–184, 187, 189–191
  synaptic patterns and, 255
  ventroposterior thalamic afferents and, 217–218
adenosine deaminase (ADA), 263, 264
α-2-adrenergic receptors, 254
Alcmaeon of Croton, 3
alkaline phosphatase (AP), 26–28
allocortex, 308
L-2-amino-4-phosphonobutyrate (AP4), 262, 264
D-2-amino-5-phosphovaleric acid (D-APV), 217–218, 225, 260, 262, 263, 274
trans-1-aminocyclopentane-1,3-dicarboxylate (ACPD), 262, 263, 264, 266–267
AMPA (2-(aminomethyl)phenylacetic acid), 217–218, 273
amygdala, in schizophrenia, 289–290
anaesthesia, 259
antennapedia gene product, 115
antipsychotic (neuroleptic) drugs, 303–304, 318
AP4 (L-2-amino-4-phosphonobutyrate), 262, 264
apoptosis *see* programmed cell death
D-APV (D-2-amino-5-phosphovaleric acid), 217–218, 225, 260, 262, 263, 274
arachidonic acid, 91, 92
area 4, 255
area 17, 205–206
  in binocularly enucleated monkeys, 251–255
  cell cycle kinetics, 112, 114, 251, 252, 253, 311
  see also visual cortex, primary
area 18, 206
  in binocularly enucleated monkeys, 251–255
  cell cycle kinetics, 112, 114, 251, 252, 253
'area X', 253, 255
areal specialization *see* regional specialization
aspartate, 43, 156
Ast-1, 87
astrocytes, 72, 76, 86, 97
  EGF-mediated differentiation, 88, 91–92
  FGF-2-stimulated differentiation, 89, 90
  LIF-β receptor-mediated differentiation, 92–93, 94–95
  precursor cells, 77–78, 82, 87–88
auditory cortex, 151, 158, 169
axons
  directions of growth, 60–61
  excessive losses, in schizophrenia, 279, 319
  fasciculation, 136, 148, 175–177, 192–198
  growth cone morphology, 148, 153, 162, 175–177
  handshake hypothesis, 192–198
  pioneer, *Otx* expression and, 105–106
  pruning of 'exuberant', 13, 215, 226–227

barrels, vibrissae-related, 177, 187, 189, 202
  activity-dependent patterning, 13, 217–218, 225–226
  plasticity, 274–275

Subject index

in thalamus-lesioned rats, 255
transplantation studies, 210, 215–217, 226, 228
BDNF (brain-derived neurotrophic factor), 125, 237, 240–242, 245
bicoid class genes, 101
binocular enucleation, fetal monkeys, 12, 251–255
birth trauma, 298
brain-derived neurotrophic factor (BDNF), 125, 237, 240–242, 245
brain electrical activity mapping 316
brainstem, 215
Brodmann, K., 4, 5–6, 7
bromodeoxyuridine (BrdU), 209–210, 252

Ca^{2+}/calmodulin-dependent protein kinase, α type II, 279, 280
Caenorhabditis elegans, 22, 23, 57
calbindin, 44
calcium (Ca^{2+}), intracellular, 170, 266, 267–268
calcium-binding proteins, 42, 44–45, 46
calretinin, 44, 46
carbocyanine dyes
 in internal capsule, 134, 140–141, 169
 in occipital cortex, 133–134, 135–136, 140
 in thalamus, 134, 135–136, 141, 196–197, 226–227, 233
 see also DiI
cell cycle kinetics, 112, 121
 area 17/18 differences, 112, 114, 251, 252, 253, 311
 laminar fate and, 211
cell death, programmed *see* programmed cell death
cell doctrine, 3
cell lineage, 322
 approaches, 23
 cell morphology and, 57–58
 genetic disorders and, 31–32
 migration patterns and, 21–40
 models, 29–31, 34–36, 37–39, 61, 77–78
 neuronal subtypes, 41–58
 precursor cells, 61–69, 86–89
 retroviral libraries, 24–28, 35–39
 retroviral studies, 23–24, 72–73
cellular diversity, generation of, 71–84

cerebellum, 121, 145
chimeras, genetic, 23
chondroitin sulfate proteoglycan, 194
choroid plexus, 105, 106
chronotopy, of thalamocortical connections, 138, 146–147, 195
ciliary neurotrophic factor (CNTF), 93, 95
cingulate cortex, in schizophrenia, 284, 305
cingulum lesions, 259
CNQX (6-cyano-7-nitroquinoxaline-2,3-dione), 262, 263, 273
co-cultures, organotypic, 174–175
 cortex/cortex, axonal interactions, 175–177, 194–195
 cortex/thalamus, 128–133, 143–145, 174–175, 231–232, 248–249
 axonal interactions, 175–177, 193–195
 growth factors in postnatal development, 235–237, 240–242
 growth permissiveness of cortex at E20, 129–131
 neural activity and, 240
 regional specificity, 131–133
 remote growth-promoting influence of cortex, 129, 175, 233, 234–235, 246–247
 stop signal in layer IV after P3, 131
cocaine, prenatal exposure, 291
collagen
 gels, 232
 type IV, 204
commitment, cell, 120
computed tomography (CT), 304
connectivity, functional cortical, 278
 in schizophrenia, 279, 282–284
corpus callosum, 319
 subplate projections, 56, 155, 170–171
 see also corticocortical projections
cortical layers *see* layers, cortical
cortical plate, 7–8, 9, 22, 151
 determination of areal specialization, 8–12
 efferent projections, 161, 162
 Emx expression, 106–108
 model for cell lineage, 29–31
 targeting by thalamic axons, 219, 220–221

cortical slice co-cultures *see* co-cultures, organotypic
corticocortical projections, 7
  growth factors in development, 237
  in transplantation studies, 208, 209, 215
corticofugal projections, 174, 194
  attraction to thalamus, 174–175
  axon–axon interactions, 175–177, 192–198
  growth factors in development, 237
  guidance of thalamocortical fibres, 133–136, 140–141
  subplate neurons as pioneers, 161–162, 170–171, 174, 177
  transplantation studies, 215
  waiting compartment, 141
cortisol, 259, 270
critical period, 258, 270
  factors affecting, 259
  NMDA receptors and, 260, 261, 273
  synaptogenesis in, 271–273
culture medium, serum-free, 232
cultures
  dissociated cell *see* dissociated cell cultures
  neocortex explants, 25, 232
cut mutations, 22
6-cyano-7-nitroquinoxaline-2,3-dione (CNQX), 262, 263, 273
cyclic AMP (cAMP), 262, 263, 264, 266
cytoarchitecture, cortical
  areal variations, 4, 5–6, 7, 41–42
  in epilepsy, 299–301
  in schizophrenia, 278–279, 305
  transplantation studies, 215–217
  see also barrels, vibrissae-related; layers, cortical; ocular dominance columns
cytochrome oxidase, 6, 7

D3 dopamine receptors, 213, 284
D4 dopamine receptors, 284
dark-reared animals, 240, 247–248, 259, 265, 270–271
dendritic morphology
  differentiation, 59–61
  spiny stellate cells, 178–181, 182–184, 186–191
Descartes, René, 1–2

determination, 322–325
  unconditional and conditional, 322
DiA (4-(p-dihexadecylaminostyryl)-N-methylpyridinium iodide), 233, 234–235
diacylglycerol (DAG), 91, 92, 266
dibromodeoxyuridine (dBrdU), 81
diencephalon
  areas and boundaries, 104–106
  Emx and *Otx* gene expression, 102, 103
differentiation, cell, 15–16
  in dissociated cultures, 124–125
  factors regulating, 85–99, 203–204, 205–206
  transplantation studies, 202–203, 311
DiI (1,1′dioctadecyl-3,3,3′,3′-tetramethyl-indocarbocyanine perchlorate), 169
  cell migration, 23, 25, 157
  cortical injections, 219
  thalamic injections, 134, 153, 159, 196–197, 226–227, 233
  thalamus/cortex co-cultures, 129, 130, 233, 234–235
  transplantation studies, 208
  see also carbocyanine dyes
dissociated cell cultures, 232
  cortex, growth factor requirements, 237
  differentiation factors, 124–125
  multipotential precursor cells, 76–77
  neuronal precursor cells, 75–76, 81–82
  precursor cell fate, 73–75
  thalamic cell survival, 236–237
  visual cortex, 252
Dlx genes, 104, 105
Dlx2 gene, 35, 233, 234
dopamine hypothesis, schizophrenia, 303–304, 318
Drosophila, 22, 37, 58, 100–101
dyslexia, developmental, 291
dysplasia, cortical, 301

EGF *see* epidermal growth factor
electroencephalography, 316
Elvax, 217
empty spiracles (ems), 101, 220
Emx genes, 100–116, 220
  in developing cerebral cortex, 106–108
  in E10 mouse embryos, 101–103
  Xenopus homologues, 110

Subject index

entorhinal cortex, in schizophrenia, 278–279, 280, 305
environmental determinants
  neuronal phenotype, 45, 81, 202–203, 205–206
  see also transplantation studies
epidermal growth factor (EGF), 88, 91–92, 203–204, 209
  receptor, 204–205, 210–211
epigenetic influences, 12–13
epilepsy, 291, 296, 297–301, 307
  aetiology, 298
  disordered cortical cytoarchitecture, 299–301
  epidemiology, 297–298
  neuropathology, 298–299
  pathogenesis, 309–310, 313, 320–321
  and schizophrenia, 301–302
  temporal lobe, 301–302
eyes, prenatal removal of both, 12, 251–255

fasciculation, axonal, 136, 148, 175–177, 192–198
FGF see fibroblast growth factor
fibroblast growth factor (FGF), 111, 237
  precursor cell proliferation and, 82
  thalamic cell survival and, 240, 242
fibroblast growth factor 1 (FGF-1), 96
  in neuronal differentiation, 89–92, 125
  second messenger pathway, 91
fibroblast growth factor 2 (FGF-2)
  neuronal differentiation and, 89, 96, 124–125, 203
  precursor cell proliferation and, 86, 87–88, 89, 90
forebrain, gene expression patterns, 102–103, 201
forskolin, 263
frontal cortex
  in schizophrenia, 282, 284–285, 289, 307
  transplantation studies, 229
functional subdivisions, 3–6

GABA receptors, 80
  γ_2-subunit, 279, 280
GABA$_A$ receptors, in schizophrenia, 284
GABAergic neurons, 15, 53–54, 273
  lineage studies, 43, 44–45, 46
  postnatal development, 47–48, 49
  in schizophrenia, 289, 291
  specification, 75, 79–83
β-galactosidase
  mosaic mice, 23, 55–56, 64–69, 309
  retroviral marker studies, 25, 26, 43–44
  variability of staining, 52–53
γ-aminobutyric acid see GABA
GAP-43 (growth-associated protein 43), 213, 259
Gbx2 gene, 234
genetic disorders, 22, 31–32, 298
genetics
  of determination and specification, 324
  of schizophrenia, 303, 317–318
glial cells, 86
  migration, 68
  radial, 8, 10, 97–98, 313
  see also astrocytes; oligodendrocytes
glial fibrillary acidic protein (GFAP), 86, 92, 94–95, 97–98
gliosis, absence in schizophrenia, 279, 281, 307, 319
glutamate, 259, 320
glutamate receptors, 217–218
  GluR2 subunit, 279, 280
  metabotropic (mGluR), 259, 261–266, 267–268
  NMDA receptor interactions, 266–267
  subplate neurons, 156–157
  see also NMDA receptors
L-glutamatergic (Glu) neurons, 53–54
  lineage studies, 43, 44
  postnatal development, 47–48, 49
  specification, 75, 79–83
glutamic acid decarboxylase (GAD), 81, 156
  in schizophrenia, 284, 291
growth-associated protein 43 (GAP-43), 213, 259
growth cone morphology, 148, 153, 162, 175–177
growth factors
  chemical nature, 240–242
  cortex-derived, promoting thalamo-cortical development, 129, 234–237
  corticofugal/corticocortical connections and, 237
  neocortex development and, 231–250

neural activity and, 240, 247–248
precursor cell differentiation, 87–93, 203–204
see also brain-derived neurotrophic factor; epidermal growth factor; fibroblast growth factor; nerve growth factor

haematopoietic cell lineage, 21–22
hamartomas, 300–301
handshake hypothesis, 192–198
head injury, 298
heparan sulfate proteoglycan (HSPG), 89–91, 96
heregulins, 204
heterotopia, periventricular, 32
hippocampus, 121, 211, 212
 in schizophrenia, 305
homeobox genes, 100–101, 234
Hox genes, 101, 123, 201
3-hydroxy-3-methylglutaryl coenzyme A (HMGCoA) reductase promoter, 68–69

ibotenic acid, 171
immunoglobulin (Ig) superfamily proteins, 213
influenza, 277, 293, 303, 316, 318
inositol trisphosphate (InsP$_3$), 261–262, 263, 266
intelligence, 317
γ-interferon, 87, 88
intermediate zone, 9, 106, 151
 Emx expression, 108
 non-radial migration, 26
internal capsule, 155, 174
 afferent and efferent axonal interactions, 193–194, 196
 carbocyanine dye injections, 134, 140–141, 169
 cortical efferent projections, 134, 161, 162
 thalamocortical projections, 135
interstitial neurons, in schizophrenia, 281–285, 289

kainate receptors, 273
kainic acid lesions, 156–161, 162, 165–166, 170, 249

lacZ gene, 43, 309
 see also β-galactosidase
laminae, cortical *see* layers, cortical
LAMP *see* limbic system-associated membrane protein
lateral geniculate nucleus (LGN)
 ablation studies, 255
 in binocularly enucleated monkeys, 251, 254
 co-culture studies, 128–133, 143–144
 cortical projections, 150–151, 153–155, 157–161, 163, 219
  see also ocular dominance columns; thalamocortical projections
 corticofugal projections, 141
 migration and rotation, 197
latexin (PC3.1), 201–202
layer I, 8, 22, 56
layer II, in schizophrenia, 305
layer III, 180
layer IV, 6, 151, 271
 spiny stellate cells, 177–181
 subplate ablation studies, 158–161, 163, 165–166
 thalamic axon stop signal, 131, 177–178
 thalamic innervation, 153–155, 163, 169, 177–181, 187–188, 236
 in transplantation studies, 228
 see also ocular dominance columns
layer V, 6, 56, 60, 215
layers, cortical, 3–5, 42, 308
 specificity of thalamic innervation, 177–181, 187
leukaemic inhibitory factor (LIF), 93, 94–95
leukaemic inhibitory factor (LIF)-β receptor, 92–93, 95
LGN *see* lateral geniculate nucleus
limbic cortex
 molecular specification, 203–204, 310–311
 transplantation studies, 202–203, 312
limbic system-associated membrane protein (LAMP), 201, 310–311, 312
 antibodies, 211, 212
 environmental influences, 202–203, 209–212
 expression in thalamus, 211
 molecular influences, 203–204, 209, 211

Subject index

molecular structure, 212–213
lumping errors, 24, 72–73

magnetic resonance imaging (MRI), 304, 317
major histocompatibility complex (MHC), class I molecules, 87, 88
MAP-2-positive cells *see* microtubule-associated protein 2 (MAP-2)-positive cells
marginal zone, 8, 9, 22
Mash1, 35
Matrigel, 204
MCPG (α-methyl-4-carboxyphenylglycine), 263, 266
medial geniculate nucleus (MGN), 151, 158, 163, 169
mediodorsal thalamic nucleus, in schizophrenia, 281, 292
membranes, cortical, 175, 177
mesencephalon
  Emx expression, 102, 103
  Otx expression, 102, 103, 104
mesocortex, 308
α-methyl-4-carboxyphenylglycine (MCPG), 263, 266
N-methyl-D-aspartate receptors *see* NMDA receptors
microdysgenesis, 291
microglia, 96–97
microgyria, 171
microtubule-associated protein 2 (MAP-2)-positive cells, 157
  in culture, 89, 90, 124, 209
  in schizophrenia, 282, 283, 289
microtubule-associated protein 5 (MAP-5), 76–77
migration, cell, 8, 9, 29–31, 220, 278, 308–310
  in β-galactosidase mosaic mice, 23, 64–69
  distances, 119–120
  in epilepsy, 313
  periodicity, 121–122
  radial, 9, 10, 36–37, 65–68, 121–122, 308–310
  retroviral studies, 23–28, 29–30, 35–39, 61–64, 68, 122–123, 220
  in schizophrenia, 279, 290, 313
  tangential (non-radial), 26, 67, 117–123, 309

mirror index, 179–180, 181, 184
MK 801, 260, 273–274
modular specialization, 8–12
molecules
  membrane-bound, guidance of thalamic axons, 168, 175, 177–178
  in positional specification, 200–213, 310–311
monocular deprivation, 13, 267
  factors reducing/abolishing effects, 258–259, 260, 273
  growth factors and, 245
  spiny stellate dendritic morphology, 180–181, 182–184, 187, 188, 190
mosaic mice, β-galactosidase, 23, 55–56, 64–69, 309
mossy fibre projections, 211
motor cortex, 112, 252

N-methyl-D-aspartate receptors *see* NMDA receptors
NADPH diaphorase-positive cells *see* nicotinamide-adenine dinucleotide phosphate (NADPH) diaphorase-positive cells
neocortex, 103, 308
  early neuron specification, 251–256
  plasticity in development, 214–230
  role of growth factors in development, 231–250
  see also specific areas
nerve growth factor (NGF), 125, 237, 268
  thalamic cell survival and, 240, 245, 246
  visual cortex plasticity and, 242, 244–245, 247, 259
nestin, 76–77, 124
netrins, 245–246
neuroepithelial cells, 76–77, 82–83, 86–87, 220
  MHC class I expression, 87, 88
neuroepithelium, 7, 85
  see also ventricular zone
neurofilament proteins, non-phosphorylated (NF-H), 282, 283
neuroleptic (antipsychotic) drugs, 303–304, 318
neuron–astrocyte precursors (N–A cells), 77–78, 82, 88

neuron–oligodendrocyte precursors (N–O cells), 77–78
neurons, 72
  apoptosis *see* programmed cell death
  cell lineage of subtypes, 41–58
  differentiation, 15–16
    factors regulating, 89–92, 96, 97, 124–125, 203–204
    transplantation studies, 202–203
  interstitial, in schizophrenia, 281–285, 289
  migration *see* migration, cell
  precursor cells, 72, 75–76, 77–78, 82, 86–87
    in adult mouse forebrain, 88–89
    early commitment, 87, 88, 251–256
    FGF-1 effects, 89–92
    later commitment, 87–88
    reduced numbers, in schizophrenia, 279–281, 290–291, 292
  retrograde transport, 13
  see also specific types
neuropsychiatric disorders, 22, 296–321
  developmental neurobiology and, 307–313
  see also epilepsy; schizophrenia
neurospheres, 92
neurotransmitters
  expression of related genes, in schizophrenia, 281–285
  neuronal subtypes, 43
  subplate neurons, 155–156
neurotrimmin, 213
neurotrophic factor 3 (NT-3), 124, 125
neurotrophic factor 4/5 (NT-4/5), 125, 245
NGF *see* nerve growth factor
nicotinamide-adenine dinucleotide phosphate (NADPH) diaphorase-positive cells
  developmental expression, 293–294
  in schizophrenia, 282, 283, 289, 293, 294, 307, 319
NMDA, 217, 266–267
NMDA receptors, 83, 217–218, 259
  antagonists, 217–218, 259, 260
  metabotropic glutamate receptor interactions, 266–267
  visual cortex plasticity and, 260–261, 262, 267–268, 270–271, 273–274
non-NMDA receptors, 217–218, 273

non-pyramidal neurons, 16, 53–55, 75
  calcium-binding proteins, 42, 44–45, 46
  cell death, 49–50
  lineage, 42–47
  postnatal development, 47–50
  progenitors, 43–44, 46–47, 62–63
noradrenaline, 259
notch, 58
nucleus accumbens, in schizophrenia, 281, 292
numb mutations, 22

occipital cortex, 150
  carbocyanine dye injections, 133–134, 135–136, 140
  co-culture studies, 128–133
  projections from, 133–134, 135–136
  thalamic afferents, 134–136, 153
  see also visual cortex
ocular dominance columns, 6, 13, 155, 177, 248, 272
  factors reducing/abolishing changes, 258–259, 270–271
  metabotropic glutamate receptors and, 265–266
  NMDA receptors and, 260, 261, 262
  role of growth factors, 245
  spiny stellate cell dendrites, 181–184, 187–191
  subplate ablation studies, 158–161, 165–166, 170
oligodendrocytes, 72, 76, 95–96
  differentiation in culture, 124
  precursor cells, 77–78
oncostatin M, 93, 95
opiate-binding cell adhesion molecule, 213
optic radiations, 153, 197
organotypic co-cultures *see* co-cultures, organotypic
orthodenticle (*otd*), 101, 105–106
Otx genes, 100–116, 201
  in developing cerebral cortex, 106–108
  diencephalon areas and boundaries and, 104–106
  downstream targets, 110
  in E10 mouse embryos, 101–103
  in mid-gestation mouse embryos, 103–104
  regulation of expression, 110, 111

Subject index

parvalbumin, 44, 46
PC3.1 (latexin), 201–202
perireticular cells, 196–197
perirhinal cortex, 202–203, 208, 209–210
periventricular heterotopia, 32
phospholipase Cγ (PLCγ), 91
planar induction, 111
plasticity, 13, 312–313
  in development of neocortical areas, 214–230
  somatosensory cortex, 274–275
  visual cortex *see* visual cortex, plasticity
pons, 121, 292
positional specification, 308–309, 310–312, 323
  Emx and *Otx* genes in, 100–116
  molecular markers, 200–213, 310–311
  plasticity *see* plasticity
  see also regional specialization
positron emission tomography (PET), 304
posterior commissure, 104
posterior thalamic nucleus (PO), 143, 144
potassium chloride (KCl), 239–240
precursor cells *see* progenitor cells
prefrontal cortex, in schizophrenia, 281, 284, 289, 304, 305
preplate, 8, 9, 22
  guidance of thalamocortical axons, 8, 133–136
  neurons, 8, 85
primordial plexiform zone *see* preplate
progenitor cells, 22–23, 34–39
  determination, 73, 74–75
  differentiation *see* differentiation, cell
  expression of molecular markers, 310–311
  expression of signalling molecules, 204–205
  generation of diversity, 71–84
  levels of commitment, 64
  lineage potential, 61–69, 86–89
  multipotential, 47, 63, 71–72, 76–77, 86–87
    in adult mouse forebrain, 88–89
  migratory, 29–31, 34–36, 37–38, 63
  neuron–astrocyte (N–A cells), 77–78, 82, 88
  neuron–oligodendrocyte (N–O cells), 77–78

neuronal *see* neurons, precursor cells
non-migratory, 29–31, 34, 63, 118
non-pyramidal neurons, 43–44, 46–47, 62–63
pyramidal cells, 43–44, 46–47
regional specificity, 30–31, 34–35, 251–256
retroviral labelling, 23, 61–62, 72–73, 79
specification of phenotype, 72–73, 81–82
working model, 77–78
programmed cell death (apoptosis), 12, 48–50, 248
cultured neuroepithelial cells, 76–77
cultured thalamic cells, 236–237, 248–249
subplate neurons, 155, 156, 249, 282–283, 289
proliferating cell nuclear antigen (PCNA), 252
protein kinase, α type II Ca^{2+}/calmodulin-dependent, 279, 280
protein kinase C, 259, 268
proto-map hypothesis, 9, 12, 111, 118, 308, 312
protocortex hypothesis, 11–12, 309, 312
puberty, 270, 272
pyramidal cells, 16, 53–55, 75, 196
  dendritic differentiation, 59–60
  lineage, 42–47
  postnatal development, 47–50
  progenitor cells, 43–44, 46–47
  in schizophrenia, 305
  visual cortex, 14, 16, 54–55, 188–190

quisqualate, 262–263

radial glial cells, 8, 10, 97–98, 313
radial induction, 111
radial unit hypothesis, 9, 10, 36–37, 67, 308–309
recombination, 323
Reeler mouse, 136–137, 138, 141–143, 145, 147
regional specialization, 6–7, 41–42
  determination, 8–12, 308–309
  epigenetic regulation, 12–13
  plasticity in development, 214–230
  progenitor cells, 30–31, 34–35, 251–256
  thalamic influences, 11–12, 13, 127

thalamocortical innervation *in vitro*, 131–133
see also positional specification
reticular cells, thalamic, 196–197
retina
  cell lineage studies, 24, 57
  Drosophila, 37
retroviral libraries, 24–28, 29–30
retroviral markers, 23–24, 72–73, 79, 95–96
  cell migration, 23–28, 29–30, 35–39, 61–64, 68, 122–123, 220
  early stage embryos, 38, 63–64
  errors, 24, 72–73
  neuronal subtypes, 42, 43–50, 56–57
  variability of staining, 52–53
rhombomeres, 123, 201

schizophrenia, 277–295, 296–297, 302–307
  aetiological hypotheses, 277–278, 292–293, 303–304, 313–314, 316–318
  cerebral cortical pathology, 278–279
  developmental *vs* degenerative pathology, 307
  disturbances of cortical development, 281–285, 289–294
  epidemiology, 302
  and epilepsy, 301–302
  neuropathology, 278–281, 305–307, 318–320
  pathology, 304
short-latency responses, 218, 225
somatosensory cortex, 112, 197–198
  activity-dependent patterning, 217–218, 225–226
  infragranular layers, 252
  plasticity, 274–275
  pruning of exuberant branches, 215, 226–227
  thalamic afferents, 8, 134, 177, 187, 219
  transgenic marker, 202, 311, 312
  transplantation studies, 12, 202–204, 207–212, 215–217, 226, 228–229
specification, 322–325
spinal cord, 215
spiny stellate cells, 16, 54–55
  dendritic morphology, 178–181, 182–184, 186–191

stop signal for thalamic axons, 177–178
splitting errors, 24, 72–73
star pyramids, 190
stem cells, 22, 31, 37–38
strabismic cats, 182–184, 187, 189
striate cortex *see* visual cortex, primary
subplate, 8, 9, 22, 55
  defects in schizophrenia, 282–283, 289–290
  neurons, 150–172, 174, 194–195, 220–221
    ablation studies, 156–161, 162, 163, 165–166, 168–171, 188, 319
    development, 155–156
    guidance of thalamic axons, 8, 133–136, 155, 156–158, 168–170, 195
    intracellular Ca^{2+} changes, 170
    ocular dominance columns and, 158–161, 165–166, 170
    persisting, 281–282, 294
    as pioneers of cortical efferents, 161–162, 170–171, 174, 177
    programmed cell death, 155, 156, 249, 282–283, 289
    retroviral marker studies, 38
    targeting by thalamic axons, 219–221
    'waiting' of thalamic axons, 8, 134–135, 153
subventricular zone, 7, 9, 79, 86, 106
  multipotential precursor cells, 88–89, 90
  oligodendrocyte precursors, 95–96
superior colliculus, 155, 215, 223–224
synaptic contacts
  activity dependence, 184
  in critical period, 271–273
  subplate neurons, 153, 156, 170
synaptic plasticity, 13

telencephalon
  corticofugal projections, 134
  Emx and *Otx* expression, 102, 103, 108
  thalamic afferents, 135, 153–155
temporal lobe
  epilepsy, 301–302
  in schizophrenia, 282, 304, 305
tetrodotoxin (TTX), 218, 259, 271

Subject index

TGF-β (transforming growth factor β), 111, 125, 203–204
thalamocortical projections
  axon stop signal in layer IV, 131, 177–178
  axon–axon interactions, 175–177, 192–198
  cell proliferation in ventricular zone and, 119, 252, 253
  columnar specificity, 181–184, 187
  cortex-derived growth factors promoting, 129, 234–237, 240–242
  cortical targeting specificity, 218–221
  development, 153–155
  guidance, 8, 127–149, 174
    by corticofugal projections from preplate/subplate, 133–136, 140–141
    regional non-selectivity *in vitro*, 131–133, 145–146
    role of subplate neurons, 8, 156–158, 168–170, 195
    by temporal sequence of cortical signals, 128–131, 137–138, 146–147
  laminar specificity, 177–181, 187
  neural activity in patterning, 217–218, 225–226
  patterning of connections, 150–172
  in Reeler mouse, 136–137, 138, 141–143, 145, 147
  regional cortical differentiation and, 11–12, 13, 127
  side branches, 167–168, 197
  in transplant studies, 208, 215–217, 227–229
  waiting period, 8, 134–135, 153–155, 193
thalamocortical slices, 226–227
thalamus, 7
  carbocyanine dye injections, 134, 135–136, 141, 196–197, 226–227, 233
  co-cultures *see* co-cultures, organotypic, cortex/thalamus
  early demarcation of nuclei, 224–225
  Emx expression, 102, 103
  LAMP expression, 211
  lesions, 255, 259
  Otx expression, 103, 104–105
  posterior nucleus (PO), 143, 144

programmed cell death, 236–237, 248–249
reticular cells, 196–197
in schizophrenia, 279, 281, 292
somatosensory, 197–198
specificity of cortical interactions, 173–191
subplate projections, 155
see also lateral geniculate nucleus; medial geniculate nucleus
thymidine, [^{3}H]-labelled, 209–210, 251, 253
time-lapse video microscopy
  migrating cells, 25–26, 220
  thalamocortical relationships, 175–177
transforming growth factor β (TGF-β), 111, 125, 203–204
transgenic mouse
  β-galactosidase integrated into X chromosome, 23, 55–56, 64–69, 309, 311
transitional field, 106–107, 108
transplantation studies, 12, 30–31, 146
  cell differentiation, 202–204, 207–212, 311
  cortical plasticity, 214–217, 226, 227–229, 312
trauma, epilepsy and, 298
TrkA receptors, 245
trkB gene, 241–242
TrkC receptor, 124–125
tuberous sclerosis, 31–32, 318

ventricles, enlarged, 279, 304, 305, 318–319
ventricular zone, 7, 9, 85–86, 151
  cell migration, 35, 37, 122
  Emx expression, 106, 107–108
  generation of cellular diversity, 71–84
  neuronal precursors, 43–44, 47, 251–256
  Otx expression, 103
  positional specification in, 310
  retroviral markers, 61, 79
  see also progenitor cells
ventroposterior thalamocortical afferents, 215–218, 219, 225–226
vibrissae-related barrels *see* barrels, vibrissae-related

155 Subplate neurons as forerunner to cortical efferents.
Express NTs long before cx (excitatory not inhibitory) 156
Subplate neurons & programmed cell death. 156

161 Subplate axons guiding cortical axons

173 Periph Inf thr Thal define funct architect of cx.